Creating Location Services for the Wireless Web

Professional Developer's Guide

Creating Location Services for the Wireless Web

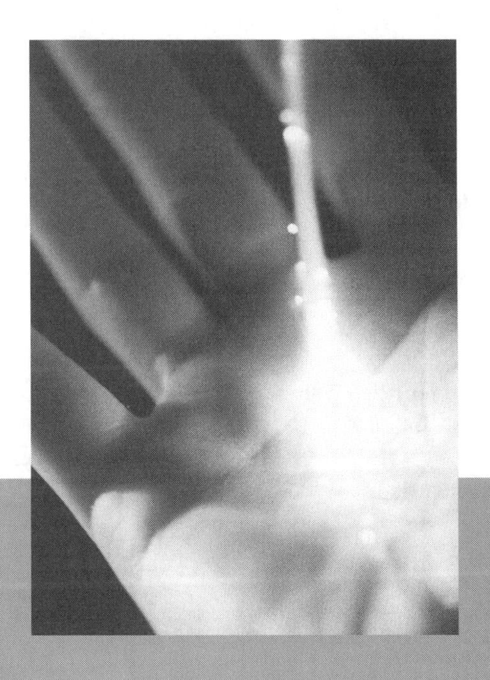

Johan Hjelm

Wiley Computer Publishing

John Wiley & Sons, Inc.

Publisher: Robert Ipsen
Editor: Carol A. Long
Developmental Editor: Adaobi Obi
Managing Editor: Angela Smith
New Media Editor: Brian Snapp
Text Design & Composition: D&G Limited, LLC

This book is printed on acid-free paper. 

Published by John Wiley & Sons, Inc., New York

Published simultaneously in Canada.

Library of Congress Cataloging-in-Publication Data:

Hjelm, Johan.
Creating location services for the wireless web : professional developer's guide / Johan Hjelm.
p. cm.
ISBN 0-471-40261-3
1. Wireless Application Protocol (Computer network protocol) 2. Web site development. 3. Geographic information systems. 4. Cellular telephone systems. I. Title.
TK5105.5865.H554 2002
005.2'76--dc21

2001006512

Printed in the United States of America.

10 9 8 7 6 5 4 3 2 1

Professional Developer's Guide Series

Other titles in the series:

Mobile Information Device Profile for Java 2 Micro Edition by Enrique Ortiz and Eric Giguere, ISBN 0-471-03465-7

Voice XML: Strategies and Techniques for Effective Voice Application Development by Chetan Sharma and Jeff Kunins, ISBN 0-471-41893-5

WAP Integration by Robert Laberge and Srdjan Vujosevic, ISBN 0-471-41767-X

XSLT: The Ultimate Guide to Transforming Web Data by Johan Hjelm and Peter Stark, ISBN 0-471-40603-1

GPRS and 3G Wireless Applications by Christoffer Andersson, ISBN 0-471-41405-0

Constructing Intelligent Agents Using Java by Joseph P. Bigus and Jennifer Bigus, ISBN, 0-471-39601-X

Advanced Palm Programming by Steve Mann and Ray Rischpater, ISBN 0-471-39087-9

WAP Servlets by John L. Cook, III, ISBN, 0-471-39307-X

Java 2 Micro Edition by Eric Giguere, ISBN 0-471-39065-8

Scripting XML and WMI for Microsoft® SQL Server™ 2000 by Tobias Martinsson, ISBN 0-471-39951-5

Contents

Preface

"I know precisely where we are," I said to my girlfriend, "but can you please slow down so I can read the map? We just passed the second turnoff, and I don't think there is another for a few miles."

Knowing where you are, where you are going, and how to get there are all position-dependent. Making a computer show it to you is creating a position-dependent application. As you will see in this book, there are many widgets on a plough, as my father used to say (meaning that a simple thing can be made very complicated). I personally learned this lesson already in 1996, when I was part of a project that built the arguably first wireless Web application and definitely the first location-based application using a mobile phone system (although at the time, you could not get the position out of the mobile phone system).

The On the Move project was intended for us content providers to meet the wireless industry, and it was a mutual culture shock. But as luck would have it, we did everything that we were supposed to do—and created applications that some claim are state of the art today. In the five years since then, I have followed this industry (I arranged a very enlightening workshop for the World Wide Web Consortium, or W3C, and WAP Forum in 2000), and now it is finally where it should have been five years ago—about to happen.

Developing location-dependent applications is the first step toward applications that are situation dependent, where parameters from the users' surroundings and personal profiles affect the presentation of the data. This knowledge will change the way we perceive and use information as much as the World

Wide Web did, and this time we have the chance to get it right before it is shanghaied by the advertising industry.

Situational services are the user interface aspect of the Semantic Web revolution, where machines communicate to exchange, generate, and format information. Application servers and location gateways do work together without human involvement, and this industry will be the first where we will see everything that the visionaries talk about happening. But I have written another book about that.

Location-dependent services also have the potential to be a major revenue spinner for many people. In Chapter 1, I touch on how that might happen, but I will not go into the payment streams and the mechanisms for handling them. Argentum non olet, but predicting business models is much more difficult than describing billing mechanisms. I do touch a little on how this function works in Japan (the only country where a mobile operator is actually making money by selling position information). But these mechanisms will be so dependent on the agreements that are struck between operators and application providers that at this time, nothing can really be said about them.

How to Read This Book

I have tried to write an introductory book, not a book about the details of programming location-dependent applications (because they will depend very much on the platform you choose). I am also writing about an area that touches many things, where there are excellent books (for instance, user interface design, data modeling, and servlet programming).

In other words, whoever you are, you have to read the whole book to get an impression about what the area is. If you are technically oriented, I have tried to move the very technical information into appendices so that they will not make reading difficult for people who want to find out about concepts, not implementation details. Some of the appendices are also introductory; for instance, discussing XML, WAP 2, and CC/PP.

Who Is This Book For?

This book is for people who want to find out what location-dependent services are and how they can apply them in the business they have today. It is not a book for programmers looking for a recipe or learning how to program an LBS language, and it is not for people who look for lofty prophecies about the future. This book is for people who want to find out what it is, and what they need to learn and do to get started.

The Web Site and the CD-ROM

We enclosed a CD-ROM with this book that contains trial versions for several useful tools for developing location-dependent applications. Where available, it also contains specifications that I found to be useful. On the CD-ROM, there is also a "surf page" with links to sites that can help.

That page is replicated on my Web site, www.wireless-information.net, where I also give some more examples and provide more links and a collection of specifications and presentations (yes, I give away the PowerPoints) and where you can also download a couple of tutorials that I have created to show you how to develop location-dependent services. There are also a few articles and other stuff, mostly things that did not belong in the core of the book. I also create an irregular newsletter, which is free to subscribe to if you can stand my opinions and want to learn something about what I have experienced.

Disclaimer

The content of this book represents the position and opinions of the author. It does not represent the position or plans of Ericsson or any of his previous or future employers, nor any companies in the LBS business that he knows of, other than to the extent that they overlap his own.

Acknowledgments

This book is dedicated to a number of people who have been helping me get everything right, none named and none forgotten, but with special thanks to Lalitha Suryanarayana, Magnus Olsson, Helena Lindskog, and Matthias Nolle. This book is also dedicated to my colleagues at Ericsson: Kuraoka-san, Yu-san, Peng-san, Zhu-san, and Jonas-san, who kept helping me without knowing it and putting up with me at my crankiest when the deadline approached and passed. But mostly, it is dedicated to my mentor and friend, Jan Martinsson, who summed up an entire industry before it was born with the words "but this is just database massage!"

CHAPTER 1

Developing Location-Dependent Services

You're on your way home from the pub, and it has started snowing. Snowing so hard you can't read the street signs. No problem: Whip out your mobile phone, press the "Taxi" button, and wait for 10 minutes until the cab comes and picks you up.

Or earlier in the evening: You and your friends are on your way home from a tennis tournament, and you decide to drop in somewhere on the way for a beer. The only problem is none of you know any good places. No problem: Take out your mobile phone from your pocket, select "pubs" from the menu, and check out the closest ones. Get a map with directions to the one you choose.

Or even earlier: It looks like it might start to rain, or maybe not. Better check the weather forecast. Just press the "weather" button, and there you are—50 percent likelihood of rain in your area.

A couple of hours before: Your wife said she wanted a jar of saddle grease. Get in the car, turn out the driveway, turn on the phone, and ask the Yellow Pages: Where do you buy saddle grease? At a saddlers'. Where is the closest one? Across town, of course. Map, please! (What the heck is saddle grease? Don't ask.)

Or, remember your vacation: You were driving through this quaint little German town, and all of a sudden discovered you had crossed the Polish border. But, you were going to spend the night in Berlin, and that is 200 miles away! How do you find a map, and even more important, how do you ask how to get from there to where you thought you were going?

These five examples have one thing in common: They illustrate more convenient ways of getting information and accomplishing tasks than the old ways.

Imagine having to find a phone booth, get out coins, remember the number to the cab company, wait on hold for 20 minutes, and then speak to a dispatcher who won't send a cab unless you can say where you are. Or, imagine trying to find a nice pub in a part of town you have never been before, without asking some shady character who looks like he will either rob you or try to sell you drugs. Or, imagine getting the weather information for the spot you are on, anytime you want. Or, imagine finding a local store that sells saddle grease, without spending half an hour in front of telephone books and making endless calls.

All the information in the examples is readily available, as well, in different databases, except one—your position. Oh sure, you know where you are. But, the taxi dispatcher does not, and if you cannot tell him, you are on the corner of Hollywood and Vine, he is none the wiser. You need a way of getting your position to him, and the taxi company needs a way of connecting your request for a taxi—with the position—with the nearest cab.

Mobile information services, including some that are location-dependent, are already deployed in Europe and Japan. According to one source, the Jnavi mapfinder already has 600,000 users per day. Even if Japan is different, it can't be so different that a good business idea won't work here, can it?

Making the position available means communicating your position someway, and there are two basic ways of doing that (apart from typing it in, which is inconvenient): getting it from the mobile phone network, and getting it from another positioning service, such as GPS. Either way, it has to be sent to the service provider to be useful.

Another reason for the surge in interest is that for once, U.S. legislation is actually driving an industry. Cellular phone operators have to provide location information to the 911 call centers. Emergency 911 call centers receive approximately 50,000 calls per day, and 25 percent of the calls come from cellular phones. The *Federal Communications Commission* (FCC) mandate requires mobile network operators to provide positional information to emergency services. The first phase of the requirements—to be able to position the user in the right cell and find his mobile dialling number (the number where the user can be called back)—was supposed to be in place in 1998. The second phase, which was to be in place on October 1, 2001, meant locating the user within 410 feet 67 percent of the time. Only two carriers had fulfilled this requirement at the time, however. They all received a reprieve and are required to make progress reports to the FCC.

Another part of being convenient—and convenience is the killer app of the wireless Web—is not to clutter your pockets with more gadgets. However, leaving all the navigation functionality in your car is out ("honey, where is the nearest ladies room? Sorry, can't tell you, I left the locator in the car"). You can have a separate display in the car, but you will want to have the information with you. That means fitting the device in a pocket, and that means it will look like a cellphone or PDA.

So, What Are Location-Dependent Services Anyway?

The concept Location-Dependent Services usually refers to wireless services dependent on a certain location, for example, the location of the requesting *mobile station*, but this kind of services does not necessarily have to include a map. Two basic kinds of location-dependent services exist: triggered and user-requested. In a user-requested service, you are retrieving the position once and using it as input for the information that is returned to the requester. This function works in the same way as any other Web system in which you fetch an external piece of information to your server. The system and the result are somewhat different, but the principle is the same.

Triggered services work differently. Here, you set a trigger that retrieves the position of a given device in the intervals you request; for instance, the time or when the user passes through the boundaries of the cells in the mobile network. Triggered services are intended for situations when the application server calculates something, such as a position of a car in a fleet to be put on a map or on a driving log.

Triggered services can get a position as long as the trigger is valid (for instance, during a time period), but they also can be one-shots in the same way as requested services. Emergency services, where the call to the emergency center triggers an automatic location request if the call comes from a cell phone, is probably a good example of how this system will work. Other ways to characterize services is from the perspective of the user, irrespective of technology. This means that services can be personal location services using WAP, SMS, or PDA; third-party services, for example tracking fleet and buddy groups; service locating, for example, finding the nearest pub; triggered services; and others. Or, you can look at whether the service is executed in the mobile terminal or uses the mobile network. We can slice it in plenty of other ways, but triggered or user-requested services cover most of the applications on the market, using Web technology or not.

User-requested services often involve a map. Some kind of indicator usually shows where the user is. The map can be panned and zoomed, and other information elements exist in the map as well. However, they do not have to. Presenting information in other formats is often clearer, for instance when it comes to weather information (even TV meteorologists can find weather maps hard to read).

User-requested services are typically information services, either personal location or service location. We will look at user-requested services in this book. Other services, such as fleet tracking or tracing containers, may be more business-oriented, but they also involve proprietary interfaces—as do all location-based services today. Navigation services are a special case of fleet tracking services in one view; in another, they may be part of the wireless Web. But, we will not discuss them separately in this book; calculating a path to be navigated is not different from calculating any other map.

At least as important as the map is other information, and the fact that it has to be tied to a position, a process called *geocoding*. Having this information is necessary for the creation of applications, and it is a major piece of work to create it. One way is to enroll users in creating the information, but so far, the services released in Europe and Japan that use geocoded information do not involve user interaction in the information creation, except one, the search engine Somewherenear, which I will tell you more about in Chapter 6, "Providing Databases and Doing Searches."

Direction information may be as important as maps and is preferred by some people (if you are a professional courier, it is easier to follow instructions than to read a map, because you can stay in a linear frame of mind and do not have to shift to the overview mode of the map reader, for instance). Having two positions—yours and your goal—the system can also calculate the route between them, either just a set of driving directions, or a route influenced by traffic conditions, weather, and so on.

In an information service, information is filtered with respect to a specific location, and the location of the user. This means providing information to the user about his surroundings—or goal—in the form of a map, and it may also mean filtering information based on other preferences, such as removing all stores that are not shoe stores from the map. If this selection is done in advance, and the phone allows it, information can be pushed to the user when the push service is triggered by the positioning. An example where this could be used is a traffic accident. This event could trigger a service, which sends a map to the user (or to all users on roads that will be affected) with the location of the accident marked. For a store, showing the location of the store in a map and simultaneously presenting current sales could be a possibility.

Different location-dependent services have different requirements for accuracy and timing. In tourism and store location services, the experiences from Europe show that an accuracy of up to 60 meters and a retrieval time from request to response of about three seconds does not feel inconvenient to the user. This puts rather strict requirements on the system, not only the positioning technology used (not all can fulfill it), but also on the way the service is designed. If you have to retrieve an advertisement that you want to include on the page, and it takes three seconds to fetch it, you have lost the users interest already. Mobile users are, if anything, impatient and do not want to wait for their mobile phones. Services like these are supposed to be possible to use on a busy streetcorner, which makes them very different from services you are supposed to use while sitting in your den.

A special problem is obtaining good cartography and a clear graphical layout of a map on a display of the size somewhere between a matchbox and a playing card. A key issue is to cut down the amount of visible information as much as possible and to show only the information of definite interest. The user is in most cases interested in just one or a few issues of geographic nature, and more information than necessary is usually seen as confusing.

How Do Location-based Services Fit with the Web?

Developing services that are dependent on position (or location-dependent services—the two expressions are interchangeable) is not very different from developing other Web services. Most of those services that exist today are based on proprietary technologies, however, and use text messages for presentation. This is bound to change, as new standards enable Web proxies and servers to get location information directly from the user or from the mobile network.

Most of the information needed for the services is already available on the Web, or in databases owned by someone. In this book, I will talk about how to take that information—or create new information—that you need to build these services. The business of geographic information is already well established and has developed both a lot of data and a very special terminology and way of looking at things over the years. Sometimes, that may be a problem if you have not come across it before. If you find the terminology intimidating, there is a glossary at the end of the book that explains the most common words and abbreviations in this specialized business.

That means that while retrieved services work more like any Web service, triggered services are more similar to traditional applications. The protocol might be *Hypertext Transport Protocol* (HTTP), but it requires that the positioning provider—the Mobile Positioning Center—provide the information at the trigger intervals, and there are currently three ways to perform this task: using a special CORBA method, using a proprietary interface, or using HTTP POST.

The position information will most likely come from a network, for reasons you will see in Chapter 3, "Position Technologies and Coordinate Systems." We can also use the *Global Positioning System* (GPS). GPS receivers will be part of the user terminal, but while it does not cost anything once you have bought the terminal, it works badly indoors and in cities (if at all). It is also controlled by the American military, which makes it contentious in some parts of the world.

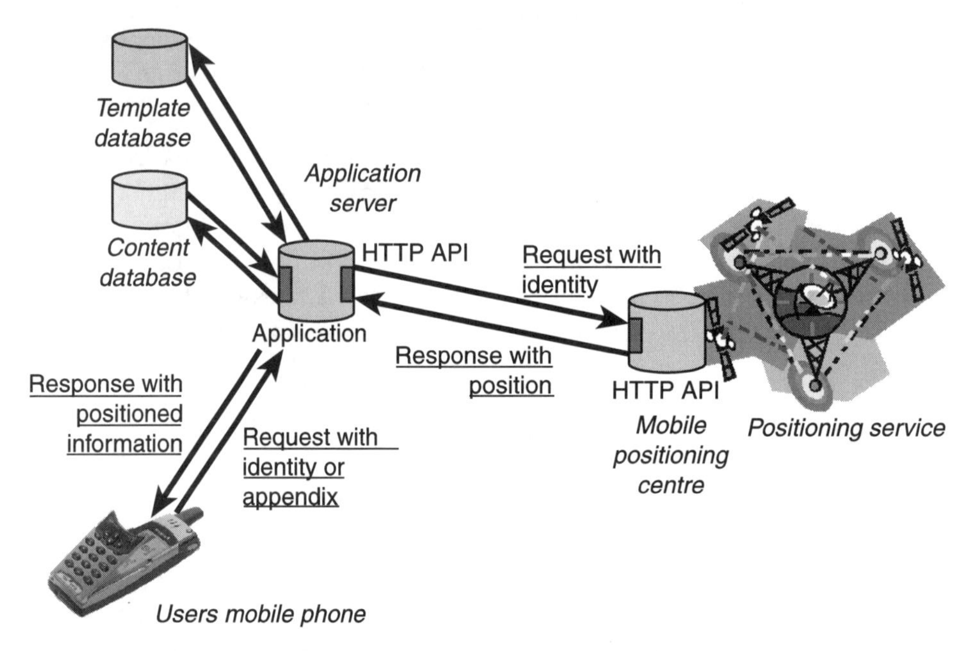

Figure 1.1 The location-dependent information system architecture.

A position-dependent service looks, in principle, something like Figure 1.1. The user accesses the application server, which coordinates the handling of requests, the retrieval of information, and so on. The position is retrieved from the positioning system and the data and presentation format comes from the database; then, they are combined according to the rules in the application and the presentation is returned to the user.

As you can see from the figure, the application server is the core of the position-dependent service. There are other ways of developing position-dependent services as well—using an *intelligent network* (IN) or letting the mobile phone do the processing of the application—and I will be talking about those in the book, as well. But the main thrust is toward application servers for a very simple reason: It is the least-different way of programming for a Web programmer.

In principle, you could build your own application server. All it takes is a Web server with three specialized programs: one to retrieve the position request from the user, one to connect to the mobile positioning center, and one to take the data and layout and apply your rules to them. They can be servlets that need to interact with the server, or they can be XSLT programs (more about that later).

While the application server is central, it is not the only part of the system, however. You must have data; the data must have a user interface; and most important of all, you must get the position information. The operators of mobile networks have realized that this is a considerable value, and are likely to charge for access to the position information. This scenario is more like credit card processing than other services, because it is likely (at least, initially) that you need to set up a relationship with the operator, which requires more than a telephone subscription. NTT DoCoMo, for instance, the Japanese operator that runs the most successful mobile Internet service in the world, requires you to install a special piece of software that talks to its service center if you want to provide information to its iArea service. You are not allowed to develop your own.

The operator is needed because the easiest way of getting a position of a user is from his or her mobile phone. Most GPS receivers, with the exception of those built into mobile phones like the Benefon Esc, do not have the capacity to get information from the network and so are forced to rely on what is programmed into them. This information can be a lot, but it is not possible to update it—and we will not discuss it in this book.

Another reason why I talk about the mobile phone as the user terminal in this book is because it is the only portable device that is in widespread use. Fewer than 10 million *personal digital assistants* (PDAs) have been sold, but the number of iMode users in Japan is topping 30 million if not more at the moment you read this book. They all use a simplified Web client that can display images and text (and run limited Java programs and play MIDI files).

In the rest of the world, *Wireless Access Protocol* (WAP) has been slow in gaining widespread use, but with the new WAP version (WAP 2.0), it is fully aligned with the standards of the World Wide Web. WAP 2.0 uses *Extensible Markup Language* (XML) as its data format, and in Appendix C of this book you can read about how XML works if you are not familiar with it. Appendix D is an introduction to WAP 2. It is important that you understand how XML works, because it is both the data format and the language in which the *application programming interfaces* (APIs) are defined. The Mobile Positioning Center (MPC) also turns out to be easiest to access using HTTP, but there are also options to get the users position by using CORBA and specialized protocols, which I mention in Chapter 4, "APIs and Protocols", as well.

The Location Server is called the *Gateway Mobile Location Center* (GMLC) in the GSM and IMT-2000 standards and *Mobile Positioning Center* (MPC), in ANSI standards. The location server is a logical entity, however, so there are other possible implementations. In this book, I will mostly talk about the MPC when I mean MPC or GMLC.

Having retrieved the position information, you will need to merge your information with it. That information is most likely stored in a database; XML documents can actually be seen as database containers. Chapter 6, "Providing Databases and Doing Searches," discusses data modeling and how you should handle your information. XML is, by the way, also a universal interface language, so I discuss how you should apply it here. The XML transformation language, XSLT, which allows you to transform data in the XML format to almost anything else, is briefly discussed (there are other books about that; for instance, *XSLT: The Ultimate Guide to Transforming Web Data* by Johan Hjelm and Peter Stark, John Wiley & Sons, 2001, if you'll pardon the plug).

Getting the position correct is only half of a location-dependent service. Without knowing what information is available about objects at that position, you will not be able to create a service. Just knowing where you are is worthless if you cannot relate it to anything. You also have to make sure that the information is in a format that you can use in your application, which means XML.

XSLT is an excellent candidate for taking the data from the database and applying your position-dependent heuristics (a fancy word for rule of thumb). But how should you apply those rules? It partly depends on what data you have and partly on how you intend to present that data. Chapter 8, "The User Interface to Location Information Services," talks more about presentation, although I do not discuss the creation of templates—there are many other excellent books about that. User interfaces on mobile phones do place some special requirements on your design, however.

By the way, when I talk about mobile phones in this book, I also mean PDAs—like the Palm Pilot or Compaq ipaq. Here in Japan, where I live, you can buy a telephone card for your PDA that fits into the memory card slot. And because the third-generation mobile telephone system works like a packet network when you are not making voice calls, you are connected to the Internet just in the same way as you would be if you were using the Ethernet connection in your office. Anything larger than a PDA, however, and you will not be able to carry it in your pocket (and some PDAs are doubtful in that regard, at that).

One special aspect of position information is that it is very close to personal. If you are using a telephone system that does not allow for anonymous connections (and there is no way of doing that if you want to retrieve the position information, as you can see in Chapter 3, "Position Technologies and Coordinate Systmes"), there is a very sensitive privacy issue. Especially for applications that track users, the service provider will (in principle) have a record of where you have been—which is very easy to combine with where you are. This scenario is a significant intrusion on your privacy, of course, and if misused, it can cause severe damage. So there are a number of things that the provider of the information needs to think about, especially how to establish and maintain the users' trust. One way is using the Protocol for Privacy Preferences; another is setting the privacy flag in the *Home Location Register* (HLR). I will talk more about the privacy aspects in Chapter 12, "Privacy and Location."

Location-Dependent Services in Japan

There are very few location-dependent services worth the name in the United States. Many services have tried to catch the buzzwave by sticking "location-dependent" in their name, but they are still databases for download. It is in places where people are using their mobile phones for something real that location-based services are happening, and it is in Japan that they really shine.

In Japan, the Vodaphone-controlled mobile operator Jphone has always been the most inventive. Delivering email on mobile phones long before the dominant player, NTT DoCoMo, got their services running, they were also early with the wireless Web, and in 2000, they became the first to launch a location-dependent wireless Web service.

Seventeen million businesses are stored in the database with their coordinates (geocoded). Despite claiming upward of 50,000 users per hour, and the fact that the maps are generated dynamically when the user requests the data, the service does not take more than two seconds to download a map of the location you are searching. Servlets in the phone handle presentation, including panning and scrolling.

However, Jphone has a range of similar services. The latest is a system that enables you to get a map of your own position, if you want to mail it to friends who have problems finding you (not unusual in the crooked and unnamed streets of any Japanese city). See Figure 1.2.

Jphone charges the subscribers 300 Japanese yen per month for the service, about 25–30 U.S. cents. Content providers can participate free of charge.

Gigant NTT DoCoMo, the operator of the wildly successful iMode service and the company credited with making the wireless Web popular (although so far only in Japan), is not staying far behind, however. In July 2001, they launched the iArea service, which is an add-on service to iMode, the wireless Web portal they established in 1999.

The user selects "iArea" from the iMode menu (the iArea selection appeared one day on all iMode handsets, because it uses the cell ID and does not imply any modifications of either the network or the mobile telephone, which has a limited Web browser in Japan, and a small Java environment). There were four functions at the start: weather forecasts, including a rain-alert; local guides

Figure 1.2 An interactive map of your location.

with shops, restaurants, hotels, and so on, including dining information arranged by specific location or type of occasion, which change as you move; maps, both detailed maps searchable by address or place name and overview maps, including a function to send maps to others via e-mail; traffic updates with 24-hour updates on traffic congestion, accidents, and estimated travel times for the national highways and expressways.

The service is free in itself, and some of the services (like weather) are free. DoCoMo charges for the traffic, of course, and also takes a transaction fee for those services who charge the users (such as the map services). In addition, they cannot get the position of the user directly from a mobile positioning center owned by DoCoMo but must integrate a piece of software in their application, which works toward the application service provider (ASP) that DoCoMo owns, who gives out the information.

It is much more convenient to get your weather information if the phone knows where you are, because the number of clicks you have to do is far less: three, compared to eight in the nonpositioned version of the service. For the user, eight clicks in a busy situation, which is when you use your mobile phone, is worth a lot. That user satisfaction is higher when getting data instantly is not just conjecture. Reliable data from different surveys in Japan as well as psychological studies on users beginning in the 1970s (the basic research behind the surveys goes back much longer) that show that user irritation increases—and hence, satisfaction decreases—if a user has to wait for more than three seconds for a request to be served.

These services are not the only ones, however. NTT DoCoMo also runs a service on the older Personal Handyphone System (PHS), called ImaDoko (which means something like "where now?"). This is a very simplified device, not really a mobile phone because it has only one button, and is suitable for anyone who cannot or does not want to use a mobile phone—children, for instance, or people with Alzheimer's disease.

When you dial a special number and give a PIN code, the system locates the device and faxes a map to you, telling you where it is. This works well in Japan, because the cells in the PHS system are only about 200 by 200 meters. Similar devices for GSM exist in other parts of the world, but we will not discuss them deeper in this book—without a user interface, it is hard to be part of the wireless Web, after all.

Business Models for Location-Dependent Services

The only thing that can be said about the business models for location-dependent services at this stage of development is this: Anything that anyone can tell you is speculation, except that the mobile operators will make money.

One problem is that once you know where someone is, it is easy to get greedy. Position-dependent advertising has been touted as a future source of income for the operators, but they also talk about charging the users for access to the positioning information. As tempting as it might seem to grab every penny you can, especially if you paid billions of dollars for a third-generation license you are not using, it is not something that builds trust, however. Because the mobile telephone is also an intensely personal device, the user will tend to regard advertisements as even more an intrusion than on the fixed Web (where there is solid research to show that users avoid looking at advertisements). The solution might be to create more personal advertisements, because you have to personalize the information that you send to the user depending on his or her position anyway.

In other words, advertising can, at best, be counted on as a supplementary source of income. If you view the creation of your position-dependent service as a marketing expense, helping users to find your nearest stores or offices, you will see that the economics become very different, especially because the effects are immediately measurable in your Web logs. You can also use additional marketing mechanisms, such as coupons, to enhance the effect (and to get an additional measurement parameter). You could, of course, also consider charging the companies for the number of times information about them has been searched (essentially charging for clickthroughs, but this model has not been very profitable on the Internet).

Advertising is basically about charging a third party for the information, and there is another third party involved in providing location-based services: the network operator. He or she stands to make a great deal of money, first from selling the service of providing location-based information and second from the traffic generated by the user when accessing the service (which, of course, he or she charges for). Many operators also look at the model of NTT DoCoMo, where the subscription fee for services is charged over the telephone bill but DoCoMo takes 9 percent of the fee for providing that service.

If you can provide sufficiently interesting information and have a strong enough brand, you might consider turning the favor around and asking the operator to pay you for generating traffic for him or her. While that is an alien thought to many operators (who believe they have unbeatable brands and own the customers), with the emergence of "virtual operators" (who do not own networks) and the current economic downturn squeezing the traffic volumes, operators might consider different models for increasing their traffic as well as increasing their income. This situation is especially true in Europe, where they overpaid horrendously for getting licenses to operate third-generation services. An interesting combination, actually, is using a dual-mode handset where the data traffic goes over third-generation systems but the positioning is done by using GSM (which gives higher precision).

Many operators also cannot let go of the idea of a "walled garden" where they provide a set of services to users that are only accessible over their network. Successfully performing this task requires a near-monopoly situation like NTT DoCoMo employs in Japan and a market where the impediments for users to change service providers is high. In most mobile markets, the churn rate (how many users leave the service every year) is more than 25 percent, which means that the differentiation is hardly substantial. Differentiating themselves, operators also risk shutting out large groups of users, and that is not something you will want to risk in the current economic climate.

The most likely way of getting paid for position-dependent services, however, is charging the users. This procedure can be done in a number of ways: charging per positioning request (essentially adding a charge on top of the operator's charge for the service); charging a subscription fee (where you have to be careful to calculate the average number of position requests so you do not end up charging your users less than you are paying the operator); or a combination of these. It is, of course, also possible to strike a revenue-sharing deal with the operator to give him or her a percentage of your revenue, not paying a fixed percentage because you are driving traffic to a new service he or she did not have before.

Possibly, the best business model is to charge the customers of those who benefit from selling the services that are located that way. Taxi companies could have priority customers who get sent a cab within 10 minutes, if they used their mobile phone to order and paid a $50 subscription fee every month.

When you make the business plan for your location-dependent service, you might come across a term that you have not heard before if you are not in the telecommunications industry: ARPU, or Average Revenue Per User. Other businesses that sell subscriptions do not calculate that way, but because the margin of the network operator is higher the more users who use more of his or her services, the ARPU has become a magic number in the telecommunications industry. It is a trap, of course, because it is liable to both of the typical errors in calculating a mean: Very few users can drive up the average by using the service a lot, and if you have very few users, the calculation is meaningless. Instead, looking at the ARPU, the median income per user, and the number of users will allow you to make a sound calculation.

Charging users for accessing your service means that you have to design a service that entices users to come back. So you not only have to build a user interface that is intuitive and simple, but you also have to set your prices in such a way that you do not scare them away. Guessing how this situation will work is an art rather than a science, and I will discuss some of the background reasoning in Chapter 8, "The User Interface to Location Information Services."

Before you develop a location-dependent service, however, you need to understand how the data should be encoded and how you can get the position information. This is what Chapters 3, 4, and 5 are about. How it all ties together is something I discuss in Chapter 9, "Maps as User Interfaces."

And with that, let us look at some of the prerequisites for positioning: coordinates and positioning services.

CHAPTER 2

Positioning Technologies

When you can measure time precisely, you can measure your position if you have a point of reference. In the mobile telephone network, the position of the base station antennas is fixed and can be used as reference points. On a planet that rotates, that procedure is not easy—so the points of reference for GPS are not on the Earth. Wherever your reference point, using radio waves at longer distances than a few meters means introducing inaccuracies, because the atmosphere disperses the radio waves, and obstacles might distort them. For this reason, any method of reporting position has to include a method of reporting inaccuracy.

Quality of Position: Describing Accuracy

In any positioning technology, there is a possible inaccuracy and a probability for error (actually, the likelihood for an error of a maximum size). How close you can measure the position determines the granularity of the technology. Actually, it is a probability value too, but if you know that the technology you are using only very rarely puts you further away from 50 m of your true position, it makes sense to talk about the technology as having an accuracy of 50 m. The granularity, the closest you can measure possible positions, will be 50 m.

The probability is not necessarily fixed, however. It can vary with certain factors, like the number of satellites, the frequency used for the radio signals, and so on. Increasing the probability means decreasing the error rate and so increasing granularity. Of course, probability is a bell curve, which means that there is always a likelihood for the accuracy to be zero, even if it is small to zero. It can be and is ignored for most practical purposes.

There is also a third dimension to relevance: age. How long ago it was that the information was collected will mean that the position might have changed. If you are traveling at a speed of 100 km per hour, and your position is one minute old, that will mean that the inaccuracy is a mile (or a bit more than 1.5 km). The faster you move and the more imprecise the measuring technology you use, the greater this inaccuracy will be. You might even have stopped somewhere for a break on the way.

For this reason, the industry talks about *quality of position* (QoP). It can mean any of the following:

- Age of the location information; in other words, when was the information actually collected?
- Accuracy of the location information.
- Confidence in the accuracy information; for example, "with 65 percent probability."

The more precise the measurement system, the lower the likelihood of inaccuracy, and the more often measurements are made, the better the quality. Of course, different applications require different qualities of position. Measuring the position of a bus on a country road might not be necessary more than every 15 minutes to know that it is getting from Edmonton to Banff, but knowing where a fighter plane is every two seconds will greatly improve its aim, or even better the ability to abort its mission. How often to measure the position, how to calculate when to do it, and how to use that measurement is up to the application designer.

Knowing the quality of position is not only a fact in tracking applications, however. You will also have to know it if you are designing applications that retrieve information, because it will affect how information should be presented. You might want to present one map if the user can be certain of where he is within a meter, but present another if he cannot know it within a kilometer.

Positioning Technologies

The most frequent way of measuring position, though, is asking the user. This method is used in almost all the applications that are available on the World Wide Web today, and because there is no way to either measure or report the position of the user on the Internet, there is no standardized way to include that information without asking.

GPS

There are two basic ways of positioning a device on the market today: using GPS (or its Russian cousin, GLONASS) or using the mobile telephone network. There are combinations of these, which increase the accuracy and decrease the cost for the user. But first, let's try to understand how GPS works.

Introduction to GPS

GPS is based on 24 satellites and 5 monitoring stations around the world that enable the satellites to broadcast a signal that can be used as a reference in determining the position of a user.

The satellites, NAVSTAR, were all in place in 1994. A further 21 satellites are planned. The system is run by the U.S. military, which has caused some concern in some countries. The European Union is planning a separate system, and there is a Russian system, GLONASS (Global Navigation Satellite System), which works in a similar way. Both transmit spread-spectrum signals at two frequencies in the L-band (1.2 GHz and 1.6 GHz) and have pledged to make a partial set of signals available for civil use without any user fees for the next 10 years or more. The full GPS signal was made available to all users in 2000, but there is still a possibility that in a situation it judges to be an emergency, the U.S. military could shut off the general availability.

GPS is not only used for navigation; it is also increasingly used for scientific studies because it gives an unambiguous reference (in time and position) for sensors. Because it takes the signal from four satellites to get a position fix, however, the system works poorly indoors, in cities, and in areas where the satellites (which are not in geostationary orbit but rather orbit every 12 minutes) do not have good coverage. The satellites are positioned to cover the entire area of the Earth, but coverage is generally better the closer to the equator you come; in northern Arctic (and southern Antarctic) regions it can be problematic to get the signal from (or acquire) four satellites, although the system is dimensioned to always have five satellites in view of each point of the Earth at all times.

The ground stations of the U.S. military have a crucial role in determining the precision of the system, but there are many other people who both track the satellites and who provide data about and from them. One such organization, just to give an example, is the International GPS Service, which provides orbit data (in the RINEX meteorological file format) through a number of different data centers around the world. The Sub bureau uses its data for Reference Frames of the *International Earth Rotation Service* (International Terrestrial Reference Frames, ITRF) to provide accurate data about the Earth.

GPS works by the user's receiver (which can now be made so small that they can be built into watches and cell phones) acquiring a sufficient number of satellites (a minimum of three) and triangulating them by using the travel time of the radio signal from the satellite (more properly, trilateration or resection because the system is not calculating angles). Because the position of the satellite in space is known (because the orbit and the time is known) and the distance is known (because the time it takes for the signal to get from the satellite is known), you can calculate the distance arc on the Earth's surface from the satellite (although you have to correct for delays and errors that might occur when the signal passes through the atmosphere). It will be an arc (think of the satellite as the point where you put one leg of a compass and the arc as the circle you draw with a pencil). Although it is not an arc, it is a sphere (well, really an ellipsoid) because we are making three-dimensional measurements. Taking the arcs from more than three satellites enables you to draw their intersection on the surface of the Earth where the arcs meet. You have to use more than two satellites because you are not just drawing an X on a flat surface. Not only are you drawing on an ellipsoid mapped out on the surface of the Earth, the distance from the satellite is actually an edge of a sphere, and intersecting two spheres gives you an ellipsoid; applying the third satellite gives you two points where the ellipsoids intersect, and the intersection with the fourth sphere will enable you to narrow it down to one point (although usually, one of the two points is not a reasonable answer and can be discarded). In principle, it would be sufficient with three satellites, but four are needed for the clock in the GPS receiver to set itself to the precise time of GPS. See Figure 2.1.

This method not only enables you to measure position with great accuracy, but it also enables you to measure time, velocity, and altitude.

The reason you can measure the distance to a satellite in orbit is because the signal from the satellite travels with a known velocity—the speed of light (as all radio

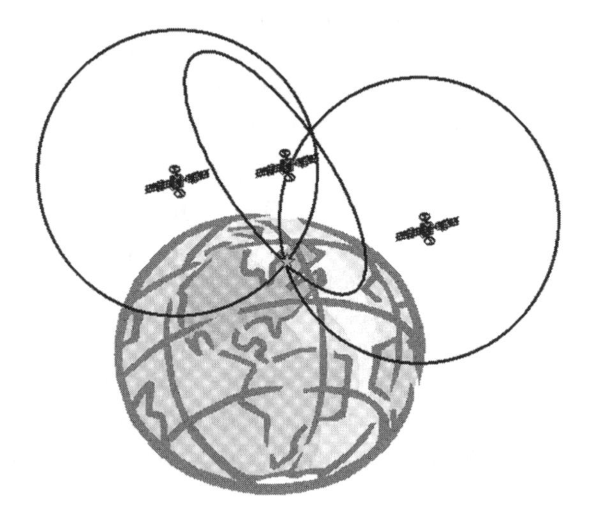

Figure 2.1 How GPS works.

waves do). Measuring the travel time from the satellite enables you to calculate the distance. The orbits are of course not mathematically exact, if for no other reason than because the Earth's magnetic field, the solar wind, and the gravitational pull of the sun and moon would cause some drift over time (although by being in orbit, they do provide a reference point where the gravitation of the Earth is constant, as required by WGS-84 to measure the ellipsoids on the Earth's surface, which we will discuss in Chapter 3). For this reason, the orbits are monitored by the U.S. Department of Defense monitoring stations, using high-precision radar to check the speed, altitude, and position of each satellite. The errors are called "ephemeris errors" because the orbit is determined by the calculation of the trajectory of a body over the sky or its ephemeris (an old astronomical term, actually).

Once the ground stations have measured the ephemeris error, they relay the information back to the satellite, which includes it in the signal it is broadcasting (in effect, correcting the signal by the error so that the precision is maintained). This error is not the only error that can occur, however. The signal passes through the atmosphere for a piece of the distance to the receiver, and the atmosphere has different optical and electromagnetic properties than a vacuum. So, the signal will be affected by the passage through the atmosphere. The angle from the satellite to the receiver will determine the distance the signal passes through the atmosphere. But once we have this information, and if we know the typical properties of the atmosphere, it is possible to compensate for the error in the receiver. Using the two different frequencies on which the signal is broadcast, you can also determine the characteristics of the atmosphere (at the point of measurement) and calculate the possible error.

Unfortunately, there is yet another problem that has to be overcome, and that is bouncing signals off reflecting objects (for instance, buildings). This situation can sometimes be seen in TV signals, where a building is close to the transmitter and catches part of the signal. Because the GPS signal is so weak, it will be harder for the receiver to determine which signal is the correct one based on the strength. But receivers have become very sophisticated at performing this task, and by using four different satellites it is possible to calculate a correction to the signal because of the way it is encoded. And, using the two different frequencies of the signal, this type of error can be minimized because the echo will have different characteristics depending on the frequency (which has to do with the properties of radio waves and reflecting surfaces, but let's not go into that now). Obstructions are a very real problem for GPS, however, and as anyone who has tried to acquire a weak satellite signal knows, even objects like trees can dampen out the reception totally. This reason is also why GPS does not work indoors.

The real problem, then, is to measure the time it takes the signal to travel from the satellite. And because the time is very short (because signal travels at the speed of light, the satellites are rather close to Earth in cosmic terms, and the time during which you measure the distance is very short as well—if a satellite is straight overhead, we are talking something like a six-hundredth of a second), not only will you need a very precise clock, but you will also need to know the time from four different signals.

The clock in the satellite is an atomic clock (using the vibrations of an atom to derive the time signals), and it is extremely precise. The clock in a GPS receiver is not an atomic clock, however, and it uses the signals from four different satellites to synchronize itself to the universal time.

If it were possible to have a perfect clock in the receiver, all the satellite range measurements would intersect at your position with extremely high precision. But because the clock is imperfect, three measurements will synchronize but the fourth will be off (done as a cross-check). The receiver's computer registers this information and calculates one single correction factor that it could subtract from all timing measurements to allow them to intersect at one single point. Applying this correction factor synchronizes the clock in the receiver with the universal time. It can then apply this timing to its measurements, and the positioning will become precise. One consequence of this principle is that the GPS receiver will need to acquire at least four satellites so that it can make the four measurements simultaneously.

If we were only measuring time, it might be sufficient with a beep like that of the Sputnik, the first artificial satellite. But we have to measure distance, as well. The way this procedure is done in GPS is by transmitting a code, or rather a string of codes, at very precise intervals. If the transmission starts at a predetermined time, you can calculate the time it takes from the satellite to you by calculating where in the code you would be if you had been transmitting from the surface of the Earth and then comparing this information to the signal you receive from the satellite. The signal from the satellite would be slightly delayed, because it had been passing the distance from the satellite to Earth. If you wanted to measure the delay, we could make the receiver delay its virtual broadcast until it was coordinated with the broadcast from the satellite. We would then know the exact time it took to travel from the satellite to the surface of the Earth, and from this information (and the errors we have also calculated) we could calculate the distance.

As we said earlier, the GPS satellites transmit signals on two carrier frequencies. The L1 carrier is 1575.42 MHz and carries both the status message and a pseudo-random code for timing. The L2 carrier is 1227.60 MHz and is used for the more precise military pseudo-random code.

The code, which the satellites transmit, is called the *Pseudo Random Code* (PRC). This code is a stream of bits that carries the timing signal in a way that allows the time to be calculated from it and includes the satellite address. The first pseudo-random code is called the *Coarse Acquisition* (C/A) code. It modulates the L1 carrier at a 1MHz rate. It repeats every 1,023 bits, and each satellite has a unique pseudo-random code. The C/A code is the basis for civilian GPS use. The second pseudo-random code is called the P (Precise) code. It repeats on a seven-day cycle and modulates both the L1 and L2 carriers at a 10MHz rate. This code is intended for military users and can be encrypted. When it is encrypted, it is called Y code. Because P code is more complicated than C/A, it is more difficult for receivers to acquire, which is why many military receivers start by acquiring the C/A code first and then move on to P code. The encryption was turned off in 2000 by order of President Clinton, which has meant that the GPS system can be used to measure positions with much higher accuracy than before that date.

In addition, there is a low-frequency signal added to the L1 codes that gives information about the satellite's orbits, their clock corrections, and other system status (where the ephemeris error is transmitted, for instance).

Because each satellite has its own unique PRC, this feature also guarantees that the receiver will not accidentally pick up another satellite's signal. So, all the satellites can use the same frequency without jamming each other (as a matter of fact, the basis for many of the CDMA patents is satellite communications techniques). Because the GPS signal is encoded differently based on which satellite it is coming from, knowing which satellite you are looking at means you do not actively have to decode the signal.

Using different codes also makes it more difficult for a hostile force to jam the system. In fact, the PRC gives the U.S. Department of Defense a way to control access to the system, which is the reason why the Russians maintain their own system and the European Union is planning a system of its own.

The GPS signal is relatively immune to interference although it is rather weak. If a GPS receiver were to start "playing" its signal, it would match the random noise from electromagnetic radiation in the atmosphere about 50 percent of the time. But if a signal is added to the noise, the pattern is enhanced and the signal will stand out from the noise and can be matched to the virtual broadcast in the receiver (basic information theory, actually).

Because the GPS signal has a cycle width of almost one microsecond, it is possible for an error to creep in here, however. A microsecond at the speed of light is almost 300 meters (which means that the point you measure is 300 meters wide). What happens is that the synchronization between the satellite signal and the virtual signal in the receiver is coordinated, but the precision is not very high (because the P signal is broadcast at a higher frequency, the bits will be shorter and the precision higher, apart from the code being more precise). The code-phase GPS, which only uses the signal to measure position, can thus give an error that can be quite a problem, although designers of receivers have come up with various tricks to improve precision most of the time to something like 5 meters.

The frequency of the carrier wave is higher than the frequency of the signal that can be used to correct some of the error, however. The 1.57 GHz GPS signal travels at the speed of light and has a wavelength of roughly 20 centimeters, so the carrier signal can act as a much more accurate reference than the PRC by itself. And, if we can get to within 1 percent of perfect phase like we do with code-phase receivers, we would have an accuracy of 3 or 4 millimeters. Needless to say, this value is theoretical. But the carrier phase can be used to increase precision considerably.

Because this extra measurement trick can be used to get us a time measurement perfectly synced to universal time, we have everything we need to measure our distance to a satellite in space. Just knowing the distance will place you on a plane, however, not a sphere, and you would get a set of infinite planes at the measured distance from each other. You also need to know where the satellite is in the sky to be able to calculate where you are.

Receivers maintain an almanac or long-term model of the satellite trajectories as well as the ephemeris data (which is short-term) for all satellites, and they update these almanacs and ephemeris tables as new data comes in (through the satellites, as mentioned earlier). Mobile phones with GPS receivers can get the almanac from the mobile network, which means they can look for the satellites when they are known to be overhead, saving the battery—which is critical for mobile phones. Typically, ephemeris data is updated hourly.

The satellites' positions (or rather, their orbits and the ephemeris error) are stored in the receiver, and which satellite you are receiving can be determined by using the codes. Because you also know the time, you can use this information to determine its position.

Basic geometry itself can magnify these other errors with a principle called *Geometric Dilution of Precision* (GDOP). Because there are usually more satellites available than a receiver needs to fix a position, the receiver picks a few and ignores the rest. If it picks satellites that are close together in the sky, the intersecting spheres that define a position will cross at very shallow angles. That increases the error margin around a position. If it picks satellites that are widely separated, the circles intersect at almost right angles and that minimizes the error region. Good receivers determine which satellites will give the lowest GDOP and use them (it can calculate the satellite angles because it has the almanac data).

Because the receiver does not know where it is when it starts measuring, it has to do a search for the satellite signals. If the receiver has an almanac, it can minimize the search pattern and do a warm start (searching in the orbits where it is more likely the satellites are). From cold start with the almanac, it typically takes two minutes to determine its position. A warm start takes about one minute, a hot start (with the last fix less than one minute old) about 15 seconds. Without almanac, however, the time it takes to acquire the satellites is approximately 12 minutes. Even 15 seconds can seem long to a mobile phone user, and two minutes will be intolerable. The cost of including the almanac can also increase the cost of the receiver. These problems can be overcome by including the almanac and the ephemeris data with the satellite positions in an assistance signal.

Differential GPS

Originally, GPS was provided with 1.5 meters accuracy for military use and 50 meters for civilians, thanks to an encoding of the signal (selective availability), and only the U.S. military and its allies had the key to the encoding. This situation meant that there was a market for a service to increase the accuracy by using a reference signal from a known position. Several such systems were in fact commercialized, for instance, for use in airports (where an error of 50 meters in approaching the runway would be disastrous, to say the least). Differential GPS is also used for geodetics; for example, in determining reference point locations on a map.

Different sources have been used for the reference signal, such as radio broadcast towers (broadcast over the FM band) and mobile networks as well. The reference signal needs to contain a time reference of the same format as the GPS signal, however, and that means that mobile systems in themselves are unsuitable as reference signal providers (because there is no synchronization of the time signals in most mobile networks).

Differential GPS involves two receivers. One is stationary in a predetermined, known location; one is mobile (like handheld or in-car receivers). Both receivers are usually on the ground (because the system works better if they are on the same altitude). The stationary receiver is the key. It ties all the satellite measurements into a solid, local reference. The stationary receiver monitors variations and calculates errors in the GPS signal and communicates the correction values to the other receiver. Because it has no way of knowing which satellites the mobile receiver will pick up, it calculates corrections for all the available satellites. Error transmissions not only include the timing error for each satellite, but they also include the rate of change of that error as well. That way the roving receiver can interpolate its position between updates. The communication is done through some other radio network than the GPS network; for instance, the FM network or the mobile phone network (see the section on Assisted GPS).

The second receiver can then correct its calculations for better accuracy, because the distance between the receivers can usually be discounted.

In the early days of GPS, reference stations were often established by private companies that had big projects demanding high accuracy or as a service by companies such as radio broadcasters. Public agencies, especially outside the United States in countries where the local military does not have access to the decryption codes for GPS, have established reference transmitters and made them available to the public. These stations often transmit on the radio beacons that are already in place for radio direction finding (usually in the 300 kHz range). Most ships already have equipment to receive this signal, and many new GPS receivers are being designed to accept corrections (and some have built-in radio receivers for the differential signal).

Sometimes, however, it is possible to correct the received GPS signal afterward. This is true if you do not need the precise positioning immediately. If you are not navigating a ship but want to position a feature of geography (for instance, a new building) on a map, you just need to record the measured position and the exact time of the measurement. Later, in the office, this data can be merged with corrections recorded at a reference receiver and corrected. Some academic institutions are also experimenting with the Internet as a way of distributing corrections.

The same principle can be used when positioning mobile objects (and is then called inverted DGPS). The moving receivers are standard GPS receivers with a transmitter that transmits their received GPS positions back to the tracking office. The tracking office can apply the corrections to the received positions.

Augmented GPS

The aviation industry has been developing a type of differential GPS called the Wide Area Augmentation System (WAAS). This system is really a differential GPS system covering a continent (in this case, the north American continent).

The system also fulfills some other requirements that the regular GPS system has problems fulfilling. The first is reliability. Although the system is very reliable as computer systems go, every once in a while a GPS satellite malfunctions and gives inaccurate data.

Normally, the GPS monitoring stations detect this information and transmit a system status message that tells receivers to disregard the broken satellite until further notice. This process can take several minutes, however, which would be disastrous if the position were being used by a plane coming in for landing.

The American Federal Aviation Authority has set up its own monitoring system that responds faster by using a geosynchronous satellite positioned over the United States that can alert an aircraft instantly when there is a problem. The transmission is done on a GPS channel, so aircraft can receive it on their GPS receivers and do not need any additional radios.

The geosynchronous satellite can also be used as a positioning reference, because it is stationary over a point on the Earth's surface, and it transmits the correction signal. The *Federal Aviation Administration* (FAA) figured that with about 24 reference receivers scattered across the United States, it could gather good correction data for most of the country. That data would make GPS accurate enough for Category 1 landings (in other words, very close to the runway but not zero visibility).

The system would guarantee that DGPS corrections would be freely available, although the need is much smaller now that the encryption of the signal has been turned off. To complete the system, the FAA wants to eventually establish

Local Area Augmentation Systems near runways. These would work like the WAAS but on a smaller scale. The reference receivers would be near the runways and would be able to give much more accurate correction data to the incoming planes. With a LAAS, aircraft would be able to use GPS to make Category 3 landings (zero visibility). Similar systems have also been installed in other countries.

This system is underway. Specifications have been drafted and approved, and field trials have been conducted.

Assisted GPS

As we noted in the section, "Differential GPS," it might take a while to track four satellites (approximately 12 minutes if you do a cold start without an almanac). So, if the information about where the satellites are can be provided to the handset before it starts searching for them, it will save a lot of time and thus battery life—something that is crucial in mobile telephone systems, where maximizing the time between the chargings is a major selling point.

Knowing where to find the satellites also enables higher sensitivity in the positioning (because more satellites can be used). If we know in which cell of the mobile telephone system the handset is (in other words, an accuracy of up to 9 km given the maximum size of a cell) and a precise time signal, it enables the calculation of the satellite orbits so that the antenna can point at approximately the right places, get a fix, and calculate the position faster. The GPS receiver does not have to scan as large a volume to acquire a satellite and can spend more time reading the signal from the satellite. It can use more time to determine the distance to the satellite, and the effect of obstructions (foliage, buildings, and so on) will be smaller.

GPS receivers can be made small enough to be integrated into a handset, and several manufacturers have demonstrated prototypes. Just having the GPS receiver in the handset means that you have to do a cold start every time you want to retrieve a position, however. You need a way of getting additional data about the satellites to the GPS receiver.

There are two ways of doing assisted GPS: either have all the functionality in the handset or transmit it via the network. Because the satellites are running in predictable orbits, it takes a major upset to require updates to the information about them. Chips with all the functionalities of an assisted GPS receiver are readily available. It does cost more and mean a drain on the batteries, however. It also means that you have to have a mapping to the cells of the operator in the GPS receiver if the cell ID is to be used as a base for calculating the satellite positions. Potentially, the library of cell IDs can also be very large if the handset is intended for global use.

The alternative is to download the satellite data over the mobile network. The GPS receiver in the handset is not a full GPS receiver, and the data needed to calculate the position is sent every time. This situation means a higher load on the network (although not much), and it gives higher precision as well as enables the operator to keep his or her cell structure confidential (because you only have to send the assistance information for the cell where the user is, not give out information about all of your cells). As we will see in the next section, this function is very important to operators of mobile telephone networks.

Assisted GPS is also a standard from the American T1P1.5 standards body and has been adopted by 3GPP, the group that standardizes the *Universal Mobile Telephony System* (UMTS), which is also more properly known as IMT-20000. That means it is a standard for the E911 services, although no manufacturer had released handsets that use it at the time of this writing.

Another advantage is that it works well with UMTS and that the E-OTD measurement can be used as differential information to further enhance the position. The big disadvantage of GPS is, of course, that it works badly indoors and in areas far to the north and south, which means that additional positioning methods might always be required in those locations.

Network-Based Positioning

GPS is globally present, and despite its being controlled by the U.S. military, it is being used all over the world. The main way of measuring and reporting a user's position, though, is likely to be the mobile network—except where you also need altitude. The mobile telephone network cannot measure that, because there is no way of measuring the angle of the mobile station to the base station.

In all mobile networks, including wireless *Local Area Networks* (LANs) such as 802.11 and Bluetooth, there is a user identity that is built into the terminal (or mobile station) that is unique. In 802.11 networks, it is derived in the same way as in the Ethernet, from a hierarchic number series where the first part is allocated to the manufacturer. A mobile telephone gets a number in a similar way, which is allocated by the *International Telecommunications Union* (ITU) through a process where it is actually allocated by other standardization bodies.

This unique identity is what is used when the terminal is positioned. Just like a network identity is allocated to an Ethernet identity (your IP address is connected to the Ethernet number for the duration of the session, so all data packets addressed to winnie_pooh.johanhjelm.com, who has the IP address 197.22.23.01, are sent to the Ethernet address 3LJ4N000465AQDO). The database that keeps track of the user keeps track of this unique number and can be used to measure the position of the terminal that has it.

All mobile telephone networks that are used now are built the same way: The network is organized in cells around the antennas of the base stations, which connect the user terminal to the global telephone (or IP) network. As the user moves around, he or she will pass from one cell to another. The network has to keep track of this information to be able to send the data or the phone call to the right receiver, so the network always has to know the base station (and which antenna in the base station) to which the user is connected.

This information is managed in a couple of databases, the *Home Location Register* (HLR) and the *Visitor Location Register* (VLR). The values in these databases—which handle millions of users at speeds where the handover has to be done at shorter times than milliseconds—are what are used to calculate the user's position. As a developer, you will not be connecting directly to the mobile telephone system. You will be connecting to a *Mobile Positioning Center* (MPC), which hides the telephone system and its positioning method from you. You do not have to know whether OTA or TOA is used; rather, you only have to care about the position and the quality of position values you get back from the MPC.

That is a very simplistic explanation, and it is somewhat different in different mobile telephone system technologies, but that is how it works.

For instance, for users of the Japanese PHS system, there has been a service since 1997 where you can dial a number and a code and find out in which cell of the network the telephone you are asking about is located. And, you can get the result faxed on a map. Because the cells are no larger than 200 by 200 meters, this result gives a good idea about the position of the user. The main users are not truckers who want to keep track of their cars, but rather parents who want to keep track of their kids. Every year at the start of school, there is a heart-gripping commercial about a little boy who goes home from school alone for the first time and gets lost. His mother, who wisely has equipped him with the system, comes (together with the police, as it happens), and the end is happy.

In most countries, the cells of the mobile phone system are a lot larger than 200 by 200 meters, and that accuracy is too low for most applications. Instead, using more than one antenna of the base station, you can triangulate the position of the user. The technologies vary somewhat between the different mobile systems used. You can see how they are built in principle in Figure 2.2.

Just to give you a brief idea, there are four mobile systems that are used globally: GSM, WCDMA, CDMA, and TDMA (to see what the abbreviations mean, see the Glossary). They are all digital (analog systems are rarely used anymore). Then, there is PDC, which is only used in Japan, and the local mobile

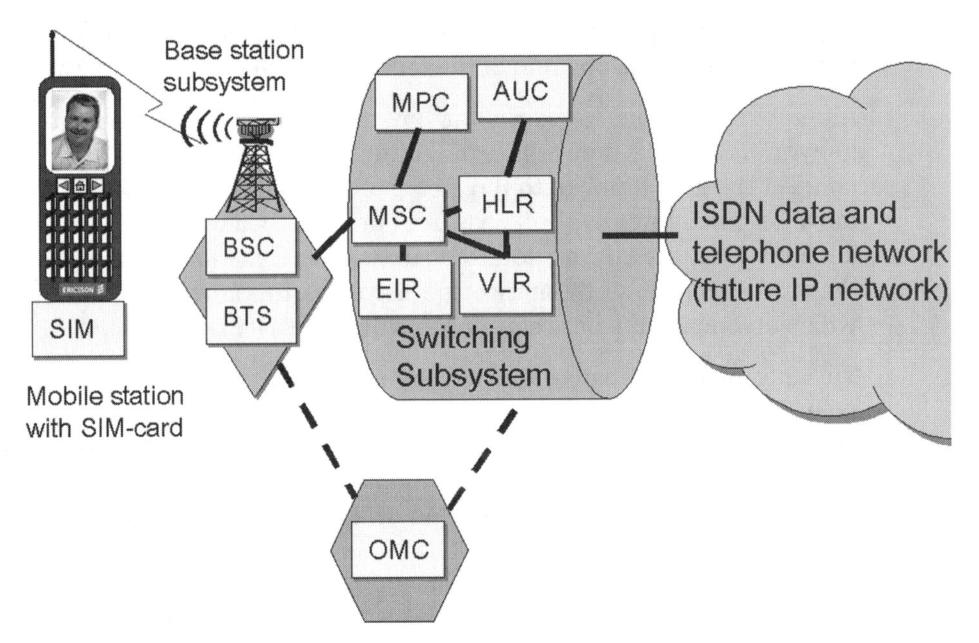

Figure 2.2 How the mobile system is built.

systems PHS and DECT, which have a hard time handling the user moving from one cell to another. Then, there are other systems like wireless Ethernet (802.11) and Bluetooth, which basically work in the same way when it comes to positioning as PHS and DECT. Blutetooth, though, has begun to be used in beaconing systems which, in principle, are the same as a system that works with cell ID but does not enable communication.

In principle, all mobile telephony systems built today work in the same way: The user connects to a base station (which handles the radio traffic and which follows one of several radio standards: GSM, CDMA, WCDMA, TDMA, PDC, or PHS) that is connected to a network (the core network), which has two main functions: It connects the user to the global network (either through setting up an ISDN connection, like a regular telephone call but digital; or to the Internet). Also, it handles signaling and user administration.

The signaling can be done in two ways: through *Signaling System 7* (SS7), a special protocol developed for telephony applications or through *Internet Protocol* (IP). That said, no IP-based mobile networks are deployed today, because the standards are not finalized. Standards also exist for transporting SS7 over IP, but they are still in development. Although this convergence is happening, SS7 functions exist to handle the positioning of a terminal, encoded in the internal system. In Chapter 4, "APIs and Protocols," we will look at how you can access those functions using APIs. The calculation and communication of the shapes described in Chapter 3, "Position Technologies and Coordinate Systems," and velocity are part of the SS7 standard. It takes 13 bytes to describe an ellipsoid point with altitude and an uncertainty ellipse.

Standards are extremely important in the mobile telephone industry, because any divergence in the radio signal between the mobile telephone and the base station will mean that they do not connect properly. For this reason, the mobile telephone industry is so keen on standards and has testing procedures that are vastly more rigorous than testing procedures for Internet equipment.

The user information is kept in a distributed database in the core network (the Home Location Register, or HLR). It keeps track of where the user is and interfaces with the billing system to make sure that the user gets charged for the services he or she is using. Charging and billing are a chapter in themselves and the subject of much contention in the mobile industry (where it is often assumed that certain billing models, like billing for the time a user is connected to a service, are a natural law). But as you saw in Chapter 1, location-based billing is one emerging model that is different from the traditional billing paradigms.

Cell ID-Based Positioning

In all of the mobile systems in operation today, when the user presses the "on" button or enters a cell, the mobile phone connects to the network through the base station and there is a signal to a central database stating that a mobile telephone (a mobile station is the correct terminology) has entered a cell. This central database is connected to a telephone system (or a packet network, in the case of UMTS, GPRS (the packet version of GSM), and CDPD (the packet version of TDMA)). In GSM and UMTS, it is used not only to keep track of where the user is, but also to indicate which frequencies, time slots, and codes the user has been allocated. There is no need to go into that level of detail here, however. In principle, an operator of 802.11 or Bluetooth networks could have the same type of database, although the cost is often seen as prohibitive—and that the operator foregoes this feature to keep access cheap or free. So, on many of those networks, this method of positioning is not possible—nor is there a standard for retrieving the information if it was collected, or a way of tying the identity of the cell to a position. In many cases, the operator is also a regular telecommunications operator, and the databases of the WLAN or Bluetooth networks can be connected to the core network of the mobile operator (and the same databases used for user administration and positioning).

The network operator always knows which cell the user is in so that data is sent to the right destination (and voice calls are connected to the right mobile telephone). This data is what is being provided when the positioning system is based on cell ID. All cells have a globally unique identity, the Cell Global Identity (CGI). Depending on the cell size, the accuracy can be between 100 meters and 35 kilometers, because this is the maximum radius for a cell.

The network operators rarely want to provide the coordinates of their cells to third parties, however. The competition in the business of mobile operators is cutthroat, and if you gave out your cell information, their thinking goes, you would be telling your competitors what coverage your network has, how many users you can handle in certain areas, how much money you have invested, and so on. In many countries, there are companies that drive around triangulating cell sites and looking at the number and type of antennas to determine this information. So, the network operator provides the coordinates for an area where the user is (actually, a point with an uncertainty area).

All mobile telephone systems work this way, and in some systems (like GSM), the user can expose the cell ID on the telephone (if the operator has chosen to provide it). It can and has been used for advertising (in the Stockholm suburb of Kista, one operator sold the cell ID position on the phone to an office supply store). In GSM, this position can be picked out by using the SIM Toolkit (which I will talk more about in Chapter 11, "Location-Based Services in Terminals"), but you still have to find a way to know that 644 990 covers the tarmac of the Frankfurt airport in Germany. Knowing the cell ID might not be very useful if the position of the cell is unknown. How the cells are numbered or named in the user display depends on the network operator; there is no standard for it.

There is also a technology for providing information to all users in a cell, called cell broadcast, which is defined in the GSM and IMT-2000 standards. Not all network operators have implemented it, however.

Because all handsets have to be in a cell to work and must know the cell information, all mobile handsets work with this positioning method. In large cells, however (the size limit for a GSM cell is 4.5 km radius), the cell information is not particularly helpful for the user, even if you know the position of the cell. To know that you are almost 10 kilometers from something is not much help, whether you are looking for a restaurant in downtown Bangkok or an eagle nesting in a tree in the middle of a Swedish forest. In the Jnavi and iArea systems of the Japanese operators Jphone and DoCoMo, the cell ID is used as the key for the information, but the cells in the Japanese mobile system are much smaller than those in other countries.

Timing Advance

In the GSM (and TDMA) system, where each mobile station is allocated a specific frequency and time slot to send and receive data, there is another way of determining position. Timing advance measures the time it takes for the signal to go from the mobile station to the base station, or the reverse. It is used in the GSM system to make sure the time slot management is handled correctly (that the data bursts from the mobile station arrive at the base station at the correct time, in the time slot allocated for them).

The resolution is one GSM bit, which has the duration of 3.69 microseconds (and because this value is a measure of the round trip from the base station to the mobile station or the reverse, the resolution is 1.85 microseconds, which equals 550 meters at the speed of light).

Combining Cell ID with Timing Advance

Timing advance only gives a distance; cell ID only gives a cell. Combining the two makes it possible to determine what distance from the base station in the cell the user is located. If the system also registers which antenna the user is coming in to, the sector of the cell in which the user is located can be determined (normally, there is not one single antenna radiating in all directions, but there are separate antennas radiating in lobes in different sectors to make a full-circle coverage around the base station).

If you know the position of the antennas (or the base station), you can calculate the position of the user (using antenna positions as circle sectors) and use the timing advance value to calculate the distance to get an arc for the user's position. See Figure 2.3, and it will probably become clearer how this system works.

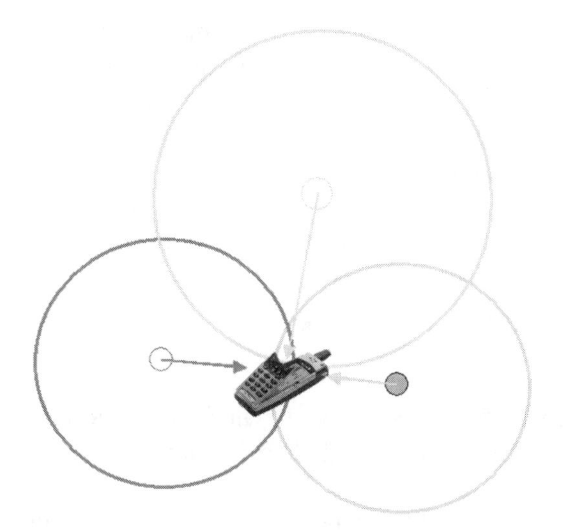

Figure 2.3 How cell ID sector with timing advance works.

Cells are not perfect circles, however, because the different antennas can send out lobes of radio waves that are of different length (different effect gives different distance, and objects in the terrain make the lobes reach different distances at the same effect). In other words, the ideal case of determining a sector of the cell hardly ever works, and in cases where the cell size is smaller than 1 km, it does not give any advantage to use timing advance (because the resolution of the cell will be smaller than the resolution of the timing advance).

If the operator is willing to provide a map of the actual coverage, the system will be able to map the cell ID to a position—and depending on the application, this accuracy might be sufficient (in rural areas, knowing that a turnoff is coming within 10 seconds will give you good time to react; if you are traveling at a speed of 80 km per hour, which means 22 m per second, you get 220 meters in which to react—and that corresponds to a resolution that is double that possible with timing advance in GSM).

Using Timing Advance (TA), the mobile system can calculate in which sector of the cell a mobile station is located, further improving accuracy. The operator always knows where his or her antennas are and what coverage the cells have, so the user can be positioned with greater accuracy. The base stations in GSM and TDMA always know which users are in their range even if they are not connected to it, so it is always clear in which area the user is located—and that area has coordinates that are registered in the network operator's database. In CDMA and IMT-2000, it is slightly different, because the user connection to the mobile network works differently. The measurement is done by measuring the delay between the beginning of a timeslot and the arrival of a burst from the mobile station. Because the speed is equal to the speed of light, the distance is possible to calculate, and because the position of the base station is known, the coordinates for the terminal can be calculated within an uncertainty area that describes the accuracy of the measurement in its size and shape. This shape depends on the configuration of the antennas. Using timing advance in itself, the accuracy is typically as low as 550 meters. It can also take a few seconds to locate the terminal, depending on the traffic in the network and the load on the Mobile Positioning Center. The information that can be calculated this way is only two dimensional and relative to the coordinates of the serving base station.

Measuring the Effect from the Handset

When a mobile station connects to the base station, it will (in GSM and IMT-2000) measure the effect needed to connect to the base stations near it, not only that to which it is connected, but also those that have lower field strength and therefore are farther away. This measurement, the RxLev, can be used to measure the distance from the base station to the mobile station.

The power measured at the handset is the same as the power transmitted from the base station plus the gain in the antenna of the base station, minus the loss during the transmission path, plus the gain in the antenna at the mobile station. This equation can be used to figure out the distance from the base station, because the pathloss is related to the distance (there is no need to go into the equations determining how right now).

You can either use a database to figure out the distance, or you can use the equation to triangulate the handset (in other words, use the pathloss distance from several cells to measure the distance and direction from each, which means obtaining a precise point where the mobile station is located). The mobile station has to transmit the data to the base station, where it can be forwarded to a system in the network which figures out the distance (it could also be collected from the network).

The precision of this method varies with the cell density, however, between 5 and 550 meters. Because it uses the intersections of several circles (the distance from the base station will be drawn as a circle on a map, because the only thing the system can know about the distance is that it is constant from a central point, not any direction), it will give a higher precision than measuring from one base station. But it does require that you either signal the value from the mobile station to the network, or that you coordinate the measurements from several different base stations.

Uplink Time of Arrival (TOA)

Measuring the effect might work in GSM and IMT-2000 (where the effect is always variable), but the effect required to connect to the base station might also vary depending on things other than the distance (such as electric noise in the vicinity, for instance). However, the time required for a signal to pass from the mobile station to the base station does not vary; it is the same as the speed of light divided by the distance.

This method is the basis for the Uplink *Time of Arrival* (TOA) method, which was developed for GSM. (See Figure 2.4.) Signals from the handset are detected at receivers (which can be the base stations but is usually separate equipment located in the same place as the base stations). The propagation delay from the terminal to the receiver is measured, and the position of the handset is estimated by triangulation.

Originally, TOA was presented to the T1P1.5 standardizations group as a solution to the E911 mandate in the United States. Theoretically, its performance complies with the E911 requirements, in other words, with a radius of no more than 125 meters and a success rate of 67 percent. The larger the cell and the slower the handset is moving, the smaller the error, which can vary from 30 to

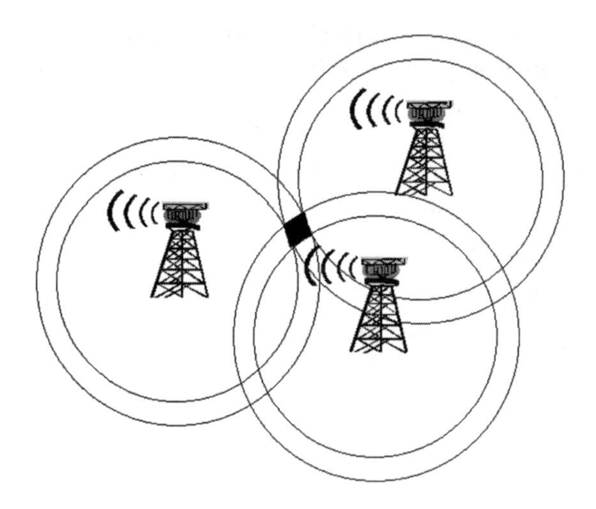

Figure 2.4 Uplink Time of Arrival.

120 meters (which is sufficient for an emergency service to locate the user). The more receivers, the lower the error and the more precise the location, but the receivers have to be positioned precisely, and their coordinates known to the MPC (which is the unit that does the actual calculations). An accurate clock also must exist in the receiver, which can be determined using a GPS clock. The better the position and the time are known, the better the precision.

The receivers must share a common clock reference (with each other and with the handset). By forcing the handset to make a handover request (which contains a time signal) although it is not passing from one cell to another, and by adding up the differences in the time from the transmission to the reception, the distance can be calculated and the intersection of the distance arcs from the two receivers will be the position of the handset.

The receivers are complex, however, and there must be one per base station site. There is a capacity limit because several receivers must be assigned simultaneously. The more traffic there is on the network, the lower the performance (mainly because of interference). The positioning itself creates interference by the positioning requests and the forced handover. In other words, for emergency services it might be sufficient, but if 1,000 tourists are making positioning requests every few minutes, the network will be overloaded.

Another problem is that while it works with legacy handsets, it might not work on the future networks. There is no standardized equivalent in IMT-2000 (or CDMA, for that matter).

The accuracy of this method depends on the accurate positioning of the base stations and the clock. Up to 50 meters in areas with little interfering obstacles and 150 meters in areas where there are obstacles to the radio signal (such as in urban areas), are the values usually given for this method. This is a two-dimensional value that can be recalculated into an uncertainty shape with coordinates. However, it may take a few seconds to locate the terminal, depending on the traffic load and the load on the MPC.

Enhanced Observed Time Difference (E-OTD)

In *Enhanced Observed Time Difference* (E-OTD), the handset measures the time from three different, geographically separate, base stations (or more). The difference from TOA is that the handset does the calculation (and signals back its position to the network). Instead of using the uplink (from the handset to the basestation), it uses the downlink (from the base station to the handset).

In practice, this signal is sent on the control channel for the handsets, and it requires some modification of the mobile station (in other words, some additional software), and the handset also needs to be able to do the computations. There is additional information that is required to make it work: The handset needs to know about reference base stations and neighboring base stations, identity signals, and timing information. It also requires the coordinates for the base stations. Some information can be provided by the network, to assist the terminal with the calculations. Alternatively, the terminal can send just the measurement values, and the MPC in the network can perform the calculations.

If the network really uses the same time signal, then E-OTD will work without anything but extra software. If it does not (which is actually the normal situation), it will require measurement receivers in the network in the same way as in TOA. There can be nothing more remarkable than handsets, however, which are used to listen for the signal.

Like TOA, E-OTD has been submitted to the T1P1.5 standards group. The accuracy is about the same as for TOA. It can be implemented in IMT-2000, but it does require extra equipment in the network—and the performance is uncertain if the cells are big and the base stations far between Typically, an accuracy of 60 meters in rural areas and 200 meters in bad urban areas can be expected. The resulting position is coordinates with an uncertain shape in two dimensions. It can take a few seconds to position the terminal the first time, however, depending on the traffic in the network and the load on the MPC.

Internet-Based Positioning

During the New Economy years, there was much hullabaloo about how the Internet did not have a fixed location, that it was everywhere and nowhere, and so on. It will no doubt come as a significant surprise that today there are companies who try to sell services where the position information comes from the Internet. This system works (sort of) because there are generic ways in which addresses can be mapped to positions. There is no way of actually querying a device about its position and no way for the device to determine its position, however.

IP Address as Position

There are companies that sell services where a user is located based on his or her IP address. This system, of course, only works for users who are not mobile, because in the current systems mobile IP is not yet implemented (with mobile IP, the data packets are redirected from your home network address to an address in the network in which you are visiting, and in the process you must of course get an address on that network).

Instead, they base their technologies on the fact that allocation of IP numbers is known and can be bound very roughly to a position—in some cases. IP numbers are allocated in blocks, and these blocks were given to *Internet service providers* (ISPs), companies, and universities. They do, of course, have a location that is known, and if the address is within a block for which the owner is known, it can be mapped to a position (at least, at city level). This knowledge is then used to personalize information, in practice, to sell advertisements.

That is the theory. The practice is that not only do companies and universities have multiple locations, they also often have access points that might or might not have anything to do with the location of the user (the traffic from my computer, for instance, is redirected to a proxy server in Sweden. I will get Swedish advertisements, notwithstanding the fact that I live in Japan). Using the mapping of address blocks to user addresses also means that users who are not from the current location, for instance because they are dialing into an ISP from another city, will get incorrect advertisements. While the theory works, it will only work up to a point and not for all users. And, of course, there is nothing that enables the system to tell which end of Dallas I am in.

Position Based on DNS

Positioning based on the *Domain Name System* (DNS) works somewhat better than positioning based on IP number. For the same reasons, however, it only works with fixed positions. First, there are two types of domains: the top-level domains, which are detached from location, such as .com, .net, and .org, and the national domains, such as .ca, .us, .se, and so on. If I use the Japanese proxy of Ericsson, my address will be in the co.jp domain, and I will look like a Japanese user (and as a matter of fact, I frequently get advertisements and other fringe information in Japanese).

There is no way of obtaining a finer granularity except, ironically, in the United States, where the states are used as suffixes to the .us domain—.ma.us, .ny.us, and so on. The .us domain is underused, and it looks likely to stay that way—outcompeted by .com, .net, and .org, unfortunately. As a curiosity, this positioning method might give users in Rhode Island better position measurements than the mobile system, because in theory Rhode Island could be covered with one single base station (being the smallest American state).

Of course, another problem is that servers and clients of different nationalities can use the country addresses and they actually would work better if it were recognized that they mapped to cultures, not countries. But that is another discussion.

Because the country is a very coarse-grained position, you need to add extra information about the position. There are a number of RFCs (RFC actually means Request For Comment, but they are in reality standards) from the *Internet Engineering Task Force* (IETF) that describe how to attach position information to the DNS record. All of them, however, are in the experimental category, which means "this is a good idea and try it if you like, but we will not make it a standard."

RFC 1876 describes one system where a new *Resource Record* (RR) for the *Domain Name System* (DNS) is used and reserves a corresponding DNS type mnemonic (LOC) and numerical code (29). The position is described as a sphere, and the coordinates are in the center of the sphere, described in northings and eastings. It also describes the altitude. The format would add a number of fields to the DNS record, which would look like the following: rwy04L.logan-airport.boston. LOC 42 21 28.764 N 71 00 51.617 W -44m 2000m (the regular information from the DNS records would of course be there as well). The fields added are the version, the size of the sphere, the vertical and horizontal precision, northing, easting, altitude, and precision.

That RFC was published in 1996. RFC 1712 was published as early as in 1994, and it foresees not just adding the position of a host, but also actually distributing position information. The record would add a number of fields to the DNS records so that the following would represent a number of hosts in the city of Perth in western Australia:

```
Authoritative data for cs.curtin.edu.au.
;
@      IN     SOA     marsh.cs.curtin.edu.au. postmaster.cs.curtin.edu.au.
```

```
                (
                        94070503        ; Serial (yymmddnn)
                        10800           ; Refresh (3 hours)
                        3600            ; Retry (1 hour)
                        3600000         ; Expire (1000 hours)
                        86400           ; Minimum (24 hours)
                )

                IN      NS      marsh.cs.curtin.edu.au.

marsh           IN      A       134.7.1.1
                IN      MX      0       marsh
                IN      HINFO   SGI-Indigo IRIX-4.0.5F
                IN      GPOS    -32.6882 116.8652 10.0
ftp             IN      CNAME   marsh

lillee          IN      A       134.7.1.2
                IN      MX      0       marsh
                IN      HINFO   SGI-Indigo IRIX-4.0.5F
                IN      GPOS    -32.6882 116.8652 10.0

hinault         IN      A       134.7.1.23
                IN      MX      0       marsh
                IN      HINFO   SUN-IPC SunOS-4.1.3
                IN      GPOS    -22.6882 116.8652 250.0

merckx          IN      A       134.7.1.24
                IN      MX      0       marsh
                IN      HINFO   SUN-IPC SunOS-4.1.1

ambrose         IN      A       134.7.1.99
                IN      MX      0       marsh
                IN      HINFO   SGI-CHALLENGE_L IRIX-5.2
                IN      GPOS    -32.6882 116.8652 10.0
```

The GPOS record contains the longitude, latitude, and altitude.

A major problem with these ideas is that the information is static. Although you can update your local DNS records, they have to be propagated through the hierarchy of DNS servers to take effect, and that will mean that it will be at least 24 hours between the time you update the information and the time the information is distributed to other users of DNS (this situation is not a technical problem, actually, but has to do with the time the central systems are updated).

That makes DNS-based positioning unsuitable for applications where the terminal is moving, and indeed, the examples imagined were all for computers that were sitting still (Unix workstations, as can be seen in the GPOS example, and those are hardly something you stick in your pocket and walk away with).

There is also a *Simple Network Management Protocol* (SNMP) variable in the *Management Information Base* (MIB). The sysLocation MIB variable would require the host to be running an appropriate agent with public read access. That MIB was also intended to reflect local management data (for example, this host is on level 5 room 74) rather than a host's geographical position.

The third experimental RFC, 2009, published in 1996, discusses a way to base routing on positioning instead of assuming that the router network is fixed. It attempts to integrate positioning on the Internet as a base technology, but it has never become more than an experimental RFC.

It assumes that there are GPS-addressable base stations with known coverage areas and that areas can be made addressable by using polygons with coordinates. It would also be possible to have a DNS-like system, mapping the polygons to place names. IP address fields map to geographical atoms, which can be aggregated to bigger atoms (not molecules), and partitions, which are larger yet (on the level of cities).

It uses a combination of unicast, geographical addressing, and multicast to enable multicast groups (which can be ad-hoc, based on the users currently in the positions specified). DNS is used to facilitate the use of GPS geographic addressing for sites of interest. The aim is to describe specific geographic sites in a more natural and real-world manner by using a postal service-like addressing method. Essentially, the DNS would resolve a postal service-like address, such as City_Hall.New_York_City.New_York, into the IP address of the GPS router responsible for that site. The GPS router would then route the message to all available recipients in the site.

The DNS would be used when a message is sent by using an addressing scheme, as follows:

```
site-code.city-code.state-code.country-code
```

The DNS would evaluate the address in reverse starting with the country code, then the state code, and so on. This method is the same used currently by the IP DNS service to return IP addresses based on the country or geographic domains.

In the protocol, users without GPS receivers in their terminals would get the address for the base station they were in and therefore would be able to participate in multicast groups and receive position information. In situations where there is no GPS coverage, the RFC foresees the use of beacons indoors, transmitting GPS data on the GPS frequency.

Microlocation

GPS does not work indoors, and the mobile phone network might be too coarse-grained. In several countries, the mobile phone network works very well in subways and other locations that are obscured from line of sight from base stations, but the cells are equal to the tunnel lengths (because the antennas are cables running in the ceiling of the tunnels). The system in Stockholm cannot distinguish whether you are in Näckrosen or Hornstull, a distance of more than 30 kilometers.

But if a mobile telephone network can give you a position based on the cell you are in, couldn't a more fine-grained wireless LAN do the same thing? Or some other kind of network—maybe Bluetooth or infrared? Maybe even dedicated beacons?

There have been experimental beacons for positioning since the experiments with ubiquitous positioning at Olivetti Research Labs and Xerox Parc in the late 1980s, when infrared diodes first become cheap enough for this kind of application. But there has never been a successful indoor system (although along some Japanese highways, there is a beacon-based positioning system).

Instead of having beacons, which only transmitted their position, the ideal would be if you could obtain this information from the local network. Experiments have been done with wireless LANs in several locations and several times—for instance, in the British city of Lancaster, where the University of Lancaster developed a tourist guide system where the position was given by the wireless LAN. The principle is basically the same as in the mobile phone network but on a smaller scale: Because the position of the transceiver of the wireless network is known and each transceiver has a unique identity, the user must be in the cell of that particular transceiver if he or she receives its identity.

Bluetooth and wireless LANs (802.11 and its variants) work very differently, but the principle at the positioning level is the same. It misses a number of features that the mobile phone network has, however, which in some cases can make it more sensitive than a wireless LAN (it is a somewhat different matter with Bluetooth, because it works at an effect of about a tenth of the wireless LAN, and the area a receiver can cover is correspondingly smaller—about 30 meters radius). And it is not possible to use the same data as in the mobile network if the network operator is not providing the system (because you cannot connect to the databases otherwise), and so you cannot do seamless handover from an outdoor position based on the mobile telephone network to an indoor network based on microlocation beacons.

How well the system works depends on the application. On the CeBIT fair in Germany in 2001, the German company Lesswire demonstrated a Bluetooth access point that also could work as a positioning device and a corresponding trade fair application, which showed what was interesting where you were.

Because there is no coordinating entity, like the MPC, it does not do much good to receive the identity of the base station even if you know that it sits on the wall 15 meters from you but do not know what coordinates that represents. For your terminal to be able to do something intelligent with the data, you need a system that translates the address into a position. That means that the owners of the networks will have to create this mapping (or allow someone else to do it). And aside from ad-hoc efforts and experimental systems like the Guide and Lesswire systems, that is not done today (and if it were, another problem would of course be that the coverage would be so spotty—there is no government mandate included in the license for 802.11 networks to cover an entire country because there is no license).

A possibility would, of course, be if the mobile operator could integrate wireless LANs or Bluetooth into his or her networks. But there is no standard for integrating positioning based on microlocation with network positioning, either. Some manufacturers talk about combining wireless LANs and mobile telephone networks and using the HLR to handle the user information in wireless LANs as well. In that case, the MPC could give the position of the user for all the different networks to which the user connects.

It does not help you much if the network knows the position, and you do not, if you want to use it in creating a positioning system. You have to have a way of getting the position, but you also have to describe it. For this, you need a coordinate system, and that is the topic for Chapter 3.

CHAPTER 3

Position Technologies and Coordinate Systems

When you are moving around, you always know where you are: right here. The problem is that you do not know how to relate "here" to other information. You need to know what your absolute position is, and you need to relate that to other information. Traditionally, that has meant holding a GPS receiver to get your coordinates and possibly looking at its built-in maps to see what is near where you are. You are restricted to the information that is in the receiver, however, and there is no way for you to put in more information (for instance, that you prefer Chinese restaurants).

What Is Position Information?

The mobile Internet communication devices commonly known as Web phones have changed everything. They have a graphical display, but they also have a two-way connection to the Internet. In other words, they can send your coordinates and receive an information set that is adapted to those coordinates—and they can also adapt to any other personal parameters you might have put into the system.

Coordinates and shapes are not only used in determining your position on the Earth and the inaccuracy with which it is given, but they are also used in spatial calculations by using spatial databases. These have specialized functions for calculating shapes and coordinates. That is another reason to understand how they work. To represent a point in space, you assign it coordinates. A coordinate system can represent a point in a two- or three-dimensional space. It can actually represent a four-dimensional space as well if you regard time as a set of coordinates. While that might sound like overkill, considering there are actually relativistic effects (the special theory of relativity describes the relation between the dimensions of space and time) in satellite orbits and that the most frequently used system for position determination, GPS, uses satellites, it is not so farfetched to remember.

Coordinate Formats

When you are working with position information, you will mainly use two different coordinate systems to describe a point in space: UTM and Lat-Lon. The coordinates are calculated in very different ways, but they are designed to do the same thing: give you a point on the surface of the Earth (or a little bit above it).

A large number of measurement systems and frameworks exist for measuring the Earth (and hence, positioning objects on it). The first system for measuring the Earth was created by Aristotle (who, incidentally, knew that the Earth was round). Coordinate systems used in the west have used the equator, the tropics of Cancer and Capricorn, and the longitudes of Paris and Greenwich as references to position an object on the surface on the Earth. Cartographers in China used rectangular grid systems as early as 270 A.D.

A large number of different coordinate systems are in use today. Table 3.1 lists some of the more frequently used ones, but there are others (as we will see later in this chapter).

Models used today range from models of the Earth as flat (used for surveying—because the Earth's curvature does not matter for distances under 10 km) and complex models used to describe the size, shape, orientation, gravitation field, and the angular velocity of the Earth. These are used, for instance, in astronomy, surveying, and navigation.

Lat-Lon Coordinate Systems

If you take a map and fold it over a sphere, you will quickly see that it is only smooth over the equator (because the largest circumference is the only place a sphere can be mapped to a cylinder). To make it fit over the poles, you have to cut it up into triangular strips. This process is actually how the UTM system works, as we will see in the next section.

If you take your paper cylinder, you will see that you do not have to use the equator as the reference for the smooth area. It can be anywhere on the globe as long as it is where the diameter is widest. Any objects can have a number of

Table 3.1 Some of the More Frequently Used Coordinate Systems Today

CODE	NAME
GDC	Geodetic
GCC	Geocentric
GEI	Geocentric Equatorial Inertial
GSE	Geocentric Solar Ecliptic
GSM	Geocentric Solar Magnetospheric
SM	Solar Magnetic
GCS	Global Coordinate System
PS	Polar Stereographic PCS
LCC	Lambert Conformal Conic PCS
TM	Transverse Mercator
UTM	Universal Transverse Mercator
LSR	Local Space Rectangular

coordinates, depending on where you start measuring and how many dimensions in which you measure. The Earth is not even a globe, but a spheroid (because it is round around the equator but flattened at the poles). In other words, its cross-section is not a circle but an ellipse.

Most shapes can be represented as segments or arcs in an ellipse. Even straight angles can be approximated. The fact that most shapes can be represented in terms of ellipses lies in the way that the Lat-Long system works.

In a Lat-Lon model, you draw a number of imaginary arcs over the surface of the Earth to measure the latitude (the vertical distance from the pole to the equator) and the longitude (the distance from a reference meridian). Latitude is relatively easy to measure with decent accuracy, because you can use the height of the sun, the length of the day, and other astronomical phenomena to determine your distance from the equator (for instance, the Tropic of Cancer and the polar circle are both determined in terms of astronomical phenomena —the spring equinox and the length of the day in summer—and were used in the past for navigation).

While the latitude was relatively easy to measure (you just need to keep an eye on the sun), the longitude was harder. It was not until chronometers, which could measure time accurately, were developed that the distance from one longitude meridian to the next could be measured accurately. Christopher Columbus assumed that the Earth's circumference was less that it actually was, so he believed that he would reach India by sailing west before his ships ran out of food and water. The longitude was not accurately measured until the 17th century.

If you always had to measure the longitude and latitude as concentric circles overlaid on the Earth, however, the system would become very complex. Instead, the longitudes and latitudes are measured in terms of sections drawn on the surface of the Earth. In other words, the longitude and latitude (which intersect where you are) always are ellipses (imagine, if you will, slicing through the Earth through your position and the center, first vertically then horizontally). Any position can be expressed as two intersecting ellipses.

The ellipsoidal model was developed in the 18th century after latitude measurements became possible and it was demonstrated that the Earth was an ellipsoid (because it is flattened at the poles). The difference, however, between the ellipsoidal model and the spherical model is only about 20 km (in other words, the ideal sphere has a radius 20 km longer than the ellipsoid); therefore, spherical models can still be used, and are used, for navigation and approximating distance.

The Earth, however, is anything but regular. If you keep a distance over the Earth's surface such that the gravitation is constant, you will find that the Earth is anything but flat. There are troughs and ridges in the surface, for instance, which are independent of mountains and seas, since they depend on the density of the mantle (the layer beneath the crust of the Earth). A geoid model attempts to represent the surface of both the land and sea as if it were determined by gravity alone. Different parameters will be the best approximations for different areas, so the United States might use one set of ellipses whereas Europe might use another. The combinations of ellipses (a reference ellipsoid is defined by its equatorial and polar radius axis) and other parameters (such as eccentricity and flattening, which vary depending on which part of the world you are in) together comprise a reference frame for a location, or a datum. When measuring coordinates, you start with a reference point. The reference system of a coordinate system is called a datum (in a polar coordinate system). The datum defines the size and shape of the Earth as you measure it.

Datums have developed from the spherical and flat Earth models to ellipsoidal models used by GPS and other systems today (based on satellite measurements of the Earth). The *World Geodetic System 1984* (WGS84) is a reference system fixed on the Earth, providing a global reference frame including a model of the Earth. The primary parameters define an ellipsoid, which defines the shape of the Earth. The mass of the Earth is included in the ellipsoid reference. Secondary parameters define a detailed gravity model of the Earth. The WGS84 geoid performs this task for the entire surface of the Earth (and a reference grid is published by the U.S. National Imagery and Mapping Agency).

The major axis of the WGS84 ellipsoid is 6378137 m, and the minor axis is 6356752,314 m.

If you have four grid points, you can use a four-point linear interpolation algorithm to determine the geoid height for any location on the Earth. This information is what is used by the GPS system to determine position, by the way.

The WGS-84 reference frame (the datum) is consistent with another such frame, the *International Terrestrial Reference Frame* (ITRF), within centimeters. Sometimes you will see references to a specific European datum, the *European Terrestrial Reference System* (ETRS89), which also is consistent with the ITRF within centimeters. But creating it meant that several European countries had to resurvey their coordinate systems.

The coordinates for any point on the surface of the Earth in the ellipsoid (and spherical) models is determined by intersecting arcs calculated from reference points. In GPS, the arc is measured by measuring the distance from the satellite and mapping that onto the surface of the Earth. If you have three satellites, you will have three intersecting arcs that will give you a sufficient precision to measure the coordinates with reasonable accuracy (we discussed how this process works in Chapter 2, "Positioning Technologies").

The Geodetic Coordinate System represents the coordinates in terms of latitude and longitude. If no ellipsoid is specified, then the WGS84 ellipsoid is normally used. Elevations are relative to the ellipsoid (or to the WGS84 geoid; in other words, mean sea level). Latitude and longitude are given in units of degrees. Coordinates are expressed in terms of longitude and latitude relevant to the WGS84 ellipsoid. The range of longitude is −180° to +180°, and the range of latitude is −90° to +90°. 0° longitude corresponds to the Greenwich Meridian, and positive angles are to the East, while negative angles are to the West. 0° latitude corresponds to the equator, and positive angles are to the North, while negative angles are to the South. Altitudes are defined as the distance between the ellipsoid and the point, along a line orthogonal to the ellipsoid. Elevation is given in units of meters above the ellipsoid (the default) or above the WGS84 geoid. For example, "37.4506 -122.1834 0" is the latitude/longitude coordinate for Menlo Park, California, related to the WGS84 ellipsoid, that is, to the center of the Earth.

Universal Transverse Mercator Coordinates (UTM)

The coordinate systems were originally developed to enable mapping of the Earth's surface. In the 16th century, when the need to create global maps became urgent, the Dutch geographer Mercator developed a method to project the flat surface of the map onto the globe of the Earth. Today, this method has been developed into the *Universal Transverse Mercator* (UTM) projection. It is probably the one most people are familiar with because it is used on most maps. The UTM coordinate system defines two-dimensional, horizontal positions.

It was Cartesius (René Descartes, 1596-1650) who introduced the coordinate systems based on orthogonal axes. Before then, there were coordinate systems that represented the coordinate by representations to points or baselines (called polar systems), which has to do with the way the coordinate system is projected on a map, as well. Cartesian systems can be two-dimensional (as on a map) or three-dimensional (including altitude). A three-dimensional coordinate system can be defined with respect to two orthogonal planes (x, y, and z; two-dimensional coordinates are usually x and y). How the coordinates are determined in a Cartesian coordinate system depends on the unit of measurement. Various length and angle measurements have been used during the centuries. Our current system of degrees and minutes are derived from Babylonian mathematics (hence the 360 degrees of the full circle), and the meter was defined during the French Revolution as one-ten millionth of the distance from the pole to the equator (today it is defined as a number of wavelengths of the light given out by an isotope of krypton at a certain excitation state).

A spherical geometry does not follow quite the same mathematics as a geometry based on angles and ellipses, however. Try again to wrap a map around a globe. If you slice it up and make the slices triangular with the tip toward the pole, you get a fairly good approximation. That is how the UTM system works. UTM zone characters designate eight-degree zones extending north and south from the equator. The world is divided into six-degree longitudinal strips extending from 80 degrees South latitude to 84 degrees North latitude (because the Earth is not inhabited in the far South, around the pole).

To make sure the northern part of the Earth is covered, there are special UTM zones between 0 degrees and 36 degrees longitude above 72 degrees latitude and a special zone 32 between 56 degrees and 64 degrees North latitude. The zones are shown in Figure 3.1.

Each zone has a central meridian. Zone 14, for example, has a central meridian of 99 degrees West longitude. The zone extends from 96 to 102 degrees West longitude. There are special UTM zones between 0 degrees and 36 degrees longitude above 72 degrees latitude and a special zone 32 between 56 degrees and 64 degrees North latitude.

UTM provides positions accurate to one meter in 2,500 meters across the entire surface of the Earth (except the poles). At the poles, the Universal Polar Stereographic projection is normally used (this value would be less accurate around

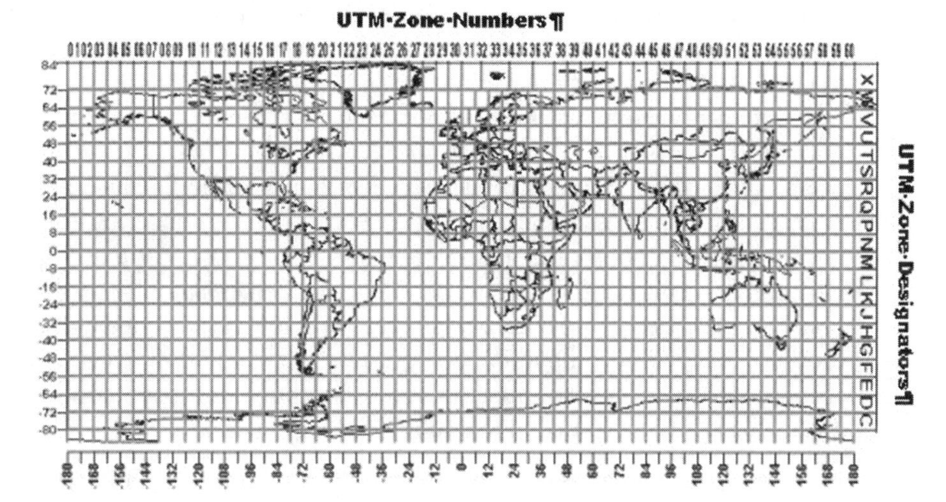

Figure 3.1 The Mercator projection zones.

the equator). For any position in the UTM grid, coordinates can be determined in eastings and northings.

Latitudes towards the east (Eastings) are measured from the central meridian of each zone (with a 500 km false easting to ensure positive coordinates; in other words, all measurements start from the eastern edge of the zone). Longitudes towards the north (northings) are measured from the equator in the northern hemisphere (with a 10,000 km false northing for positions south of the equator, so the values are always positive).

In the UTM system, coordinates are given as <northing> <easting> <elevation>". Eastings, northings, and elevation are all given in units of meters. For example, "4145173 572227 0" is the zone 10 northern hemisphere UTM coordinate for Menlo Park, California.

Local Reference Systems

If you reference the coordinates to the wrong datum, it will result in errors of hundreds of meters. Different nations have created their own measuring and modeling systems to make sure that the measurements fit their countries and to make surveying easier (because the curvature of the Earth does not matter for distances smaller than 10 km). There are a large number of datums available, and for any given point on the Earth, selecting the right datum can mean a difference in accuracy of less than a meter. Selecting the wrong datum can mean a mispositioning of several hundred meters.

Figure 3.2 shows how Sweden is ordered into a polar coordinate system. In reality, the country is tilted to the northeast, and of course the curvature of the Earth does not make the meridians straight lines. The UTM projection also distorts the proportions of the northern part of the country, because it is close enough to the pole for this to matter (the northern part of Sweden is as far north as northern Canada). Instead, Sweden is arranged into a network of straight lines, creating a national grid. Several other countries have the same type of network. Points can then be described by using x and y coordinates with a fixed point in the area as a reference. Before satellite positioning, surveying authorities in most countries created grids of reference points, typically mountains, by which the distance could be determined by triangulation. Conversion between datums must be made carefully, and the national surveying authorities provide a special formula for this conversion.

A planar rectangular reference coordinate system like a national grid can be useful for a small area (small in global terms, like a country). It cannot, of course, be used over a larger area because that would badly distort the measurements. The creation of a national grid is often colored by non-technical aspects (Sweden and Finland, despite being neighbors, have different national grids, for instance).

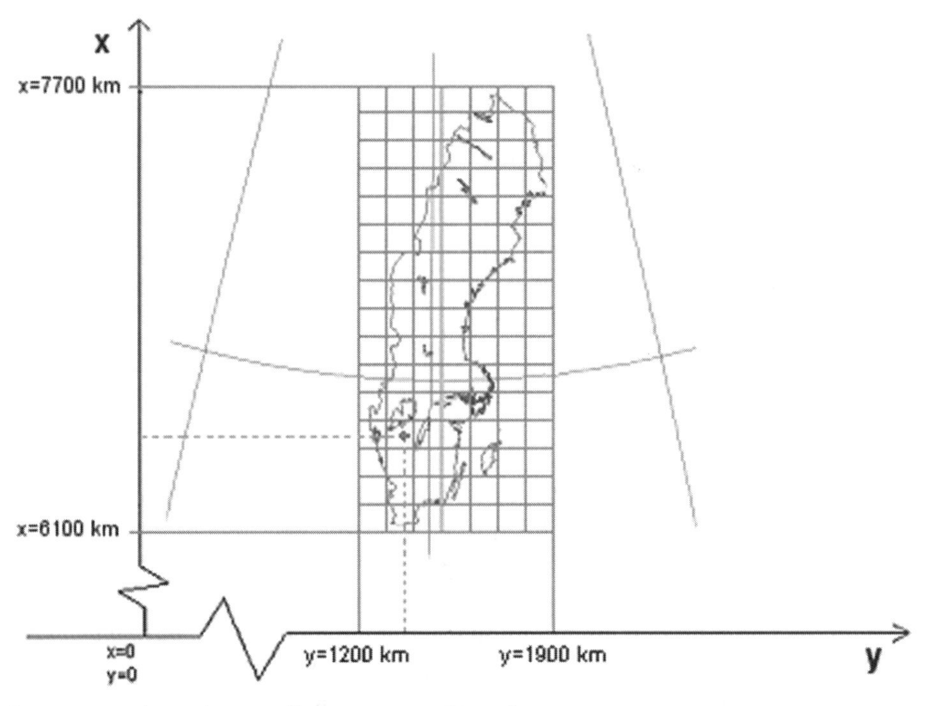

Figure 3.2 The polar coordinate system of Sweden.

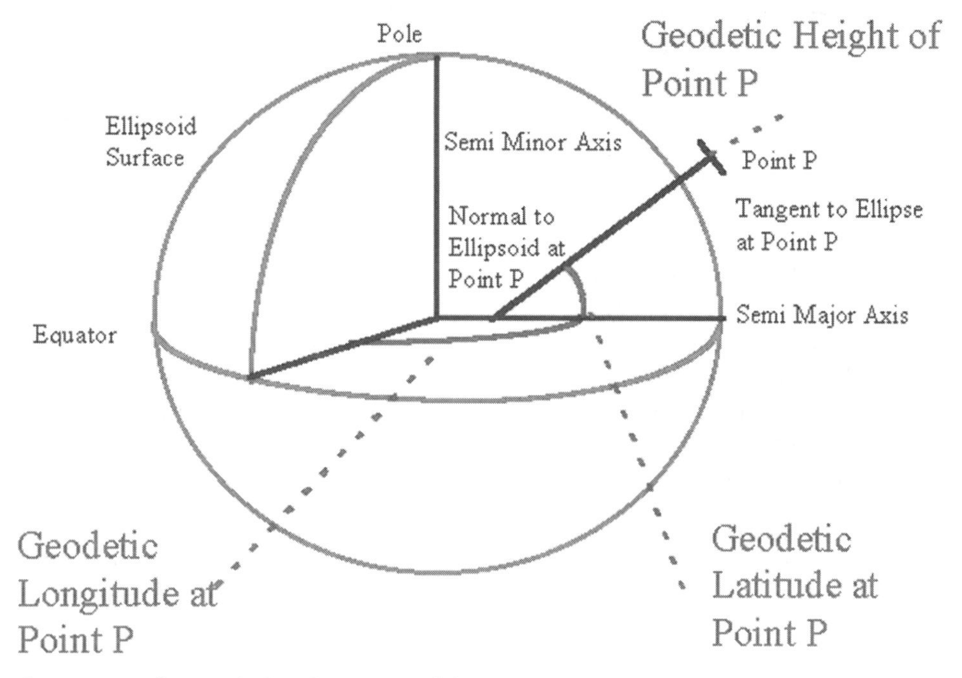

Figure 3.3 The geodetic reference model.

Strictly speaking, the geodetic latitude (as opposed to a number of different other defined latitudes) of a point on the Earth's surface is the angle from the equatorial plane to the vertical direction of a line normal to the reference ellipsoid, as you can see in Figure 3.3. The longitude of a point in the geodetic model is the angle between the reference plane and a plane passing through the point, where both are perpendicular to the equatorial plane. The height is the distance from the reference ellipsoid to the point in a direction normal to the ellipsoid. Today, the latitude/longitude/height relative to the constant gravitation system is still the most frequently used coordinate system. The reference plane used to define the latitude is the prime meridian (formerly known as the Greenwich Meridian), and for longitude, it is the equator.

Other Coordinate Systems

The Swedish national grid is a planar, rectangular reference coordinate system that uses a network of fixed reference points that are referenced to the WGS-84 geoid. There are coordinate systems that do the same thing for the entire Earth, however. Earth-centric Cartesian coordinates (Geocentric Coordinates or GCC) can be used to define three-dimensional positions. They define the three dimensions with respect to the center of mass of the reference ellipsoid (in other words, the center of the Earth). The Z axis goes from the center through the North pole; the X axis is defined by the intersection of a plane defined by the prime meridian and the equatorial plane; and the Y axis is a plane 90 degrees east of the X axis, with its intersection through the equator. The values in the coordinate system are all in meters, and the coordinate represents an (x, y, z) offset from the center of the planet, based upon the WGS84 ellipsoid. For example, "-2700301 -4290762 3857213" is the geocentric coordinate for Menlo Park, California.

Another worldwide system is the *World Geographic Reference System* (GEOREF), which is used in aircraft navigation. It is based on latitude and longitude and divides the world into 12 bands of latitude and 24 zones of longitude (each 15 degrees wide). The grid makes it possible to designate which grid square you are in by two letters, and the grid can be further subdivided by more letters and numbers. Different polygons and circles can be described, altitudes can be added, and so on.

Several countries have also developed their own military grid reference systems that map the countries onto an orthogonal grid, which is not orthogonal to the UTM system (especially in cases of countries far to the north). In the United States, there is a State Plane Coordinates system.

There are other coordinate systems as well, such as the Maidenhead Grid Squares, which were designed to help radio amateurs designate geographical position. The grid identifies "Fields" consisting of areas 20 degrees of longitude by 10 degrees of latitude with two alphabetic characters. An additional set of two numeric digits locates a specific two-degrees of longitude by one degree of latitude "grid square" area within the field. Two additional alphabetic characters can be used to refer to a 5.0 minutes of longitude by 2.5 minutes of latitude "Sub-Square" within the grid. In each case, the longitude character precedes the latitude designator. Ham radio operators use these grid designators to communicate transmitter positions to each other. Several utility programs are available to convert between latitude and longitude and the Maidenhead Grid Square system. Some of these also allow computation of distance and azimuth between stations.

Some GPS receivers display positions in an extended Maidenhead system that appends one or two additional sets of numeric and alphabetic pairs, increasing the precision with which a location can be specified. Some amateur radio operators use other terms for the Maidenhead system, such as *World Wide Locator* (WWL) squares or QTH locator squares.

Topology

Topology is a way of defining the relationship between geographic objects. These relationships do not change if the underlying map is changed in scale or reach (for example, if it was made of rubber and stretched). Typically, a limited number of relationships are used, such as:

Object A contains object B

Object A is adjacent to object B

Object A intersects object B

Absolute relationships can change, such as saying "Object A is located 200 feet from Object B." If the rubber sheet is stretched, the distance could become 400 feet. This means that this relationship is not a topological relationship; however, "Object A is located 3 of the diameters of object A from object B" will not change, if the stretch is the same in all directions. Rotate the sheet, and the direction also changes.

Topology is a mathematical discipline, which originated in geographic computations, and mathematical topology can be very different from how we think about relationships in the real world.

Topological data can best be expressed using a vector structure, which contains the distance and direction of the object. Algorithms can be applied directly to the vector structures. The mathematics of topology can be counterintuitive (for instance, a coffee cup and a donut have the same shape in topological terms), and we will not go into it more here.

Geographic Shapes

Many formats that are used to describe position (for instance, the LIF and WAP Forum APIs) describe shapes that can be used to describe positions. Geographic shapes are actually defined in an ISO standard, as well as a standard from ETSI, which means that the mobile network can use them. Only a few shapes are needed to describe a position. As we could see earlier, most shapes can be described in terms of ellipses—especially if there is an uncertainty around the edges. The shapes can also be used for defining triggering criteria to initiate a positioning when the mobile subscriber enters or leaves the geographical area that is described. These shapes are:

Ellipsoid point

Ellipsoid point with uncertainty circle

Ellipsoid point with uncertainty ellipse

Polygon

Ellipsoid point with altitude

Ellipsoid point with altitude and uncertainty ellipsoid

Ellipsoid arc

Ellipsoid Point

An ellipsoid point is a point on a surface of an ellipsoid. In other words, it can be used to refer to a point on the surface of the Earth (or close to it) and can be described by using a number of coordinate systems. You can also see how the Earth spheroid is flattened toward the poles.

In a polar coordinate system, the latitudes and longitudes must be described as straight-angle references toward a plane. The latitude is the angle between the equatorial plane and the perpendicular to the plane tangent to the ellipsoid surface at the point. Positive latitudes correspond to the northern hemisphere. The longitude is the angle between the half-plane determined by the Greenwich Meridian and the half-plane defined by the point and the polar axis, measured eastward.

Ellipsoid Point with Uncertainty Circle

Often, the point cannot be measured precisely. There will be an ellipsoid with a radius that is the same as the uncertainty around the point. The "ellipsoid point with uncertainty circle" is characterized by the coordinates of an ellipsoid point (the origin) and a distance r (the radius). It describes formally the set of points on the ellipsoid, which are at a distance from the origin less than or equal to r, the distance being the geodesic distance over the ellipsoid—that is, the minimum length of a path staying on the ellipsoid and joining the two points.

An ellipse where the semi-major and semi-minor axes are both equal to the radius, r, and the angle of orientation is immaterial, can represent a circle. Other shapes also can be approximated using ellipsoids, but circles are by far both the easiest to calculate with and the easiest to measure. They are also easy to relate to the two-dimensional plane of the map, so circles will often be used in applications that need to determine a location in terms of how close it is to something.

Ellipsoid Point with Uncertainty Ellipse

The ellipsoid point with uncertainty ellipse consists of an ellipsoid point (the origin), the distances r1 and r2, and the angle of orientation A. This information describes the set points of the ellipsoid, which fall within or on the boundary of an ellipse. This ellipse has a semi-major axis of length r1 oriented at angle A (0 to 180 measured clockwise from north) and a semi-minor axis of length r2. In principle, it is a point with an ellipse that describes the uncertainty around it.

The typical use of this shape is to indicate a point when its position is known only with a limited accuracy but the geometrical contributions to uncertainty can be quantified.

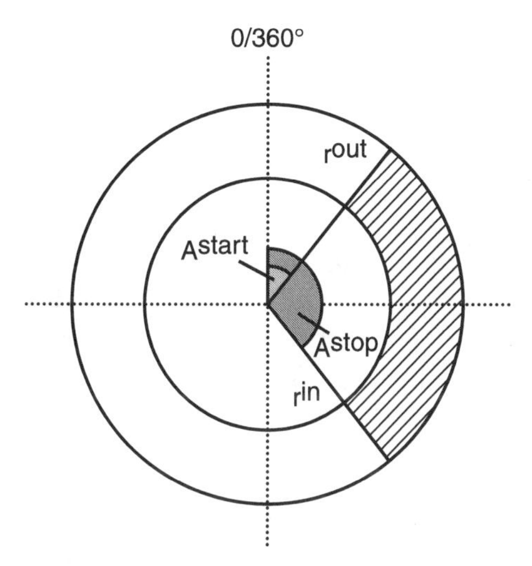

Figure 3.4 The definition of an ellipsoid with an uncertainty arc.

Ellipsoid Point with Uncertainty Arc

An ellipsoid point with an uncertainty arc is the same thing as an ellipsoid point with an uncertainty ellipse, but the uncertainty is only in a part of the ellipse (corresponding to a circle sector at a certain distance from the center of the circle). The ellipsoid point can be defined by using the coordinates of the ellipsoid point (the origin), the inner and outer radius of the arc (rin and rout), and the start and stop angles, Astart and Astop. It will look like Figure 3.4.

An arc is defined by a point of origin with one start and one stop angle plus one inner radius and one outer radius. In this case, the striped area describes the actual arc area. The smaller circle defines the inner radius, and the outer circle defines the outer radius.

Polygons

Most shapes can be described as polygons (ellipses and circles, as well). Polygons are arbitrary shapes that consist of a series of points with lines connecting them. In the LIF system, the minimum numbers of points are three and the maximum is 15. In reality, there can be an unlimited number of points, of course.

Just like the connect-the-dots puzzles, the points are connected in the order they are given. The connecting line cannot cross another connecting line, and two successive points must not be diametrically opposed on the ellipsoid. Because it must be possible for the points to map onto an ellipse on the surface of the Earth, points cannot be directly separated by more than roughly 20,000 km. Because the calculation is so difficult (and the mapping onto curves in an ellipsoid is so difficult), systems like the LIF API accept an inaccuracy of less than three meters.

Ellipsoid Point with Altitude

The description of an ellipsoid point with altitude is that of a point at a specified distance above or below a point on the earth's surface. This is defined by an ellipsoid point with the given longitude and latitude and the altitude above or below the ellipsoid point. In practice, it means that the altitude coordinates are added to the coordinates for an ellipsoid point.

Ellipsoid Point with Altitude and Uncertainty Ellipsoid

The *ellipsoid point with altitude and uncertainty ellipsoid* is characterized by the coordinates of an ellipsoid point with altitude, but instead of two uncertanities, it has three: the distances $r1$ (the semi-major uncertainty), $r2$ (the semi-minor uncertainty), and $r3$ (the vertical uncertainty) and an angle of orientation A (the angle of the major axis).

This term describes the set of points that fall within or on the surface of a general (three-dimensional) ellipsoid centered on an ellipsoid point with an altitude whose real semi-major, semi-mean, and semi-minor axis are some permutation of $r1$, $r2$, $r3$ with $r1 \geq r2$. The $r3$ axis is the vertical axis; the $r1$ is the semimajor in the horizontal plane; and the $r2$ is the semiminor in the horizontal plane. The horizontal plane is oriented at an angle A (0 to 180 degrees) measured clockwise from north. It becomes a little clearer if you look at Figure 3.5.

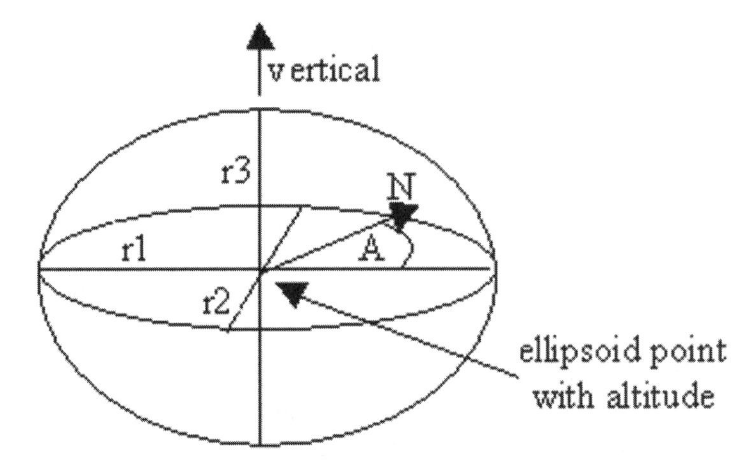

Figure 3.5 An ellipsoid point with altitude and uncertainty ellipsoid.

Transforming Geodetic Data

There are various ways of converting coordinates based on one datum to those based in another. Complete conversion is based on the transformation of seven parameters, which include three translation parameters, three rotation parameters, and a scale parameter.

A simple three-parameter conversion between latitude, longitude, and height in different datums can be done by using a reference datum (with the Earth-centered or Earth-fixed XYZ Cartesian coordinate set) and three origin offsets that are an approximation of the difference in rotation, translation, and scale. There is a set of conversions, the Standard Molodensky formulas, that can be used to convert latitude, longitude, and ellipsoid height in one datum to another datum if the Delta XYZ constants for that conversion are available and ECEF XYZ coordinates are not required.

A number of reference ellipsoids defined by different mapping agencies exist in different parts of the world. They all can be translated to the WGS-84 format, however, which is the most frequent reference. The U.S. National Imagery and Mapping Agency publishes parameters for simple XYZ conversion between many datums and WGS-84. If you take a value based in one datum and use it in another datum, you might have an error of up to 1 km (in three dimensions).

The role of the reference system in distance calculations is important. Generally, at least three types of distances may be defined between points (and, therefore, between geometric objects): map distance, geodetic distance, and terrain distance.

Map distance is the distance between the points as defined by their positions in a coordinate projection (such as on a map when scale is taken into account). Map distance is usually accurate for small areas in which scale functions have well-behaved Jacobean transformations that represent the local Gaussian metric (for those of us who are nonmathematicians, that means when the distances are so small we can ignore the curvature of the Earth).

Geodetic distance is the length of the shortest curve between those two points along the surface of the earth model being used by the spatial reference system. Geodetic distance works well for wide areas of coverage and takes the earth's curvature into account. It is especially handy for air and sea navigation, although care should be taken to distinguish between rhumb line (curves with a constant geodetic bearing) and geodesic curve distance.

Terrain distance will take into account the local vertical displacements (hypsography). In other words, you also take the elevation into distance and measure the real distance that a mobile station travels (a line on a sloping plane is longer than a line on a flat surface, because it also has a vertical distance). Terrain distance can be based either on a geodetic distance or a map distance. In practice, this measurement is usually measured on the ground (that is, by wheel rotations). If you are not calculating distances (short or long), this is the measurement you will use. Note that the terrain distance can be quite different from the map distance in mountainous areas, as any skier will tell you.

Time

If you know the time when two positions were measured as well as the value of those positions, you can calculate the velocity of an object moving between them (assuming it was moving in a straight line between the points, of course). Direction can be calculated from the velocity if one can assume that the mobile station is moving along the line that was used to calculate the velocity. If velocity is 0, there can not be any direction, and if the measurements are done with long time intervals respective to the velocity, there is no way of knowing the level of confidence that one can have in the direction (the user might have made any number of twists and turns, during which the position was not measured). So, the direction measured this way is an abstraction at best if the time interval between measurements is not very short.

To estimate velocity is only one reason for measuring time precisely. Determining time precisely is an important factor in positioning, as you saw in Chapter 2. Indeed, latitudes could not be measured properly until there was a precise way to measure time.

The measurement of time is done by using atomic clocks (driven by the number of oscillations of different types of atoms), and the International Earth Rotation Service in Paris coordinates the reference clocks; up to 2001, the *Universal Time Coordinates* (UTC) were calculated by coordinating a number of different clocks, but now, physicists in Boulder, Colorado have developed a clock that measures time down to the femtosecond (which is two steps below nanoseconds). A femtosecond is approximately to a second as the width of a human hair is to the distance from the Earth to the sun.

A single leap second 23:59:60 is inserted into the UTC time scale every few years as announced by the IERS to keep the UTC from wandering away more than 0.9 s from the less constant astronomical time scale UT1, which is defined by the actual rotation of the Earth. An example time is 23:59:59, which represents the time one second before midnight.

Distributing time is not a matter of just beaming out the time over the Earth, however. Because the accuracy is so high, the Earth is actually too big for the time measurement to be distributed in time for it to be valid (the speed of light is simply too slow for the signal to reach everywhere)—something that is further complicated by the fact that the satellites used in the GPS system are sufficiently far from Earth for a signal to take a few nanoseconds to reach its surface. In other words, systems that require this precision (some mobile phone systems, for instance) actually must take relativistic effects into account. Also, for instance, they must start transmitting before their time slot begins so that the signal reaches the base station at the exact instant when the time slot opens.

Presenting Time

Measuring time is one thing; presenting it is another. Dates and times are defined in an international standard, ISO 8601. This standard describes a large number of date/time formats. For example, it defines Basic Format, without punctuation, and Extended Format, with punctuation, and it allows elements to be omitted. A particular problem with ISO 8601 is that it allows the century to be omitted from years, which has caused some trouble now that we have entered the 21st century.

The international standard notation for dates in the standard is YYYY-MM-DD, where YYYY is the year in the Gregorian calendar, MM is the month of the year between 01 (January) and 12 (December), and DD is the day of the month between 01 and 31. For example, the fourth day of February in the year 1995 is written in the standard notation as 1995-02-04. In some countries, like Japan and Moslem countries, the official year is not the year according to the Gregorian calendar; rather, it is counted according to some other zero point (in the Japanese case, the year since the ascension of the latest Emperor—2001—is Heisei 13, and in Moslem countries from the year Mohammed left Mecca for Medina). In Moslem countries, the months are also different, the month of Ramadan, for instance, begins at the first sighting of the new moon in what is November in the Gregorian calendar.

ISO 8601 is only specifying numeric notations and does not cover dates and times where words are used in the representation. It is not intended as a replacement for language-dependent worded date notations such as "24. Dezember 2001" (German) or "February 4, 1995" (American English). ISO 8601 should, however, be used to replace notations such as "2/4/95" and "9.30 p.m."

Times in the ISO standard format are expressed in UTC (Coordinated Universal Time) with a special UTC designator, Z. UTC is what was previously known as GMT, but GMT is Greenwich Mean Time and actually changes when Daylight Savings Time is in force during the summer in Europe. UTC is constant. As an alternative to UTC, times can be expressed in local time, together with a time zone offset in hours and minutes. A time zone offset of +hh:mm indicates that the date/time uses a local time zone which is hh hours and mm minutes ahead of UTC. A time zone offset of -hh:mm indicates that the date/time uses a local time zone that is hh hours and mm minutes behind UTC.

One big advantage of the ISO format is that all the information is numerical, which means that any information can be compared by using simple string comparison techniques (and you can also subtract and add dates, times, and so on). Week 01 of a year is per definition the first week that has a Thursday in this year, which is equivalent to the week that contains the fourth day of January. In other words, the first week of a new year is the week that has the majority of its days in the new year. Week 01 might also contain days from the previous year, and the week before week 01 of a year is the last week (52 or 53) of the previous year even if it contains days from the new year. A week starts with Monday (day 1) and ends with Sunday (day 7). For example, the first week of the year 1997 lasts from 1996-12-30 to 1997-01-05 and can be written in standard notation as 1997-W01 or 1997W01. A number indicating the day of the week can also extend the week notation. For example, the day 1996-12-31, which is the Tuesday (day 2) of the first week of 1997, can also be written as 1997-W01-2 or 1997W012. Possible ISO week numbers are in the range 01 to 53. A year always has a week 52. (There is one historic exception: The year in which the Gregorian calendar was introduced had fewer than 365 days and fewer than 52 weeks.)

The international standard notation for the time of day is hh:mm:ss, where hh is the number of complete hours that have passed since midnight (00-24), mm is the number of complete minutes that have passed since the start of the hour (00-59), and ss is the number of complete seconds since the start of the minute (00-60). If the hour value is 24, then the minute and second values must be zero. (The value 60 for ss might sometimes be needed during an inserted leap second in an atomic time scale such as Coordinated Universal Time [UTC]).

As with the date notation, the separating colons can also be omitted (as in 235959), and omitting the seconds or both the seconds and minutes (as in 23:59, 2359, or 23) can reduce the precision. It is also possible to add fractions of a second after a decimal dot or comma; for instance, the time 5.8 ms before midnight can be written as 23:59:59.9942 or 235959.9942. Because every day both starts and ends with midnight, the two notations 00:00 and 24:00 are available to distinguish the two midnights that can be associated with one date. In other words, the following two notations refer to exactly the same point in time: 1995-02-04 24:00 = 1995-02-05 00:00.

The 24h time notation in the ISO standard is used all over the world except in some English-speaking countries (and Japan), where notations with hours between 1 and 12 and additions like ante meridiem (A.M.), before noon; and post meridiem (P.M.), after noon, are widely used. Most other languages do not even have abbreviations like a.m. and p.m. Even in the United States, the military and computer programmers have been using the 24h notation for a long time. Even in the United States, the widely respected *Chicago Manual of Style* now recommends using the international standard time notation in publications.

The world is divided into time zones, which serve to make sure that noon comes when the sun is in zenith. That means there is about as much daylight before noon as after. As the daylight moves around the world, so does the day, and when the world rotates around the sun so that the Pacific time zone has midnight, the next day starts. Time zones are a purely administrative division, which is why they can bend in places. Some time zones are half-hours, and some countries like China are one single time zone.

There are no standard abbreviations or designations for time zones, they are calculated from UTC by giving an offset. The time zones are shown in Table 3.2, and the alphabetic abbreviations for the time zones is shown in Table 3.3.

Time is normally measured in *Universal Time* (UTC), which is shown by appending a capital letter Z to a time (as in 23:59:59Z or 2359Z). The Z stands for the zero meridian, which goes through Greenwich in London, and it is commonly used in radio communication where it is pronounced Zulu (the word for Z in the international radio alphabet). Universal Time (sometimes also called Zulu Time) was called *Greenwich Mean Time* (GMT) before 1972; however, this term should no longer be used. Since the introduction of an international atomic time scale, almost all existing civil time zones are now related to UTC, which is slightly different from the old and now unused GMT.

If you want to designate a time zone, you add the offset from UTC. +hh:mm, +hhmm, or +hh can be added to the time to indicate that the used local time zone is hh hours and mm minutes ahead of UTC. For time zones west of the zero meridian, which are behind UTC, the notation -hh:mm, -hhmm, or -hh is used instead. For example, *Central European Time* (CET) is +0100 and U.S./Canadian *Eastern Standard Time* (EST) is -0500. The following strings all indicate the same point of time: 12:00Z, 13:00+01:00, and 0700-0500.

- YYYY = 4 digit year ("0000" . . . "9999")
- MM = 2 digit month ("01"=January, "02"=February . . . "12"=December)
- DD = 2 digit day ("01", "02" . . . "31")
- hh = 2 digit hour, 24-hour timekeeping system ("00" . . . "23")
- mm = 2 digit minute ("00" . . . "59")
- ss = 2 digit second ("00" . . . "59")

Table 3.2

TIME ZONE ABBREVIATION	TIME ZONE NAME	OFFSET FROM UTC
UTC	Coordinated Universal Time	0
UT	Universal Time (used in astronomy)	0
TAI	International Atomic Time	0
GMT	Greenwhich Mean Time	0 (In summer, UTC +1)
BST	British Summer Time	+1
IST	Irish Summer Time	+1
WET	West European Time	0
CET	Central European Time	+1 (in summer +2)
CEST	Central European Summer Time	+2
EET	Eastern European Time	+2
EEST	Eastern European Summer Time	+3
MSK	Moscow Time	+3
MSD	Moscow Summer Time	+4
AST	Atlantic Standard Time	–4
ADT	Atlantic Daylight Time	–3
EST	Eastern Standard Time	–5
EDT	Eastern Daylight Saving Time	–4
ET	Eastern Time (EDT or EST)	Depends on place and time of year
CST	Central Standard Time	–6
CDT	Central Daylight Saving Time	–5
CT	Central Time	Depends on place and time of year
MST	Mountain Standard Time	–7
MDT	Mountain Daylight Saving Time	–6
MT	Mountain Time	Depends on place and time of year
PST	Pacific Standard Time	UTC –8
PDT	Pacific Daylight Saving Time	–7
PT	Pacific Time	Depends on place and time of year
HST	Hawaiian Standard Time	–10
AKST	Alaska Standard Daylight Saving Time	–8
AEST	Australian Eastern Standard Time	+10

TIME ZONE ABBREVIATION	TIME ZONE NAME	OFFSET FROM UTC
AEDT	Australian Eastern Daylight Time	+11
ACST	Australian Central Standard Time	+9.5
ACDT	Australian Central Daylight Time	+10.5
AWST	Australian Western Standard Time	+8

Table 3.3

LETTER	WORD	DIFFERENCE
Y	Yankee	UTC - 12 hours
X	Xray	UTC - 11 hours
W	Whiskey	UTC - 10 hours
V	Victor	UTC - 9 hours
U	Uniform	UTC - 8 hours
T	Tango	UTC - 7 hours
S	Sierra	UTC - 6 hours
R	Romeo	UTC - 5 hours
Q	Quebec	UTC - 4 hours
P	Papa	UTC - 3 hours
O	Oscar	UTC - 2 hours
N	November	UTC - 1 hour
Z	Zulu	same as UTC
A	Alpha	UTC + 1 hour
B	Bravo	UTC + 2 hours
C	Charlie	UTC + 3 hours
D	Delta	UTC + 4 hours
E	Echo	UTC + 5 hours
F	Foxtrot	UTC + 6 hours
G	Golf	UTC + 7 hours
H	Hotel	UTC + 8 hours
I	India	UTC + 9 hours
K	Kilo	UTC + 10 hours
L	Lima	UTC + 11 hours
M	Mike	UTC + 12 hours

International Date Line is between time zones M and Y (24 hours/one day differs)
The letter 'J' for 'Juliet' is used to refer to local time of the observer.

Note that T and Z appear literally in the string. Example: 06.40 in the morning UTC on 30 April 1999 would be as follows:

```
1999-04-30T06:40:00Z
```

The offset follows the following syntax: [+|-]hhmm.

A positive offset value indicates a time zone east of Greenwich. Time zones as such are not possible to specify (10 AM ET is a meaningless string in this context; it must be translated to UTC, where it is 14.00Z). Note that this offset will change when countries go to Daylight Savings Time, so GMT (which is the Greenwich Mean Time in England) is not equivalent to UTC during Daylight Savings Time.

As an example, the following times are equivalent:

```
<time utc-off="+0200">2001-06-30T18:28:10Z</time>
<time>2001-06-30T16:28:10Z</time>
```

There is no international standard that specifies abbreviations for civil time zones like CET, EST, and so on, and sometimes the same abbreviation is even used for two very different time zones. Rules for Daylight Savings Time in different time zones can also vary with political decisions, which means that you have to keep track of them if you want to use them. All positioning systems described in this book are based on using UTC.

Knowing the time and the position of a reference can be used to calculate your position if you have a way of receiving the position of the reference point. However, to get your position from the network, you need an API. And, that is what we will talk about in Chapter 4.

CHAPTER 4

APIs and Protocols

As you could understand from Chapter 2, you do not have to connect directly into the mobile telephone system to get a position. You connect to the MPC, or some other system, which has a connection directly into the mobile telephone system's core network.

Some positioning solutions (that is, GPS and network-assisted E-OTD) do not use an MPC. The positioning data is calculated directly in the terminal. Some APIs, like the LIF API, can be used to represent this information as well. Whichever method is used, the MPC is the unit to which the position request for the mobile station is directed. Except in cases where the mobile station itself has to be queried for the position, it is also the entity that calculates the position and that makes the retrieval of the position independent of the positioning technology. Today, no standard is deployed for the interfaces to the MPC, and each manufacturer has its own interface. Ericsson, for instance, has developed a protocol called the Mobile Positioning Protocol, which uses XML documents transported in HTTP to transmit the request and response data. The responses can be in different data formats, so if you require Lat-Lon coordinates, that is what you will get.

The MPC also has a connection to the Mobile Switching Subsystem (MSS) and through that to the authorization database of the network. This means that it can shut out requests that do not come from authorized parties. This mechanism is likely to remain basically the same in the future, which means that you as an information provider or application developer must set up a relationship with the mobile network to get the position information. It is likely, however, that there will be intermediaries who will take care of this part of the business for you.

The Three Main APIs: an Overview

What the service can do with the position information really depends on how the system gets it. There are three main methods to get position information from a network today: using the Parlay, LIF, or WAP Forum APIs (there is also a fourth, the Magic API, but no provider of mobile positioning centers has committed to implement that). Apart from these, there are APIs to get GPS information in text format (essentially expressing how the serial port communication is managed), but since few devices implement them, we will not discuss them here. All three require that there is some kind of intermediary that has a SS7 signaling interface (for example, an SS7 card) that can communicate with the network. In the WAP Forum and LIF cases, the assumption is that this interface is the MPC, but the WAP gateway can also perform this signaling and in effect double as the MPC. In the case of Parlay, the interface to the mobile network is in the special Parlay application server in the network. In fact, the WAP Forum and the Location Interoperability Forum have also developed one API each. While they mainly work in the same way, the WAP approach is somewhat more constrained than the LIF approach, which attempts to address all the problems that obtaining the location of a mobile phone brings (the Location Interoperability Forum does not address the underlying positioning method, and neither does the WAP Forum). The WAP method is more elegant in that it has the ability to send the positioning request and response piggybacked on top of the request for information.

There are two basic levels on which a software developer can use an API in a location information system: between the mobile station and the positioning gateway, and between the positioning gateway and the application. The first level is taken care of by the industry itself, essentially providing proprietary interfaces from the mobile network system to the MPC.

Traditionalists might not want to call the LIF and WAP Forum APIs Application Programming Interfaces (because they do not contain functions you can call within your software program) but rather protocols (because they contain sequences of messages). But the specifiers have termed them APIs, so that is the term we will use.

The use cases for the two APIs overlap, but there are differences. Both have two main functions to retrieve location information: immediate and deferred. The immediate request is intended to retrieve information directly, whereas the deferred request is intended to deliver the information later—for instance, triggered by a timer (which can be triggered several times). With the WAP system, you can send an attachment with a request containing your location information. Within the LIF API, there are functions for emergency service use (as in Parlay), because that API is more tied to the telephony function. The WAP API is a part of WAP 2.0, which is vastly different from the old WAP (see Appendix C).

Both assume that the location information is delivered to the client (or the application server) as a result of a query. The query can be done by the device of the end user or by an intermediate entity. A query can also set up a triggered response. The trigger criteria are different in the two APIs: the WAP API triggering on time only and the LIF API triggering on position as well (in other words, when a user enters or leaves the area).

Both use HTTP POST to transport the data (the WAP Forum system can also use WSP, the WAP Session protocol, which works like HTTP except that it can retain state—we will not discuss it in this book, however). Also, both systems use XML documents to encapsulate the requests and responses and have functions to report the quality of the position (or rather, the confidence in the position information). The LIF system also has a function to set a privacy flag (at least in some versions).

Despite the similarities, there are differences that make it impossible to map them to each other. The WAP Forum API is not the same as the LIF API. For instance, the encapsulating XML documents have different names. In other words, application servers will have to implement both APIs, or you will have to know which system is generating your position information.

While the WAP Forum has had to bear much criticism for its proprietary protocols, at least some WAP gateway vendors will implement the WAP Location Framework APIs because the WAP Forum has a well-established system for testing the interoperability of different components, and the buyers and distributors typically request that the mobile phones they buy for the stores of the operators have passed interoperability tests, because it would not do to have handsets that do not work with the network they are providing. It is also relatively easy for an application server provider to build the WAP gateway functionality into the application server or to take an existing WAP gateway and build an application server from it.

The Location Interoperability Forum API

The *Location Interoperability Forum* (LIF) was set up in 2000 to make sure that the location industry did not fragment into a number of incompatible islands. This function was especially urgent when it came to the question of how application servers could address different mobile position gateways (or work directly with the HLRs) from different companies. Otherwise, an application server might need totally different ways of connecting to a Motorola network and an Ericsson network, despite their both being GSM or IMT-2000 networks, and following the standards in all other regards.

The LIF has produced a specification for a *Mobile Location Protocol* (MLP), which is an application-level protocol for the positioning of mobile terminals. It is independent of the underlying network technology (and thus, of the positioning method), so it can also use position data from GPS—if there is a way to get it from the GPS receiver to the *Location Enabling Server* (LES). The MLP serves as the interface between a Location Server and a Location Enabling Server, which in turn is interfacing to the application server. It defines the core set of operations that a Location Server (essentially an MPC) should be able to perform. How it works is shown in Figure 4.1. This book describes the version of the API that was current in December 2001, but the API went through several major changes during the year before that, and while the main functions probably will work as described, several things have been changed in this version compared to the previous ones. Changes may continue to be made until a final version is released. So this description should be read as a description of principles, not details. You can find detailed descriptions on the companion CD-ROM.

The assumption is that the location-enabling server initiates the dialogue by sending a query to the MPC, and the MPC responds to the query. There is also a second-use case for triggered requests, where the request is responded to during a period or while the user is in a certain area.

The protocol used is HTTP, and in the context of MLP the client is the LCS Client and the server is the Location Server (GMLC/MPC). There are two TCP ports, port 700 for secure transactions (using SSL/TLS) and port 701 for transactions without encryption (for cases where the system is operating within a trusted domain). The standard port for HTTP is port 80, but as these ports are unreserved, in an address area whwere any function can be mapped to any port number.

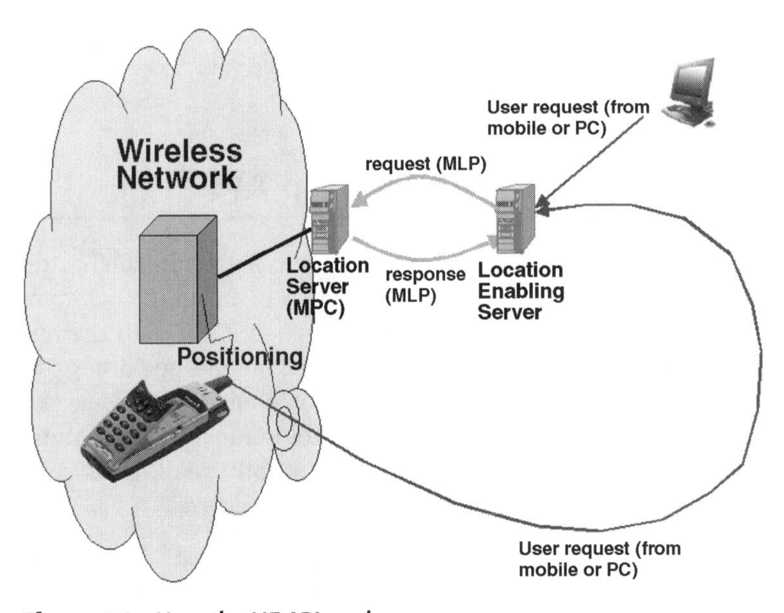

Figure 4.1 How the LIF API works.

There are two basic ways of requesting a position: an immediate request that is serviced immediately, and a triggered/deferred request that can be triggered by a timer (perhaps several times or periodically).

The application server requests a position (called a Location Query Service in the standard) by issuing an HTTP POST request toward the Location Server. The Location Query Services are invoked by sending a request using HTTP POST to a URI. It is possible to send both immediate requests and deferred requests, where the result can be delivered later, and periodic requests, where the delivery of the location information is triggered at a predetermined time period (or position). The difference between a deferred request and a triggered request is that the trigger can be set off at a later occasion, whereas the deferred request should be fulfilled as soon as possible. The deferred request is also fulfilled only once, and the triggered request is fulfilled only when the trigger criteria are fulfilled.

If the request is a deferred request (triggered or periodic), the result is delivered to the client through an HTTP POST operation issued by the MPC. This situation implies that the client must be able to receive HTTP POST requests and be able to give a valid response. The response to the invocation of a Location Query Service is returned by using an HTTP response.

The POST message contains a document, which in turn contains the requirements for the position information and the information about who triggered it. Triggers can be both the user requesting a position and an SMS being sent or received, as well as moving into another cell or making an emergency call. There are also possibilities for users outside the mobile network (for instance, a fleet management center) to request positioning not just for one mobile terminal, but also for a number of them (either by specifying their numbers or specifying a range of numbers).

The LIF API also has functions to manage the privacy flag in the Home Location Register. If the flag is set, the user has requested not to be positioned, and it can be overridden only by an emergency request. The status of the privacy flag is determined by using the PFLAG_STATUS element, which can be set in the status attribute to three different values: ON, which means no location requests can be made (although this setting can be overridden by emergency location requests); OFF, which means that the privacy flag is off and all location requests can be performed; and UNDEFINED, which means that the operator can determine the status depending on the local legal requirements (for example, in some countries, the flag can be set to OFF by default except for emergency positioning). It is a unclear how this will be implemented in the 3G systems, however.

The LIF API is formally defined in a number of XML DTDs (Document Type Definitions). They give the type definition for the XML elements that are to be sent in the different documents that comprise the messages. Because there are a number of common structured elements among the different services, the DTD that defines a single location query service is composed only by the definition of the root element and the inclusion of the necessary common DTD. The *Mobile Location Position* (MLP) is distributed over a set of common DTDs that define the core elements. In effect, the documents and DTDs together define data structures for the HTTP methods.

It is possible to trigger a location report when a mobile station enters or leaves an area. The area has to be defined by using the shape elements (more about that later) or a cell ID; because the user is registered when he or she enters or leaves a cell anyway, it is more likely that is how it will be used (if the area is a polygon inside a cell, it will still require the position to be measured regularly to determine where in the polygon the mobile station is or if it has left it).

Triggered reports can also be set based on timing. There is a special message element for both start and stop time for the request, which describes when the tracking of the user should start but can be sent at any time before the start (or stop) of the request.

The LIF API defines a number of shapes used in defining a geographical position. These go back to Chapter 3, where I described the different shapes and coordinate systems used to describe the Earth. When requesting and reporting a position, the result can be a point, but it is very rare that there are such precise positions. Instead, the position inaccuracy can be described as a circle (or some other type of shape, as we discussed in Chapter 3) around the point with a radius that describes the inaccuracy. Requests can be made for different accuracies; the response will contain the accuracy that can be delivered.

It is possible to set quality requirements on the position information by using the LEV_CONF attribute, which indicates the probability in percent that the mobile station is located in the position area that is returned (this information can also be used by responding entities, it seems). It is a percentage value. This element might sound spurious, but it has to do with the accuracy. If a system can determine that a user is in a circle sector that is long and narrow, and by measuring the circle arc in which he or she is located and finding that it is narrow and broad, or determining that he or she is in a cell, there is a tradeoff between the size of the inaccuracy area and the accuracy required. It is not very useful to know that a user is within a certain cell if it is large, but it will help the user know whether he or she is somewhere in a circle sector. At least he or she will know where he or she is going. Needless to say, the accuracy requirement will depend on the application. If the requesting system requires an accuracy that is higher than the best value it can give by using any other positioning method, it will have to return the cell.

Quality of position is not just dependent on the position, however, but also on the time when the position was measured. The RESP_REQ element can have three different attributes (in the resp_req_type attribute) that will set the quality level (nothing is said about what quality level will apply if this attribute is missing). The three attribute values are NO_DELAY (where the initial or last known location of the user will be sent back); LOW_DELAY (where the current location with minimum delay is returned with a best effort to fulfill the requirements); and the DELAY_TOL value (where the response is delay tolerant). The default value is DELAY_TOL.

The location can be one of four different types: current, last known, current or last known, or initial emergency call location. Which is defined in the LOC_TYPE element, and which type is set in the loc_type_type attribute. It can have four values: CURRENT (which means the current location); LAST (the last known location); the CURRENT_OR_LAST (current or last known location); and the INITIAL (which means the initial location when the user makes an emergency call). All these have to be qualified by the time and the accuracy of the position, of course.

To know what is current, last known, and so on, you need to have a rather precise idea about the time of your measurement. You can perform this task by using the TIME element, but there are several other elements that are used to describe timing in the LIF API. The RESP_TIMER element allows you to set the time in minutes and seconds for an interval during which the location should be obtained and returned to the requestor. The time of the requisition of a position measurement is declared by using the REQ_TIME element. This time does not have to be the same as the time the position was measured (which can be true, for instance, in the case of triggered positions).

Calculating velocity and direction is, as we noted in Chapter 2, rather tricky. There are elements in the LIF API, however, that enable the positioning system to return exactly those values. The DIRECTION element expresses position in degrees (with north as 0). It is only present if the positioning method used can be used to calculate it. VELOCITY is the speed of the mobile station in meters per second (so this value might have to be recalculated to kilometers per hour, miles per hour, or knots as required). It is also present only if it can be calculated by using the positioning method (positioning using a cell ID does not make it possible to calculate the velocity, for instance).

Altitude can also be measured by some positioning methods (for example, GPS, as we noted in Chapter 2, but not the pure network-based positioning methods). The altitude is expressed in the ALT element, which gives the altitude of the mobile station in meters (above mean sea level). The ALT_ACC element expresses the accuracy of the requested altitude in meters.

The LIF API has a special element for position information: the GEO_INFO element. Position can be in UTM or LL (Longitude Latitude) and XY coordinate formats, and it is possible to define different datum. In addition, there are a number of elements and attributes to describe shapes, as we discussed in Chapter 3.

Table 4.1 The Different Types of Mobile Identity Number in the LIF API

ABBREVIATION	TYPE OF IDENTITY NUMBER
IMEI	International Mobile station Equipment Identity
IMSI	International Mobile Subscriber Identity
MSISDN	Mobile Station International ISDN Number
MIN	Mobile Identification Number
MDN	Mobile Directory Number
EME_MSID	Emergency MSID

The LIF API relies heavily on the mobile system to identify the user. There are no alternative functions to provide an identity for a mobile station (you could, for instance, imagine IP numbers being used for this purpose—as they indeed are in the WAP Location Framework). There are six different types of identities for mobiles, as shown in Table 4.1.

Which of these number types applies depends on which network technology is used and whether the mobile is making an emergency call or not.

Among the identity-related functions in the LIF API is also CELLNO, which is a unique cell number that will identify the cell within a network (remember that there is a global numbering plan for mobile telephony that allocates numbers to networks so that all networks can really have unique numbers). The number is a string of characters; for instance, the *Location Area Code* (LAC) plus the *Cell ID* (CI) in GSM.

The VLRNO is another element that identifies a network entity; in this case, the visited location register number (which in GSM and UMTS is known as MSIN in the Global Title). The Global Title is the same format as an MSISDN; in other words, CC + NDC + MSIN.

Because the LIF API is supposed to work with the mobile positioning center, the identity is somewhat restricted. From the point of view of the mobile system, the only identities that exist are the identities of mobile phones (although there is an ID element in the LIF specification, which is a string defining the name of a registered user performing a location request. In an answer, the string represents the name of a location server). There is, however, a password element in the standard as well. The identity of a mobile phone resides on the *Subscriber Identity Module* (SIM) card of the mobile station. The system also has a function to position a range of identities, however (which is possible because the MSISDN is a number)—something that can be useful when working with applications like fleet tracking, where the phones in the cars of the fleet can have numbers in the sequence. The MSID can be of several types that are declared in the msid_type attribute of the MSID element. The msid_enc attribute contains the encoding of the identity, ASCII (which is the default value), base 64, or encrypted.

In some countries, the Network Operator (where the Location Server is assumed to be placed) is not allowed to send to the private information of a mobile station (like the MSISDN) to an MPC over the network. It might be allowed to send encrypted information, though, because only the network operator will be able to decode this information. The msid_enc attribute is intended to designate that this task is done, and the CRP value shows that the address has been encrypted (there are two other values, B64 for base 64 and ASC for ASCII).

Mobile Country Code (MCC) and *Mobile Network Code* (MNC) are two numbers that are part of the global numbering plan and are used to identify the country and the network where the call originates. The *Network Destination Code* (NDC) specifies where the call is directed.

The ESRD number is a special number that is used for emergency services. The *Emergency Services Routing Digits* (ESRD) is a telephone number in the national Numbering Plan that can be used to identify an emergency services provider and its associated Location Services client. The ESRD also identifies the base station, cell site, or sector from which an emergency call originates. The *Emergency Services Routing Key* (ESRK) is another telephone number in the national Numbering Plan that is assigned to an emergency services call for the duration of the call. The ESRK is used to identify (for example, route to) both the emergency services provider and the switch that is currently serving the emergency caller. During the lifetime of an emergency services call, the ESRK also identifies the calling subscriber. ESRD and ESRK can be of two different types: NA for North American and EU for European, plus Other for other expected formats (which regional or national standards bodies might define).

Emergency positioning is different from other positioning requests in that it can be triggered by an emergency call being made. When it is made, the EME_EVENT element is triggered and the eme_trigger attribute determines whether the emergency call started (the eme_org value) or if the call is disconnected (the eme_rel value).

When an emergency call is made, you know it is urgent. But other positioning requests than those that are emergency-call related might also need to be served urgently. To that end, there is a PRIO element in the API that can be used to determine whether the current request should be served before others. This element can be of two different types: high and normal, which are expressed with the prio_type attribute value. The default value is normal.

Each of the different methods is expressed in a separate XML document, which is sent as a POST to the MPC from the client. Here is an example of a request for immediate location information (for a range of mobile phones) in the

IDMS3 format. It contains a request for positioning of a number of mobile stations (the MSID_RANGE element), and the quality of position should be accurate within 1,000 meters altitude and low delay:

```
<?xml version = "1.0" ?>
<!DOCTYPE SLIR SYSTEM "MLP_SLIR.DTD">
<SLIR ver="1.0">
  <CLIENT>
    <ID>TheUser</ID>
    <PWD>The5PW</PWD>
  </CLIENT>
  <ORIGINATOR>
    <ID>TheASP</ID>
  </ORIGINATOR>
  <MSIDS>
    <MSID>461011334411</MSID>
    <MSID_RANGE>
      <START_MSID>461018765710</START_MSID>
      <STOP_MSID>461018765712</STOP_MSID>
    </MSID_RANGE>
  </MSIDS>
  <EQoP>
    <RESP_REQ resp_req_type="LOW_DELAY" />
    <HOR_ACC>1000</HOR_ACC>
  </EQoP>
  <GEO_INFO>
     <FORMAT>IDMS3</FORMAT>
  </GEO_INFO>
  <LOC_TYPE loc_type_type="CURRENT_OR_LAST" />
  <PRIO prio_type="HIGH" />

</SLIR>
```

The response to the request is sent in a Standard Location Immediate Answer document. Each document can contain a number of responses with positions for different mobile stations:

```
<?xml version = "1.0" ?>
<!DOCTYPE ELIA SYSTEM "MLP_SLIA.DTD">
<SLIA ver="1.0">
  <POS>
    <MSID>461011334411</MSID>
    <PD>
      <TIME>20000623134453</TIME>
      <SHAPE>
        <CIRCLE>
          <POINT>
            <LL_POINT>
              <LAT>301628.312</LAT>
              <LONG>451533.431</LONG>
            </LL_POINT>
          </POINT>
```

```
          <RAD>240</RAD>
        </CIRCLE>
      </SHAPE>
    </PD>
  </POS>
  <POS>
    <MSID>461018765710</MSID>
    <PD>
      <TIME>20000623134454</TIME>
      <SHAPE>
        <CIRCLE>
          <POINT>
            <LL_POINT>
              <LAT>301228.302</LAT>
              <LONG>865633.863</LONG>
            </LL_POINT>
          </POINT>
          <RAD>570</RAD>
        </CIRCLE>
      </SHAPE>
    </PD>
  </POS>
  <POS>
    <MSID>461018765711</MSID>
    <PD>
      <TIME>20000623110205</TIME>
      <SHAPE>
        <CIRCLE>
          <POINT>
            <LL_POINT>
              <LAT>781234.322</LAT>
              <LONG>762162.823</LONG>
            </LL_POINT>
          </POINT>
          <RAD>15</RAD>
        </CIRCLE>
      </SHAPE>
    </PD>
  </POS>
  <POS>
    <MSID>461018765712</MSID>
    <POSERR>
      <RESULT resid="10">QOP NOT ATTAINABLE</RESULT>
      <TIME>20000623134454</TIME>
    </POSERR>
  </POS>
  <GMT_OFF>+0200</GMT_OFF>
  <RESULT resid="0">OK</RESULT></SLIA>
```

There is much that can go wrong when retrieving the location of a mobile terminal, however. When an error message is returned, the position gateway will return a result code with it to describe what went wrong. The RESULT element

Table 4.2 The Result Codes in the LIF API

RESULT CODE	SLOGAN	DESCRIPTION
0	OK	No error occurred while processing the request.
1	SYSTEM FAILURE	The request can not be handled because of a general problem in the location server or the underlying network.
4	UNKNOWN SUBSCRIBER	Unknown subscriber. The user is unknown, i.e. no such subscription exists.
5	ABSENT SUBSCRIBER	Absent subscriber. The user is currently not reachable.
6	POSITION METHOD FAILURE	Position method failure. The location service failed to obtain the user's position.
101	CONGESTION IN LOCATION SERVER	The request cannot be handled due to congestion in the location server.
102	CONGESTION IN MOBILE NETWORK	The request cannot be handled due to congestion in the mobile network.
104	TOO MANY POSITION ITEMS	Too many position items have been specified in the request.
105	FORMAT ERROR	Parameter in the request has invalid format. The invalid parameter is indicated in ADD_INFO.
106	SYNTAX ERROR	The position request has invalid syntax. Details may be indicated in ADD_INFO.
107	PROTOCOL ELEMENT NOT SUPPORTED	The element specified in the position request is not supported by this implementation. The element is indicated in ADD_INFO.
108	SERVICE NOT SUPPORTED	The indicated service is not supported in the Location Server.
201	UNKNOWN SUBSCRIBER	The user is unknown, i.e. no such subscription exists.
202	HLR PRIVACY FLAG SET	The privacy flag of the MSID in the HLR is set to ON meaning that location requests cannot be performed on the MSID.
207	MISCONFIGURATION OF LOCATION SERVER	The location server is not completely configured to be able to calculate a position.

will contain the result code in the resid attribute. In Table 4.2, the result codes and the descriptions of what they mean are listed. The numbering is not consistent because several codes have been removed in the draft version.

The WAP Location Framework

The much-maligned *Wireless Application Protocol* (WAP) architecture has been entirely revised in version 2.0, using and extending Internet standards such as HTTP and XML (although the WAP markup language, WML, was an XML application from the start). A description of WAP 2.0 is included in Appendix C of this book.

The WAP Forum has developed a location architecture to fit into WAP that interacts with the MPC and the application server (or that is the MPC and application server in itself). It provides a data format and a header for HTTP, which means that any application server, as well as any client,that intends to use this system to communicate location information has to implement extra HTTP headers. The system also integrates with the WAP Push system, enabling information to be pushed to clients (using HTTP, you can only pull information). In practice, though, this process is equivalent to a POST operation.

The WAP Location Framework overlaps the LIF API in many ways, and while that is partly intended for communication from a mobile station directly to the MPC, you can easily imagine how a Web server (or WAP gateway) could use the WAP Location Framework in connecting to users' systems while the mobile positioning can be made by the server through calls through the LIF API (or through a second gateway). While it is unlikely that a Web site developer will use the LIF API directly (it being used by an application server), the WAP API seems much more adapted to the needs of that developer. It is also more stable.

You can use the WAP location framework in three different ways:

Immediate Query Service. The Immediate Query Service allows an application to query WAP Location Query Functionality for the location of a WAP client and get an immediate response. This application could, for example, be a tracking application (tracking a fleet of cars) that wants to request the position of an entity periodically and receive a response immediately.

Deferred Query Service. The Immediate Query Service allows an application to send a query to the WAP Location Query Functionality and to get back the location of a WAP client with (possibly multiple) deferred responses. Deferred response means that there is no immediate response; for instance, sending a request and getting responses every two minutes. Tracking applications (for example, for fleet management) that want to track devices periodically are use cases for this type of query, as well.

Location Attachment Service. When a user sends a request to a service and the information can be personalized based on the position, it is very convenient to include the location of the user in the request so that the response can be returned within the scope of the same transaction. One way is to use the WAP UAPROF terminal location attribute; another is to use the WAP Location Attachment Service. An example application that could use the Location Attachment Service is the classic "find the nearest restaurant" application.

Note that the framework only describes how to send messages containing location information. It does not describe what happens when a location request arrives at a client or an intermediary, and it does not say anything about the data that is to be presented to the user (other than some technical matters). Figure 4.2 is a diagram outlining how it works. Because the WAP 2.0 architecture uses the XML-based XHTML format, however, it is possible to generate documents to be presented that follow both the rules of thumb for database design in Chapter 6 and the heuristics for user interface design in Chapter 8.

The framework consists of a number of logical entities. The WAP Location Network is a convenience placeholder for intermediaries, such as the WAP gateway, which can produce WAP location queries and attach the location of the device to a request. The application server in this picture is not an application server processing location queries but rather the server to which the query is directed, such as a Web server.

The framework is formally defined in a series of DTDs, or Document Type Definitions, which essentially describe the functions of the API.

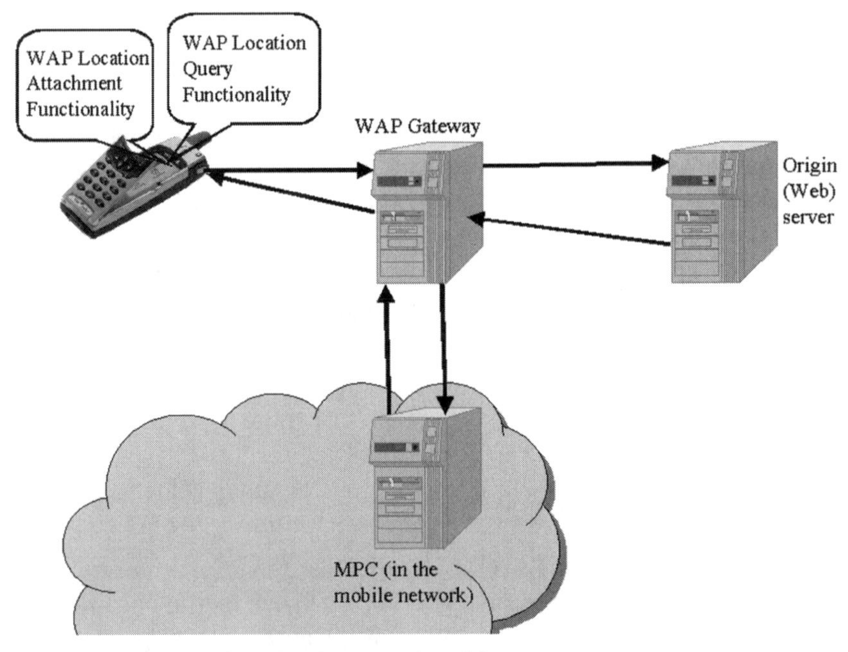

Figure 4.2 The WAP location framework architecture.

The framework does not take the creation of the location information into account, and it does not describe how other entities (for example, the push proxy) should interact with the WAP location query or the WAP location attachment functionality.

The location query can reside in the WAP client or in a supporting server (Performance Enhancing Proxy), such as the WAP gateway. This situation is possible because it has access to the identity of the user and can look up the information from the network in the same way as an application server would. Indeed, there is very little to distinguish it from an application server when it comes to the location functionality, although it does have a number of other possible functions.

The WAP Client can be a WAP browser, but it can also be a device that does not have a user interface; for instance, a telematics device in a car. The location attachment functionality can reside in a separate server (which could be a regular application server), the WAP client, or the WAP gateway.

The three different services provided by the WAP location framework vary a great deal in scope. For the immediate and deferred query, you have to have a unique identity in the client, such as an MSISDN, an IP address, and a HTTP server because the message that invokes the service is sent by using an HTTP POST, and that can per definition only be received by an HTTP server. The requestor might also provide a transaction ID, which is used to identify this particular request, as well as a short free-text description of itself (this description can be used to filter requests if desired).

The difference between the immediate and deferred services is that with the immediate service, an answer is required right away; with the deferred service, you specify a period or other way of making queries at specified intervals. The deferred query service can be initiated by someone other than the client (for example, a fleet management server). Immediate and deferred requests can be combined (for instance, "Give me your position now and every 10 minutes for the next two hours").

It is possible to request the location of several clients at the same time by having multiple client addresses in an invocation. This situation is the same as specifying multiple request messages where all parameters are identical except for the client address. It is also possible for the application to request different kinds of location information in the same query (within the same geo-info element). This situation is the same as sending multiple request messages where all parameters are the same except the requested location information.

There is also a *Quality of Position* (qop) element that can be used to request information of a certain accuracy or maximum age. If the time when the position was measured is known, it should be specified by using a time stamp. Otherwise, the system is supposed to assume that it is a general best-effort position.

The lev-conf element in the pos element is used to express the confidence in the position information. It expresses the quality in percent, and the level of confidence applies to all accuracy elements within the pd element.

The location attachment service, however, can be attached to any query, and how the identity and session are handled (in HTTP, because there are sessions in the WAP Session protocol, or WSP) is dependent on implementation, although the intention seems to be that nobody but the originator of the request for information should be able to initiate the attachment of location information (for example, by making a location-dependent request for information). The location information is always appended to a WSP/HTTP request initiated by a WAP client within the scope of that request. This situation also means that it is not possible to request attachment of location for more than one client at the time.

The request and response for all message types are both encapsulated in XML messages. There are two types of encapsulation: the Location Invocation document, which is used for requesting location information (in other words, to send invocation messages to the location service) and the Location Delivery document, which is used for returning delivery messages in response to invocations. The invocation and delivery root elements can contain one or more message elements—in other words, it is possible to send multiple request messages in a single invocation document or to receive multiple answer and report messages in a single delivery document. It is also possible to specify multiple WAP client addresses and/or multiple types of location information within a single request message.

A Location Attachment Functionality that receives the attachment-request message might reissue the WAP client's request with the attachment-answer message attached. If there is no location information available or it is not allowed to be delivered (because of privacy reasons, for example), the Attachment Functionality might ignore the attachment-request message and the WAP client can then use any content that came as a response. The document can contain a script that can be used to manually request a location from the user or a document that contains information that can be displayed instead of the location-related information.

The HTTP header used for attachment requests and responses is a new HTTP header (which is fine because the HTTP specification allows for the definition of new headers), X-WAP-Loc-Invocation-value. The header contains the header name and the data field.

The POST method is almost identical to the method used for POST to a WAP Push service. Both the immediate-query-request and the deferred-query-request can work the same way as WAP Push, although in a different direction. It integrates with the push service as illustrated in Figure 4.3 (the push initiator invokes a delivery of position information).

In the first version of WAP, there was only one type of identity that a client could have: MSISDN. In WAP 2.0, there can be a number of different methods

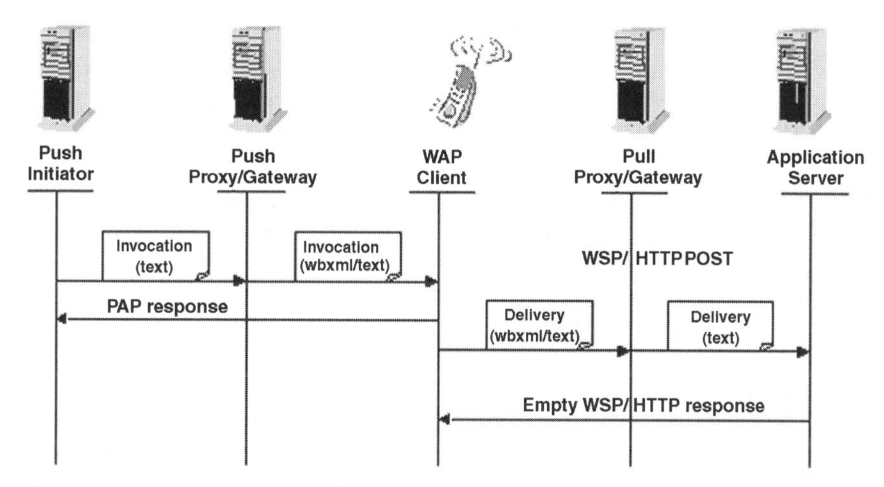

Figure 4.3 How the deferred query integrates with WAP Push.

to address a client, and the addressing also includes different types of resource servers that can provide functions for the system. There are several additional methods to handle identities; for instance, the Wireless Identity Module, which is also used in the IMT-2000 system.

Table 4.3 lists the identity formats that are used in the WAP location framework.

It is also possible to specify a range of identities within which query requests are to be made (for instance, if all cars in a fleet have sequential IPv6 addresses, it is possible to specify the range in a deferred query and the system will query each address within the range with the specified interval, essentially creating a continuous query of the entire fleet).

As we saw in Chapter 3, time stamps are very important—both for quality of position and for deferred queries. Time in the WAP Location Framework is normally given in the UTC format. If the time is not known, the element is not

Table 4.3 The Identity Formats of WAP Clients

IDENTITY TYPE	ADDRESS FORMAT
PLMN	A phone number, e.g. +4479680254567, format as specified below (global-phone-number).
IPv4	An IP version 4 address [RFC791], format see below (ipv4), e.g. 123.456.789.123.
IPv6	An IP version 6 address [RFC2373], format see below (ipv6), e.g. ABCD:6785:F65D:56F4:D687:F7DC:372F:F4D5
PAP-USER	A PAP (Push Access Protocol) user type. Examples: john.doe @wapforum.org/TYPE=USER@ppg.carrier.com; 47397547589 +15551990730/TYPE=USER@locserv.carrier.com

used. The format is YYYY-MM-DDThh:mm:ssZ. The Z designates the universal time zone.

The time element contains an optional attribute specifying the offset (hours and minutes) of the time in relation to UTC (the default is 0) to express time zones.

The WAP location framework specifies that latitude and longitude coordinates using the WGS-84 datum are always supported.

In addition, there are a number of optional information items:

- UTM coordinates using the WGS-84 datum.
- Spatial reference systems such as national grids and geocodes (for example, postal address).
- Speed, altitude, direction, and heading

The geoinfo element is used to handle the position information and describe which datum and reference system is used. Only the WGS-84 data is required by the WAP location framework, however, so the other datums will most likely be used only for specialized applications. The coordinate system used within the datum is given by the coord-datum element, which specifies the coordinate system in the coord-sys element, and the datum used. So far, only the WGS-84 datum has been specified and only UTM coordinates and LL (lat-long) coordinates (which are the default). Specific reference systems can be specified by the ref-sys element. There are a vast number of reference systems (as we discussed in Chapter 3), and the specification actually points to other specifications to define where they can be found.

A pd element can only contain information regarding a single WAP client, but it might contain multiple types of location information for the same client and be related to the same time stamp. A time stamp must always be present, if it is known, and applies to the entire pd element.

The pd element can also contain accuracy elements for the different position elements. Applications of the WAP location framework can also request, as well as describe, the quality of position (see Chapter 2). Immediate or deferred queries (although not attachments) can request a certain age or accuracy by specifying constraints in invocations; for example, "within 50 meters" or "no more than 30 seconds old." The applications can use this information, for example, to select an appropriate location method to use. This information is expressed with special elements, as described in Table 4.4. The query contains a value that the requestor wants to use, and the response is the value that the originator can provide.

Note especially the must-be-satisfied element. If the system cannot deliver this information with the level of confidence specified in this element, it should not deliver position information at all.

Table 4.4 The Quality of Position Elements

NAME	FUNCTION
maxage	Maximum requested age of location information, in seconds, as a non-negative decimal number. Only used in requests, not in responses.
lev-conf	Percentage value of confidence in position value. Applies to all accuracy elements within a pd element.
hor-acc	Horizontal accuracy is given in meters, as a non-negative decimal number.
alt-acc	Altitude accuracy as a positive value in meters
speed-acc	Accuracy of speed given in meters per second, as a non-negative decimal number.
dir-acc	Direction accuracy in degrees, as a non-negative decimal number.
heading-acc	Heading accuracy given in degrees, as a non-negative decimal number.
must-be-satisfied	Indicates that the described level of confidence is required. Possible values are yes and no

It is not necessarily the case that the requested quality of position can be honored (either because it is not possible or because the application does not choose to do it). What QoP is actually provided depends, for example, on available location methods for a particular WAP client at a particular point in time (for example, a client might be temporarily out of coverage), different billing models (for example, GPS might be free, differential GPS might be charged for —and the client might not be willing to pay for the position in this instance), and so on. There is no function to express billing parameters, however.

There is a special element, shape, to describe the shape of a polygon surrounding the position (it is even, to a limited extent, possible to describe the classic problem of a polygon with holes in it).

As an example of a circle with a radius of 240 meters around a point is described like the following, using the WAP Location Framework shape element:

```
<shape>
     <circle>
          <point>
              <ll-point>
                   <lat>30.347692</lat>
                   <long>45.437628</long>
              </ll-point>
            </point>
          <rad>240</rad>
```

```
        </circle>
    </shape>
```

The WAP location framework allows WAP location functionalities to supply the actual age, accuracy, and confidence of the delivered location information (as far as it is known). The specification recommends that applications that depend on a certain quality of position check the actual delivered accuracy. There are a number of subelements for the different message types that describe different properties of the messages themselves (they are not necessarily organized in the way described in the following tables in the WAP location specification), and presenting them as tables can be a little confusing because these are XML elements and can be nested within each other. Here, to illustrate how it works, is an example.

This coding is what a request for position-dependent information in an attachment would look like. The intention is that this document should be attached to the GET request:

```
<?xml version = "1.0" ?>
<!DOCTYPE WAP_LOC_INV PUBLIC
     "-//WAPFORUM//DTD LOC INV 1.0//EN"
     "http://www.wapforum.org/DTD/wap-loc-inv-1.0.dtd">

<invocation>
     <attachment-request>
          <application>ACME Driving Directions</application>
     </attachment-request>
</invocation>
```

The request is embedded in the invocation document (either a POST with trigger information, or presumably what is intended in this case, a GET, which means the document is empty), and it is done by an application called the ACME Driving Directions. The response is embedded in a delivery document, and it gives the position in the shape of a circle in the LL coordinate system:

```
<?xml version = "1.0" ?>
<!DOCTYPE WAP_LOC_DEL PUBLIC
     "-//WAPFORUM//DTD LOC DEL 1.0//EN"
     "http://www.wapforum.org/DTD/wap-loc-del-1.0.dtd">

<delivery>
     <attachment-answer>
          <transaction-id>12345@app.acme.com</transaction-id>
          <pos>
               <pd>
                    <time utc-off="+0200">2000-06-23T13:44:53Z</time>
                    <coord-datum coord-sys="LL" datum="WGS-84"/>
                    <shape>
                      <circle>
```

```
                    <point>
                <ll-point>
                    <lat>30.347692</lat>
                    <long>45.437628</long>
                </ll-point>
            </point>
            <rad>240</rad>
          </circle>
        </shape>
      </pd>
    </pos>
  </attachment-answer>
</delivery>
```

An attachment can contain multiple types of location information, but it can only concern one client and only relate to one single transaction. Deferred queries and responses can be much more complex because they can involve both multiple clients and multiple responses from those clients.

If there is a problem, the system must be able to report an error. There is an element called poserr, which is intended for instances when the system can not report the position for some reason; in those cases, a timestamp and the error message is sent. There are also a number of specific error codes, which are listed in Table 4.5.

Table 4.5 The Error Messages in the WAP Location Framework

RESULTATID	SLOGAN	DESCRIPTION
0	OK	No error occurred while processing the request. In the deferred query case, this error code can only occur in the deferred-query-answer and the deferred-stop-answer messages. In the case of transport mappings not supporting these messages (e.g. Push) this error code cannot be returned.
1	SYSTEM FAILURE	The request cannot be handled because of a general problem in the location server or the underlying network.
3	UNAUTHORIZED APPLICATION	The application is not authorized to obtain the location of the WAP client.
4	UNKNOWN SUBSCRIBER	Unknown subscriber. The subscriber is unknown, i.e. no such subscription exists.
5	ABSENT SUBSCRIBER	The WAP client is currently not reachable.
6	POSITION METHOD FAILURE	The WAP location functionality failed to obtain the position of the WAP client.

Table 4.5 continued

RESULTATID	SLOGAN	DESCRIPTION
7	CONGESTION IN LOCATION FUNCTIONALITY	The request cannot be handled due to congestion in the WAP location functionality.
8	CONGESTION IN MOBILE NETWORK	The request cannot be handled due to congestion in the mobile network.
9	INSUFFICIENT RESOURCES	The WAP location functionality was unable to complete a request due to insufficient resources, e.g. lack of memory, the invocation document was too complex, it contained too many request messages, etc.
10	SYNTAX ERROR	The position request has a syntax error, e.g. malformed or nonvalidated XML syntax, invalid values in some elements or attributes, invalid format of values of attributes and elements, etc. In the deferred query case, this error code can only occur in the deferred-query-answer and the deferred-stop-answer messages. In the case of transport mappings not supporting these messages (e.g. WAP Push) this error code cannot be returned.
11	PROTOCOL ELEMENT NOT SUPPORTED	An optional protocol element specified in the position request is not supported by this implementation, e.g. msid-range. 12 SER-
VICE NOT	A requested service, e.g. deferred queries or a SUPPORTED	periodic trigger, is not supported in the WAP location functionality.
13	TYPE OF LOCATION INFORMATION NOT SUPPORTED	A requested type of location information is not supported, e.g. speed, altitude etc.
14	TYPE OF LOCATION INFORMATION NOT CURRENTLY SUPPORTED	A requested type of location information is temporarily unavailable, e.g. due to some temporary internal problem.
15	QOP NOT ATTAINABLE	The requested quality of position cannot be provided.
16	QOP NOT CURRENTLY ATTAINABLE	The WAP location functionality is temporarily unable to provide the requested quality of position, e.g. due to a WAP client being out of coverage.

RESULTATID	SLOGAN	DESCRIPTION
17	REPORTING WILL STOP	A deferred query request has been cancelled, and further reports will not be produced. This error code can only occur in a deferred-query-report message.
18	TIME EXPIRED	The start time or stop time of a deferred query has expired.
1000-		Vendor-specific error codes.

The Parlay API

Both the LIF and WAP Forum APIs borrow many ideas from the Parlay API, which is not so strange because they are all using the same underlying system: the mobile telephone network. The Parlay API is very oriented toward GSM (which is not just a mobile telephony system, but rather an entire telephone system), but there are mappings from the Parlay API to functions in other networks, as well.

Parlay is an API for *Intelligent Networks* (IN), which was a big item of development in the telecommunications industry a few years ago. It was developed to enable programmers to write software that used the existing services in an intelligent network (which was an effort by the telephony industry to create a system to enable the telephone network to have more functions, which would reside in the network, not in the terminal). This situation would mean that the network structure would be maintained as it is today, using resource servers that controlled the switches. Terminals would still be stupid, and the services would be executed in the network, not in the terminals (as the Internet model would prescribe in similar cases).

The Parlay APIs work toward a server, the Parlay Capability Server, on the mobile operator's network, which in turn is connected to the different parts of the network through an SS7 interface. In that way, the idea is very similar to the LIF and WAP Forum APIs. The functions called by the API execute within a framework, which isolates it from the network (somewhat in the same way as Java works, but on a different scale—the Parlay Capability Server being the sandbox here). The interfaces between the network functionalities and the Parlay Capability Server are not defined by the Parlay specification.

Parlay gives programmers more control than the LIF and WAP Forum APIs over how position information is retrieved, and what position information will be retrieved, for applications that rely on the network. This scenario means that for applications that use the network more frequently, such as tracking applications, Parlay might be a better choice (if available and not overpriced).

The API is based not on HTTP but on CORBA, the Common Object Request Broker Architecture from the Object Management Group. In other words, the definition is more formal, and the execution of the applications is more controlled than in the LIF and the WAP APIs. You invoke methods on interfaces and get back result messages. There is one method per function in a CORBA system, but if you use HTTP, you only have four (GET, PUT, POST, and HEAD).

Parlay has a generic result message, TpResult, which is always returned by all methods. There are also a few methods that have to be present in the application that is requesting the information to receive the result of the API call.

This functionality brings more flexibility to the development, but on the other hand, it also means that as a programmer you are required to conform to a development model that might not always follow your style of working. It also means that all definitions in the API are in UML, so if you are used to reading formal specifications, it is a much easier read than the WAP or LIF APIs (it is longer, though).

Intelligent networking is very much about enhancing the value of a telephone call, and the Parlay API is very oriented toward the traditional telephony model of working. This situation also means that the Parlay API contains a number of functions that are not of concern to us in this book and that we will only look at the ones described in the mobility part of the specification.

Positions in Parlay are defined by using the TpGeographicalPosition data structure, which is a Sequence of Data Elements that specifies a geographical position using WGS84 as the reference datum. An ellipsoid point with uncertainty shape defines the horizontal location. TypeOfUncertaintyShape describes the type of the uncertainty shape, and Longitude/Latitude defines the position of the uncertainty shape. Shapes in the Parlay API are defined in terms of arcs on an ellipse, as we discussed in Chapter 3.

The TpLocationUncertaintyShape defines the type of shape within which an uncertainty is allowed. The shape can be described in terms of the primitives of the Parlay API: a circle, a circle sector, a circle arc stripe, an ellipse, an ellipse sector, and an ellipse arc stripe.

The TpLocationRequest data structure defines the Sequence of Data Elements that specifies a location request. It allows you to specify the requested accuracy, altitude, type of location, and the priority of the location request. You can also request a specific positioning method.

Using the TpUlExtendedData data structure, you can define an extended format for the location data. The optional vertical location is defined by the data element Altitude, which contains the altitude in meters above sea level, and the data element AltitudeAccuracy, which contains the accuracy of the altitude. Note that the TpGeographicalPosition data structure can be included in the

TpUlExtendedData if that data structure is used. There are a few additional user-related data elements in the TpUserLocationExtended data set that can be used to further determine the user's position (as well as a number of other things). This data structure can contain the TpUlExtendedData dataset. TpUserLocation consists of the user ID, the TpGeographicalPosition, and TpMobilityError data structures.

The TpLocationResponseIndicator lets you request a response time for the positioning. That can vary between the last known location, which is to be returned immediately; or within a timer value, without any regard for the timer. There are four different values of different accuracy. It is used together with the TpLocationResponseTime data structure, which defines the Sequence of Data Elements that specifies the application's requirements on the mobility service's response time. This data structure includes the TpLocationResponseIndicator and a timer value.

TpLocationType defines the type of location requested. There are three different values: current location, current or last known location, and the initial position of an emergency call.

The TpLocationPriority data structure defines the priority of a location request. It can be either high or normal (1 or 0).

The Parlay API allows for triggered requests for location. The triggering is set by using the TpLocationTrigger data structure in an area that is assumed to be an ellipse. These elements are numbered in the TpLocationTriggerSet data structure. The triggering criteria (whether the trigger should be set to when a user leaves or enters an area) are set by using the TpLocationTriggerCriteria data structure, which can be embedded in the TpLocationTrigger data structure.

There are corresponding data structures for the Camel and emergency methods, but we will not describe them here.

Parlay not only works for mobile users, but also for fixed users (in case their position is known or knowable), the reason being that it is intended to be used for emergency applications (which of course can be made from a fixed as well as a mobile phone). It is also defined to work with IP telephony. The TpTerminalType data structure defines whether the terminal is a fixed, mobile, or IP telephone (as mentioned before, in the world of Parlay, terminals can only be telephones. There are no other types of terminals). As I talked about in Chapter 2, there is no way you can automatically deduce the position from an IP address, so you need some kind of mapping between those addresses and the position. In principle, the routers in the network could perform this task because the routers know which addresses are attached to which interfaces. It is not defined in the Parlay standard, however, but which position methods can be used is defined.

Apart from the position, it is possible to get more information about the user's status. The status indicators include the user's terminal type and whether the user is busy, reachable, or not reachable. Remember that the Parlay API ties into the telephone network in a different way from the other APIs discussed in this chapter and can get information from the signaling in the network.

Like the other APIs I have discussed in this chapter, there are error messages for the Parlay API that can be sent in case problems occur (see Table 4.6).

Apart from the diagnostic messages, there is also a separate data structure that contains error messages. TpMobilityError defines errors that are reported by the mobility services. These are different from the diagnostic values, which can relate to other things than the pure network aspects (such as "disallowed by local regulatory environments"). The network error messages are listed in Table 4.7.

The UL service provides the IpUserLocation and IpTriggeredUserLocation interfaces. The methods in these interfaces are usually asynchronous and do not lock a thread into waiting while a transaction is performed. This situation means that the client machine can handle many more calls than can a service that uses synchronous message calls. Two interfaces especially intended for responses and reports must be implemented in the application: IpAppUserLocation and IpAppTriggeredUserLocation. They are the receivers of the result messages.

As an example of what services could look like, here two fictive services, one basic (FindIt) and the other more advanced (locate.com), are described in Table 4.8. They are described in terms of the different subtypes, and as you can see, the locate.com example uses both the IpUserLocation method and the IpUserLocationTriggers method.

The IpUserLocation inherits from the generic service interface and provides the management functions to the user location service. The application programmer can use this interface to obtain the geographical location of users. It contains four different methods, as outlined in Table 4.2. In principle, this interface in combination with the IpAppUserLocation interface follows a simple response-request model: You send a request for the location of one or more users, and you will get back a response.

The Magic API

The Magic Services API was created by a loose industry group, led by Microsoft and biased toward the automotive industry, with the goal of creating a Web services API for location information. The protocol used between the mobile sta-

Table 4.6 The Different Diagnostic Values for Error Reports

NAME	VALUE	DESCRIPTION
P_M_NO_INFORMATION	0	No diagnostic information present. Valid for all type of errors.
P_M_APPL_NOT_IN_PRIV_EXCEPT_LST	1	Application not in privacy exception list. Valid for 'Unauthorised Application' error.
P_M_CALL_TO_USER_NOT_SETUP	2	Call to user not set-up. Valid for 'Unauthorised Application' error.
P_M_PRIVACY_OVERRIDE_NOT_APPLIC	3	Privacy override not applicable. Valid for 'Unauthorised Application' error.
P_M_DISALL_BY_LOCAL_REGULAT_REQ	4	Disallowed by local regulatory requirements. Valid for 'Unauthorised Application' error.
P_M_CONGESTION	5	Congestion. Valid for 'Position Method Failure' error.
P_M_INSUFFICIENT_RESOURCES	6	Insufficient resources. Valid for 'Position Method Failure' error.
P_M_INSUFFICIENT_MEAS_DATA	7	Insufficient measurement data. Valid for 'Position Method Failure' error.
P_M_INCONSISTENT_MEAS_DATA	8	Inconsistent measurement data. Valid for 'Position Method Failure' error.
P_M_LOC_PROC_NOT_COMPLETED	9	Location procedure not completed. Valid for 'Position Method Failure' error.
P_M_LOC_PROC_NOT_SUPP_BY_USER	10	Location procedure not supported by user. Valid for 'Position Method Failure' error.
P_M_QOS_NOT_ATTAINABLE	11	Quality of service not attainable. Valid for 'Position Method Failure' error.

Table 4.7 Network Error Messages in the TpMobilityError Data Structure

NAME	VALUE	DESCRIPTION	FATAL
P_M_OK	0	No error occurred while processing the request.	N/A
P_M_SYSTEM_FAILURE	1	System failure. The request cannot be handled because of a general problem in the mobility service or the underlying network.	Yes
P_M_UNAUTHORIZED_NETWORK	2	Unauthorised network, the requesting network is not authorised to obtain the user's location or status.	No
P_M_UNAUTHORIZED_APPLICATION	3	Unauthorised application. The application is not authorised to obtain the user's location or status.	Yes
P_M_UNKNOWN_SUBSCRIBER	4	Unknown subscriber. The user is unknown, i.e. no such subscription exists.	Yes
P_M_ABSENT_SUBSCRIBER	5	Absent subscriber. The user is currently not reachable.	No
P_M_POSITION_METHOD_FAILURE	6	Position method failure. The mobility service failed to obtain the user's position.	No

Table 4.8 Two Example Location Services

PROPERTY NAME	PROPERTY VALUE	
Service 1	Property Value	Service 2
Service instance ID	0x80923AD0	0xF0ED85CB
Service name	UserLocation	UserLocation
Service version	2.1	2.1
Service description	Basic user location service.	Advanced high-performance user location service.
Product name	Find It	Locate.com
Product version	1.3	3.1
Supported interfaces	"IpUserLocation"	"IpUserLocation", "IpUserLocationTriggers"
Permitted application types	"Emergency service", "Value added service"	"Emergency service", "Value added service", "Lawful intercept service"
Permitted application subtypes	Not defined	Not defined
Priorities	"Normal"	"Normal", "High"
Altitude obtainable	False	True
Location methods	"Timing Advance"	"GPS", "Time Of Arrival"
Max interactive requests	2000	10000
Max triggered users	200	2000
Max periodic users	300	2000
Min periodic interval duration	600	30

tion and the server is Simple Object Access Protocol (SOAP), and the document format is XML. Each of the Client API functions that an application can call is converted into a serial stream of XML, stored in a SOAP envelope for transport via HTTP or SMTP. See Figure 4.4.

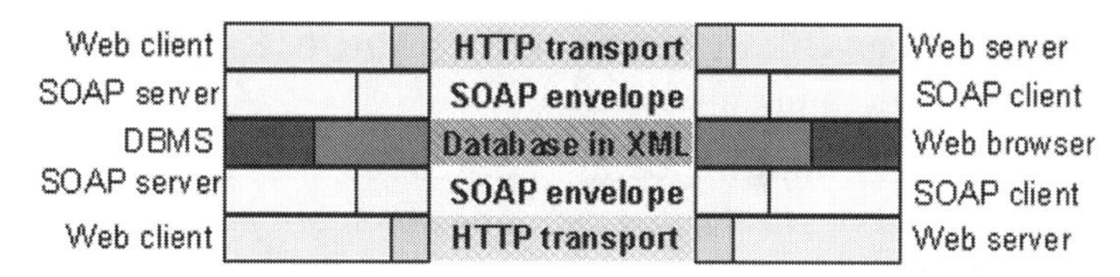

Figure 4.4 The principles of SOAP over HTTP.

Their idea is to create a set of core services, which can be called as Web services in the emerging W3C architecture, that could be used by Application Service Providers (ASPs) who intend to set up location-dependent services. The scope of the services are somewhat broader than that of the APIs we discussed previously in this chapter, because they also include route planning and geocoding, two things other APIs assume will be applications handled by some other entity. MAGIC, which does not have anything to do with Harry Potter, is an abbreviation of Mobile Automotive Geo-Information Services Core.

The MAGIC functions include:

- *geocoding:* The conversion of human text or speech defining an address or other location expression to corresponding geographic coordinates.
- *reverse geocoding:* The conversion of mobile device coordinates (latitude and longitude) to an address expressed as text or speech that can be directly understood by a human.
- *spatial query:* The location-sensitive retrieval of information based on geographic location with respect to geometric or travel-time proximity to a position, region, or route.
- *travel planning and guidance:* The specification of travel destinations and intermediate waypoints as well as the location and time-sensitive delivery of information and instructions to a traveller moving toward a destination.

The intention is to move the functions from the client to the server, which would make the requirement on the client device lower—and correspondingly cheaper. They also position themselves one step higher up in the value chain than the LIF, WAP Forum, and Parlay APIs, which means that their niche is somewhat different—not companies who primarily want the information, but companies who want to provide services based on the information. The system has a positioning API of its own, but this has not yet been developed, and the intent is to reuse a suitable API from some other organization

The MAGIC Services architecture includes seven primary components (see Figure 4.5):

1. The **MAGIC Services Client API** is the interface between a Client Application and the MAGIC Services.
2. The **MAGIC Services Client** implements the MAGIC Services Client API for a particular Client implementation.
3. The **Client-Server Communication Protocol** enables an information exchange via a wired or wireless network between MAGIC Services Clients and MAGIC Services Servers.

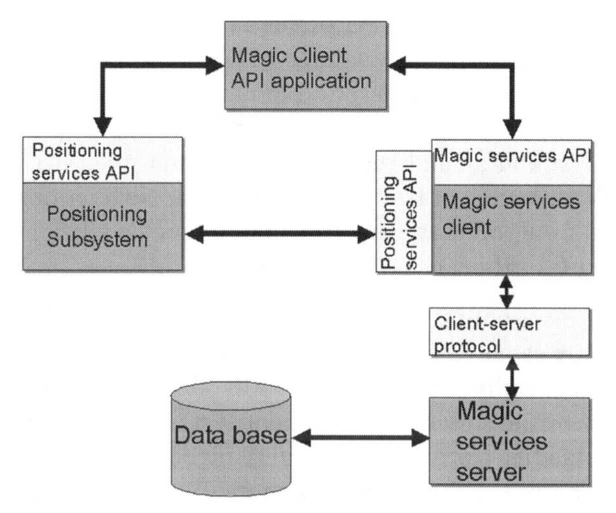

Figure 4.5 The Magic API.

4. The **MAGIC Services Server** implements the MAGIC Services to be supplied to MAGIC Clients.
5. The Data **Upload Protocol** enables a MAGIC Server to access Geographic Data via a wired or wireless network.
6. The **Positioning Services API** provides a standard interface toward **Positioning Software** and Navigation Sensors to enable efficient locating of the client device's position in real-time. This is essential for devices having limited system resources.
7. Via the **Positioning Support API**, access is granted to locally stored or cached navigable data. This is very important for real-time positioning.

As it is intended to fit into an automotive environment, the Magic API supports monolithic data delivery, that is all the data and applications coming on one CD-ROM (or DVD). Being modular, the API also supports environments with thick clients (for example, a Web browser on a PC or PDA), or thin clients (for example, a mobile phone). Because the application architecture is object-oriented, it can be divided into modules, which can be executed in different parts of the system.

The focus of the work is initially the client services, which are intended to hold the core functionality required to develop and deploy navigation, telematics, and location-based services. The client API can be implemented in a number of ways: As a library connected to a local fixed database; as requests to a remote server; and an XML document that contains the functions and parameters. Presentation is managed by the man-machine interface in the mobile client device.

The system can be implemented as one environment (monolithically) or as four different functional modules: Semantic Services, a query translation service that manages the interpretation and generation of natural language references to geographic entities; Query Services, which perform retrieval of spatial information based on location and spatial relationships; Mobility Services, which provide information about routes of travel to one or more destinations as well as real-time guidance during travel toward a destination; and Positioning Support, which gives support for the navigation subsystem.

Because no inherent session management exists in HTTP (and HTTP is a stateless protocol, that is, any request is assumed to have no previous history), a session management function is provided in the API. Data used during a session is handled as views by the session management system. Sessions also hold profiles, which can contain user preferences at very high levels (that is, "avoid roads without tolls"). There is also a policy component to manage the allocation of services and functions to sessions (and a system using numerical scales to assess the popularity or otherwise of a preference).

Although sessions are manageable (and have proven a useful concept, for instance in WAP Push), the Semantic Services part of the Magic API takes on an enormous problem: Multiple objects can have the same name. The idea is to allow for conversion between references in natural language, such as names for locations and services; associated geographic entities; and model elements in the the system. Natural language references are a pair of linguistic elements (name and class) made in speech or text strings in a specific human natural language, using a particular system of notation (script) and a specific encoding as computer representation. This should result in a one-to-one binding between the object and the representation.

According to the specification, a graph of the representations should be connected to the objects, and this can be used both for adaption of the services to the information provided and for the user to set constraints on which data is used, and how. Because geographic entities are often interrelated, and this relationship is not obvious other than in a presentation-oriented representation (for example, parts of a city having names that are not officially defined but culturally current, like "Upper Manhattan"). One way of regarding this is that semantic services provide a naming service for feature collections (more about those in Chapter 7, "Data Formats for Geography-Related Information").

The mobility services part of the API provides the basic functions of navigation systems built on the Magic API—functions to calculate the arrival of the model element, to get a geometry (as a set of model elements) for the route within the view, and to get the guidance elements for driving on the current route within a range of arrival times or a range of distances along the route (which is required when for mobile services). Route guidance also has to be able to determine when the mobile station has deviated from the route and to send a message to the man-machine interface (that is, the driver). The application can specify the starting point, the endpoint, and the waypoints. It operates on the assumption that there can be only one route from the starting point to the endpoint (and when the mobile station deviates from the route, it should be rerouted back to it). It can also include information about roadwork, traffic situations, weather, and other traffic-related data elements.

The class definitions and Document Type Descriptions (DTDs) are available on the Magic Services Forum Web site. They are not listed here, because of the detailed for an API that is not in widespread use yet.

SMS and Cell Broadcast Applications

In Europe, sending short messages to and from mobile phones has become extremely popular, especially among teenagers (this usage is exactly the same as that of Japanese users who use the mail functions in iMode). It is the same in other countries with GSM coverage—users in the Philippines send more SMS messages than all other users in the world. A type of instant messaging that is always with you, SMS has gone from being a system used to provision the handset with settings from a central database to a major source of network traffic (and revenues) for the mobile operators.

SMS works over the control channel in GSM, and the "short" in Short Message Service is quite literal: a message cannot be longer than 2156 characters (although it is possible to chain messages). The messages passed using SMS are rarely that long, however.

The reason why SMS is interesting for readers of a book about location-dependent services is that it can be used to create services that use the location of the user to provide something. For instance, there are companies that provide taxi services where you only need to send a message to a central taxi gateway and show the driver the booking number when he or she comes. Your location is identified by looking it up in the network of the mobile operator (because each SMS message contains the telephone number of the sender, this process is very easily done).

The position of the user can be obtained from the MPC or through the Parlay API. The application server in this case will not return a Web page, but rather an SMS message. It is not very easy to create an SMS application, however. You have to access the SMS Center of the operator, and the interfaces to that are not public but instead are private for each SMSC vendor.

Working directly with the SMSC to develop location-dependent applications is becoming unnecessary because many application servers have this functionality. As an applications programmer, you address the SMS functions through the API of the application server, and it connects to the SMSC, forwarding and receiving messages. Because the SMSC does not have a standardized interface (other than against the core network), it depends a lot on the SMSC you are using and how you will deliver the message, and describing the programming of an SMSC is out of the scope of this book.

In the case of SMS broadcasts to a number of users in the same cell, essentially using the cell ID, it is possible to use another function of the mobile network: cell broadcast. This function is standardized in GSM and IMT-2000 but not in other mobile telephone systems. In cell broadcast, the process is reversed. Instead of your sending a message to a service and it looking up your position, a message is sent to all users in a cell.

Unfortunately, cell broadcast suffers from the same problem as SMS: The interfaces are not standardized, but are instead proprietary. We will not cover them further in this book.

Both cell broadcast and SMS are standardized in the GSM specifications (and the standardization has carried over to the standardization of IMT-2000). The standard does not comprise the APIs, however; only the interfaces to the network.

All these APIs require that there is an application server where the actual application resides. And that is what we will talk about in Chapter 5.

CHAPTER 5

The Application Server

You can build a position-dependent application system from scratch (and I do not doubt that this exercise will be popular at many universities), but you do not have to perform this task. Several commercial companies have solved that problem for you. As soon as your system contains more than a few elements, as you will see in Chapter 9, "Maps as User Interfaces," the amount of data and rules for how the objects will interact become very large and complex very fast. Keeping track of the objects and the rules for them becomes an application in itself—and the application that handles it is the application server. Integrating the different functions that are required in an application server is not an easy task, however. As a builder of location-based services, you need to integrate (at least) five different elements into the application providing the services. These elements are as follows:

- Location determination systems
- Mapping capabilities (rendering, routing, and geocoding)
- Content and information providers
- End user handsets (with different capabilities)
- Billing mechanisms

Most application developers are not familiar with all of these segments. Some are also developed by specialists, who may not have an interest in providing a wide variety of applications, but only a specialized subsystem in their field—route calculation engines is one example in which this might apply. Such companies will want a standardized API to work toward, so they can provide their systems to as many customers as possible. In the absence of a standard API provided by the various vendors and to the various services, such integration takes time, effort, and requires specific knowledge. The application server providers want to be the ones to provide this functionality. Application servers are essentially connection points for the different components of the position-dependent service.

Who Needs an Application Server?

Application servers sit between the mobile network, the databases, and the end user. The position is derived from the mobile network (or directly from the terminal) by using a standardized API as described in Chapter 4, "APIs and Protocols" (or by communicating directly with the mobile station, as I will describe in Chapter 10, "Pulling It All Together: LBS-Enabling Your Web Site and Developing New Applications").

Today, however, we do not have the standardized APIs. What we have are network-specific APIs. If you were to develop your own application server, you would have to provide specific interfaces to all of them, which is what the application server providers do today. The application server interfaces to the network of the mobile operator in positioning systems such as assisted GPS and network-based positioning. As a developer, you use the API to call the position information. You do not have to write applications that interface directly to the system of the operator (it is, indeed, unlikely that you would ever be allowed to do so, because the operators are none too keen on letting people into their networks—and for most applications it is more practical to use an application server anyway). The application server probably does not interact directly with the core network infrastructure, nor does it include SS7 interfacing and the like. Querying the proper MPC retrieves the user location.

Your application resides in the application server. Some application server providers have created proprietary or semi proprietary APIs, which your application can call. The most bare bones application server imaginable is actually the Web server, using HTTP to call the MPC and retrieve the position document described in Chapter 4—but application servers incorporate Web servers and provide database interfaces, as well. How this process is managed is different between different application server providers, and architectures can vary a great deal. So can business models. The application server is housed in a central location, like an ISP, and the service providers can house their applications in it, and it takes care of the interfaces to the network. Some of the network operators who have deployed location-dependent services see themselves as ASPs; others plan to sell the data to companies that want to provide services. The services would then be provided in the same way as Web services today—by companies that essentially are publishers but run their own infrastructure.

An example is the API developed by French firm Webraska. The company intends to work as an ASP, making functions available to the programmer through an API on its own server. The functions that it makes available are fairly typical and comprise the following:

- Geocoding: address validation and conversion into latitude-longitude coordinates
- Reverse-geocoding: conversion of precise latitude-longitude coordinates (as provided by GPS) into an address
- Positioning: conversion of approximate latitude-longitude coordinates (as provided by network positioning technologies) into a range of possible addresses
- Mapping: customized dynamic generation of maps in gif, bmp, or wbmp format to developer-defined scale, zoom and size
- Routing: optimized route and journey-time calculation for transport by car, on foot, or by public transport
- Enhanced Spatial Searching (Ranking): provision of ranked lists of potential destinations in terms of journey-time from departure point

The Web server typically acts as a front end to the application server, using the mechanisms of the Web and WAP (HTTP and HTML, WML, or XHTML) to deliver the presentation. It is often integrated into the application server. Because Web servers are so prevalent, I will not go deeper into development using HTTP or other functions of Web servers. If there are developers who are not familiar with the Web today, there are many good books you can buy.

The Web server senses what the display capacities the requesting terminal has and delivers content accordingly. It can perform this task by using proprietary means (which is by far the most common today), or it can use a standard such as CC/PP (more about that in Appendix E of this book).

The application server also shields application developers from changes in the back end, such as the type of location determination technology used, the selection of content providers or the mapping engine used, and the databases used. Most application servers are integrated with a database with spatial processing functions (see Chapter 6, "Providing Databases amd Doing Searches"), which is used to handle the conditionality of the queries (in other words, drawing conclusions based on the coordinates and areas presented and determining how certain data should be presented). Personalized data is managed in the same way.

Most application servers contain some kind of directory function that enables developers to add applications and data sources and manage them. Depending on the architecture, this situation can be handled very differently. In the Pacific Ocean server, for instance, the directory is managed by using Enterprise Java Beans; in other systems, it can be anything from a proprietary database to an LDAP directory. You call other modules—created by you or created by the application server developer—by calling functions through this directory service (which of course means that it is done very differently in different servers). In the rest of this chapter, I will discuss some of the functions you can call, but typically they are of a number of different types:

Locator. The component that gets the user's position.

Billing Client. Used to bill a user for the service access and the positioning that is done.

Management client. Connected to the management system used by the operator.

Logger. Used for logging different events for statistics, alarming, and tracing.

Data sources. Mapping, information (such as restaurant guides), geolocation, and so on.

Generic location-dependent functions. Routing, geolocation, and so on.

Depending on the architecture, the services can be triggered in different ways. If the system includes an event handler, traps can be created that are triggered at different events. In principle, all location-dependent applications can be seen as based on events (including those that are purely request-based), because an incoming request is an event that is trapped by the database. Tracking applications will create events when the periodic or triggered position reports come in. The trick is to direct them to the right places and perform the right operations on them.

Another question is whether it has to be located on one single machine or whether the functionality can be distributed. While there might be some slight security advantages of having everything in one place in a normal data center, there are several other systems with distributed functionality, and they will work in the same way. Performance will be better if functions can be offloaded to different CPUs (especially if this function can be automated), and tuning of the operations is vastly improved. If the processes can be distributed, it is also more likely that the system can provide failover (in other words, when one machine fails, another takes over). If there are load-balancing functions, it can also help the data center operator to maintain a smooth operation. But depending on philosophy, this function can reside in the application or in the operating system. It depends on how you have designed the data center, and the application server has to consider that.

Questions to Ask When Selecting Application Servers

First, the application server has to have interfaces to the user information and positioning system used by the core network. Then, there are additional functionalities you can use to select one:

Is it integrated into a management framework, such as Tivoli or OMAP?

Does it implement standard APIs?

Does it have interfaces to the user information databases of portal engines?

What interfaces does it have to provide position information?

Does it have an integrated Web server?

Does it have an interface to a map engine or a GIS system?

Does it run on a single machine only, or can the functions be distributed? Does it provide load-balancing functions, or is that delegated to the operating system?

Does it provide for redundancy and failover?

Does it use a standardized data format?

Is there an interface to a billing system, and how difficult is it to customize?

Which version of the servlets specification does it implement?

The application servers increasingly use a standardized data format, both for their internal communications and for the communication with MPCs. This feature is increasingly based on XML, and XML libraries are available in most programming languages and are so relatively easy to implement. Depending on how your data center is built, the application server should be able to divide over several CPUs (probably using some clustering solution) to remove the risk of a single point of failure.

Different content providers (for instance, map providers) will have different specializations and abilities to provide content (for instance, being specialized in maps of one area, or only restaurant information). To connect to them, the emerging mechanism is XML channels; some might be so established that they have developed their own APIs, which the application server will implement. Content routing is a different problem altogether and can be done over the Internet.

How far distribution should be carried depends on how the data is created and organized. In principle, having all data in one data center will make access much faster, and you will not be at the mercy of the Internet to retrieve data. That might not be practical, however, if the data you are working with are updated frequently, are proprietary, or otherwise cannot be placed on your site. The simplest way is to reference the data by using a URI, but then you will have the same latency as you would have over the Internet in addition to the latency that is native to the application. That might be detrimental in an environment where the user demands that all data be no more than three clicks and three seconds away.

The Application Data Flow

How the data flows in the application depends on whether it is a triggered application, in which case the MPC will return periodic positioning messages and the application server has to forward them to the appropriate application (which will take the appropriate action). Or, if it is a request-response application, the user would generate a request that starts the data flow. The following example is adapted from the LocatioNet application server, where the different modules are called gateways. A typical request could be, "Give me John's location on an aerial photo. Show the nearest gas station and all the relevant advertisements and coupons within one mile according to John's personalized profile."

The request is sent to the application gateway. The gateway checks the user authentication and sends a location request to the location determination gateway.

The location determination gateway checks which network is the user's home network and sends a location request to the MPC of the network. Once the location information returns, the location determination gateway directs it to the application gateway.

The application gateway then sends simultaneous requests to the mapping gateway and the information gateway, which hold the location data that was just retrieved by the location gateway. If multiple information or mapping providers need to be accessed, the answers from all providers are aggregated and returned as a single response.

The application gateway then calls any personalization information included in the application, such as personal preferences (selecting a certain subset of the information held in the information gateway, for instance). Other personal information can be included or used for the structuring of the information, as well.

Once all information has been returned from the information and mapping gateways, the application gateway formats the answer according to the original request (for example, drawing the content and locations on a map if the application requires this function to be performed). It then builds an XML response.

The XML response is then formatted according to the device capabilities that were reported with the user's request (included in an HTTP header field) or based on a database of device characteristics (if there is a different way to determine the characteristics of the device). If the response is to be sent as an SMS, the formatting is negligible.

As the data is forwarded to the Web server of SMSC to be passed on to the user, the application server also triggers a billing event that is sent to the billing system. The originating application then returns the response to the user, using the appropriate channel for the format (it sends the data to the SMSC with the phone number in case the request came in over SMS and to the Web server in case it came in over HTTP). This is shown in Figure 5.1.

Because any Web request is serviced with a unique response, the adaption of each response does not represent any significant additional load on the system. The data has to be composed from the database anyway, so adding the personal adaption does not make much difference in terms of performance—but it enhances the user's experience quite drastically.

Personalization

When you adapt content to a set of user preferences, you are performing personalization. Location is just one example of personal profile data; the device characterization and the user's preferences for how it should be used is another. The service can collect a lot of information, such as how the user prefers to receive information and from where. Making the choices unobtrusive will mean that the user provides the information without thinking about it.

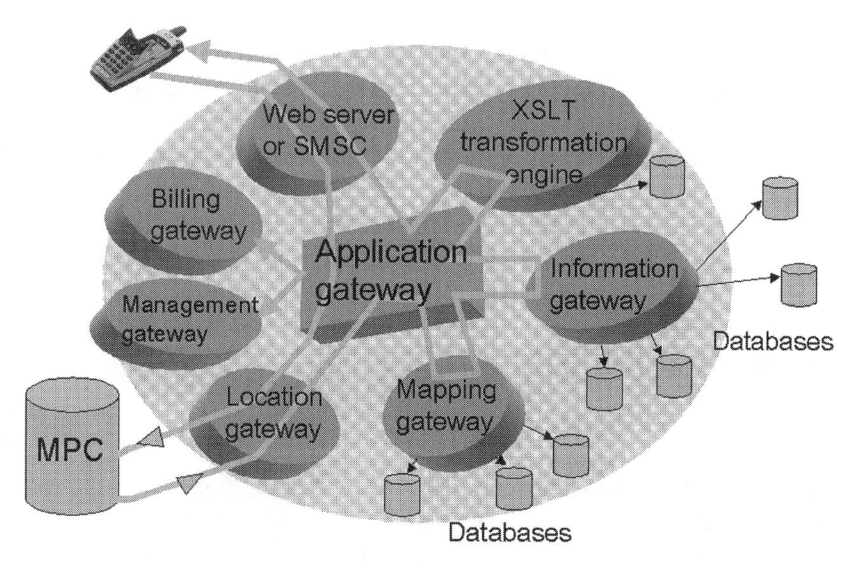

Figure 5.1 The data flow of a typical position-dependent application.

Personalization, especially combined with location data and logging by the service provider, has the potential to be extremely contentious. In many countries, it is illegal to use data for purposes other than those for which it was collected; in any case, you will seriously damage the users' trust in you (and your future business) if you try to use data for purposes other than those which the user has approved. The line is fine, and users are willing to give up quite a great deal of privacy in the interest of convenience, provided that they can trust you. The last chapter of this book, Chapter 12, "Privacy and Location," is dedicated to a discussion of privacy in the location-dependent environment.

An example of how the personalization database is integrated with the location determination is the LocatioNet server. Each location determination vendor is identified and registered in the platform database with a different ID. The same Location Determination ID is kept at the subscriber's personalization database. Once a location request is generated for a subscriber, the location determination gateway directs that request to the appropriate location determination system. This mechanism enables multiple location determination technologies support per subscriber, such as Cell-ID versus A-GPS (and so on).

The location determination channel is integrated by using the protocols of the various location providers. Because multiple location determination systems can be connected to the systems simultaneously, a unified protocol is implemented (essentially a precursor to the API of the Location Interoperability Forum, or LIF). This protocol enables location systems to input subscriber location to the platform upon request or synchronous to any request. The location determination channel needs to support the most usual coordinate and datum formats.

There are two major problems with the personalization mechanisms in most application servers. The first is that they build up a very extensive bank of personal data without any privacy guarantees. The second is that they use proprietary formats. If you have one profile on Yahoo! and another on the Netscape portal, you cannot merge them today. Actually, you cannot even look in them because they are inside the database systems of the services. But there is no interchange format, and the efforts that have been done are all hampered by the fact that the service providers regard the user's profile as something that they own (and can sell). No application server I have looked at even tries to alleviate this problem.

Of course, once you have characterized the data that you want to collect and defined interfaces and database fields, it is trivial to gather them. Taking data from the log files into a database is barely a starting point, because you can collect information about where the user is when he makes requests and if he has any preferences for what should happen with the data in different positions (for instance, does he usually look for beer on the way home on Fridays?). Combined with other personal data, such as credit card data, this situation can be a major threat to the user's privacy.

As database problems go, it is relatively simple, though. Any relational database can be used to handle it, and information can be collected by using filters. The problem is not collecting the information; rather, it is deciding how to adapt the information when it should be outputted. For instance, if there are conflicts, such as the user wanting advertisements for handbag shops when he or she passes shops that sell dresses but never advertisements after 5 P.M., how do you resolve the conflict?

This example is simple, and when you start collecting more information, you will continue to run into conflicts that most application servers will try to solve ad hoc. There is no prioritization filter in the service, and even if there were, how would you know which rule the user wanted to be applied to the content in which order (essentially a rule in itself)? You would have to ask, because the more information you collect the more complicated the process becomes until you would end up implementing a minor *artificial intelligence* (AI) system just to keep track of how different users want their personalization applied (because, of course, you can never assume that users will want the rule sets to be applied the same way or even be able or willing to provide the same information).

It is not enough that you have to adapt the data that is sent in the scope of the transaction. In mobile data networks, the network usually maintains a database containing the services that the user has selected. This information can be synchronized with a database on the handset, which makes it possible to have a personalized presentation. This process is called provisioning, and it can be done over the air—at least in GSM and IMT-2000—by using an SMS (actually, this function is what SMS was invented for). Usually, the typical process is as follows.

The operator sets up the default provisioning profile for groups of users (young users, business users, and so on) in batch mode by associating those users with a profile number from a list of defined profiles in the database. Access is typically from the operator Customer Care Center (and its specific databases) or a dedicated application in the application server, which interfaces to the computers of the customer care center.

Enterprise customers can potentially customize their users' profiles by accessing the database and modifying their users' provisioning attributes. This procedure can be done by using a Web-based application, a WAP phone, or by calling the customer care center.

End users can access the system anytime over WAP, SMS, and Web interfaces and personalize their profile and service behavior according to their preferences. This process requires the operator to allow this action, of course. Some operators will not let users into the configuration files for fear of either giving away a lot of data about their networks or for fear that the user will require an impossible configuration. Given the amount of data that has to be configured in a normal mobile phone for it to work simply, this situation is probably justified (most people do not realize how many parameters can be set and set them wrong in a mobile phone). Still, a mobile phone is very simple compared to a more advanced client such as a PDA, which is almost as complex as a PC (when I got my latest PC, it took our corporate IT department almost half a day to configure it, and that was without the Japanese settings).

Many application server providers have provided starting points for personalization, but none has understood the problem even if they claim to. This situation will be a major problem in the next generation of application servers, especially when considering the need to provide provisioning information.

Device Characterization and Content Adaption

At least initially, most terminals will not have the communications capacity to handle advanced traffic but will be Web clients, and that means that personalization of information and adaptation to file formats the terminal can display will take place in the application server.

Then, there is the problem of end user handsets, which might have very different capabilities. Some developers have created their own representation languages, but there are standards for describing terminal capabilities (and the user preferences for how they should be used). The one that is emerging as the most important is called CC/PP, and I happen to chair the working group. You will find an introduction to how it works and how you can use it in Appendix D of this book.

Strictly speaking, of course, the adaptation of content to terminal capabilities has nothing to do with location information. Neither does the adaption to other personal information, but if you see "Welcome to Paris, Johan. Here, we can recommend the following restaurants which we believe you would like," you are seeing an example of how position-dependent data can interact with personal data. Location-dependent applications are nothing more than a special case of situational applications that also adapt the information to other variables from the user's environment (including the logical environment; that is the profile of the user that is held by information providers).

The user interface to the application is discussed in Chapter 8, "The User Interface to Location Information Services," but it can be provided in two ways: directly constructed and rendered on the end-user device or constructed in the Web server and rendered on the user's client (for instance, in the previous example, when the WML deck sent to my WAP phone does not contain anything but the page—no additional formatting, no FONT tags, and no Javascript). The latter has by far become the more prevalent, especially because more computers have Web browsers than any other component these days. In environments such as WCDMA and GPRS, which have a relatively limited data capacity (as well, of course, as the original WAP and iMode), sending too much data is a concern. Formatting has to be limited. The overhead created by Microsoft FrontPage, for instance, is between 50 and 20 percent, depending how many font tags and the like you decide to use. That means at least 50 to 20 percent slower transmission of the data.

The application server is often integrated with a map engine that handles and presents maps (such as the LocationNet engine), or the maps can be retrieved separately. However, as we will see in Chapter 9, "Maps as User Interfaces," you cannot just take the map and throw it out on the user's device; you have to do more with it. In other words, the presentation that goes out from the Web server has to include the added user interface components as well as the map (or, in the case of applications that do not contain a map, the user interface as described by the application software).

It is not necessarily the case that the application has a user interface, though. If the application is a money transport security system, it might poll the transport at regular intervals to check its position against the predetermined route. Or, it might provide statistical information about the number of cars in a certain area to a traffic management system (in many countries, the assumption can be made that everyone in cars has cell phones; in other cases, it is relatively easy to derive an algorithm for how the number of cell phones represents the number of cars. But that is precisely what the application development is about, right?). In applications that are critical (time critical or business critical), such as the emergency call system (E911, E112, or E119—Enhanced emergency numbers for the United States, Europe, and Japan, respectively), you might not want the traffic to go over the Internet because this situation might literally be a matter of life or death. If the application is a positioning service for a money transport, that might be true as well (and you would probably use the ISDN signaling, which guarantees that the signal gets through). But for most applications, when it is not critical that the information gets there at an absolute time—and when the occasional delay might be OK—Internet-based systems will probably be used. In practice, that means HTTP today and probably *Simple Object Access Protocol* (SOAP) tomorrow.

In other words, the data has to be adapted in several different dimensions, and the description of the device needs to be compatible with the other adaptation mechanisms that are to be applied to the data. It is not as simple as having a number of device classes and writing information that fits precisely; as a designer, you have to give up some control to get a more efficient transfer of data (and thus creating a more attractive user experience).

This situation affects the way that the data should be handled. Pragmatically speaking, though, the question is how you apply formatting to your data. The simplest way is to create a default XML representation for each request that contains all the personalizations and then to apply the formatting afterward by using an XSLT transformation (or, if you have control over the architecture of the application server, generate appropriate XSLT transformation sheets that include the personalization information and device characterization). It is relatively easy to create one transformation sheet per device, at least for the most frequent (there are, after all, not more than some 5 or 10 PDAs and some 30 different models of telephones out there). Having one transformation sheet per device is the price you have to pay for that level of control; if you instead were to accept that the adaptation was done at the device and tried to create a device-independent representation of the data, you would not have any transformations. If the devices were using WAP 2, you could send a style sheet along with the data and just include a link to that in the transformation. You would then, of course, have to design one style sheet per device (or trust the devices to do their jobs and represent data properly).

Databases and Internal Interfaces

Application servers are the connection point between the different applications, the databases, and other resources. How they are connected to the application server varies with the implementation. Some use CORBA; others use XML. Still others use Enterprise Java Beans, JDBC, or ODBC interfaces to the databases. For communication with the user, they use the Web server, generating Web pages by using *Active Server Pages* (ASP), *Java Server Pages* (JSP), or similar techniques. As an application developer, you have to create a template that contains the generic data and the hooks for the information that will come from the database, using Java beans or any of the other means of connection to the internal databases. There must be one JSP per output format, and it is probably easiest to have formatting functions in whichever system you use. Remember, though, that an SMS message can only be 160 characters long and can only be plain text.

A query by a user can combine multiple data requests, such as the subscriber's current location, content and information of multiple kinds, and mapping information of any kind. The application developer needs to know not only how to address the databases but also which data is available and how it can be presented. Depending on the architecture, all databases that have been used previously and are registered, or only a selection of databases, can be made available.

Databases are called in the standard way, by using ODBC or JDBC. A problem is that if they implement special functions, such as spatial functions (which we will look at in Chapter 6, "Providing Databases and Doing Searches"), that have not yet been standardized. In that case, there might be vendor-specific extensions that you need to use in your programs.

Using Filters in Servlets

As the vendors of application services provide APIs, you as a programmer can create your application by creating a program that runs in the application server and provides what is unique for your service.

This service typically means a Java program, because Java seems to be on the way to establishing itself as the primary programming language for position-dependent services. Most application servers implement the servlet's APIs in addition to the proprietary APIs, which call the different services. Java programs in a Web server mean servlets, because servlets have established themselves more or less as the unique standard for Java Web server programming since it was first launched a few years ago.

Servlets run in the server in much the same way as applets run in a browser, but of course they have access to the operating system and other APIs and system functions in the server and run on a real processor.

With version 2.3 of the servlet's specification, a new function was created that dynamically intercepts requests and responses to transform or use the information contained in the requests or responses. Called filters, they typically do not themselves create responses, but instead provide universal functions that can be attached to any type of servlet or JSP page.

The filter mechanism provides a way to encapsulate common functionality in a component that can be reused in many different contexts. Filters are easy to write and configure as well as being portable and reusable. In summary, filters are an essential element in a Web developer's toolkit. Filters can be mapped to one or more servlets and vice-versa. Filters can be chained indirectly via filter mappings. The order of the filters in the chain is the same as the order that filter mappings appear in the Web application deployment descriptor.

Filters are useful if you create location-based services, because they provide a simple way to encapsulate recurring tasks in reusable units. They also enable you to transform the response from a servlet or a JSP page. A common task for the Web application is to format data sent back to the client (for instance, SVG or VRML instead of GML, as we will see in Chapter 9). Filters can be written in XSLT as well as Java. But filters can be used for anything you can do with a Java program in a Web server.

Programming the filter is only half the job of using filters—you also need to configure how they are mapped to servlets when the application is deployed in a Web container. This decoupling of programming and configuration is a prime benefit of the filter mechanism, because nothing has to be recompiled and you do not need to change the input or output of your Web application. You just edit a text file or use a tool to change the configuration. For example, adding compression to a PDF download is just a matter of mapping a compression filter to the download servlet.

The filter API is defined by the Filter, FilterChain, and FilterConfig interfaces in the javax.servlet package. You define a filter by implementing the Filter interface.

There are many ways for a filter to modify a request or response. For example, a filter could add an attribute to the requests or it could insert data in or otherwise transform the response. A filter that modifies a response must usually capture the response before it is returned to the client. This filter can be, but does not have to be, an XSLT filter. You can, for instance, set it to use the style sheets to transform the response depending on the value of a request parameter. The filter sets content type of the response according to the request parameter. The response is then wrapped in a CharResponseWrapper and passed to the doFilter method of the filter chain. The last element in the filter chain is a servlet that returns the inventory response. When doFilter returns, the filter retrieves the response data from the wrapper and transforms it by using the stylesheet.

To map a filter to a servlet, you must declare the filter by using the <filter> element in the Web application deployment descriptor. This element creates a name for the filter and declares the filter's implementation class and initialization parameters. You must also map the filter to a servlet by defining a <filter-mapping> element in the deployment descriptor. This element maps a filter name to a servlet by name or by URL pattern.

Interfaces to External Services

We already discussed the location determination systems in Chapter 4 "APIs and Protocols." You will use an API to get the location information and define how your application will use it. The application server uses the API to send location requests (for single or multiple subscribers) from the platform to a location, sends a subscriber(s) position from the location system to the platform, and a subscriber(s) status from the location system to the platform as well as sending commands from the platform to the location system.

One significant problem occurs if you want to use location information in the client and the client is not generating that information. This situation is, for instance, true if you have a periodically triggered application that generates position information and sends it to the application server. If the client does not have a Web server, there is no way to get the information to the client—except possibly WAP Push or another protocol like SMTP, which is used for e-mail.

The application server can take care of that for you, essentially working as an interface between the MPC and the terminal. That is the way it is intended to work in the LIF API, which we discussed in Chapter 4. In theory, this situation means that the clients can connect to the application server by using any protocol (for instance, DNS, SNMP, SMTP, or FTP). It could function as a gateway toward the positioning system.

The application server can also direct the request for positioning from the user to the right network. If your site is on the Internet, there is no direct connection between the user's identity and the mobile network, which can provide the position. You can receive requests from many users in many networks. That is not necessarily true in the "walled garden" model that many network operators prefer, where the user can access only information provided by the network he or she is accessing—a strategy that has been proven wrong time and time again.

The application server can also serve as the interface to other services; for instance, the billing system and other services in the network, such as portals. There is no standardized interface for user information, however, nor for billing information. How you connect the portal to the application server depends on the interfaces of each. This area could deal with standardization, but because the portals are based on their proprietary interfaces, that is unlikely to happen in the near future. It will be true a lot sooner for application servers, where there are now a number of standardization efforts going on (too many, actually). I will describe them later in this chapter. Some application servers provide for personalization beyond the use of position information as well as the transcoding of information. But neither of these are standards, and I will not discuss them in any depth—although it is clear that personalized services are the key to the future, and location-based services are just a branch of situated services where the context of the user in terms of preferences, previous experiences, and so on are used as base for the personalization.

If the application server sits in a data center, it is an advantage if it is integrated into a systems management framework, such as Tivoli or a similar system from an operator. Here, as well, the standard interfaces are lacking. But if your application server is integrated into the management framework, it will be able to send alerts, give performance measures, and generally act as any of the other application servers that are situated in the data center. If you are thinking about setting up as an ASP, having this type of functionality is a must.

Management System Interfaces

If the application server is used in a central location, such as in an ASP, then the management of the server becomes a crucial issue. This statement is especially true if the system sits in the data center of a network operator who typically has higher requirements on up time (99.99 percent is not unusual) than the average PC user. This requirement might seem excessive on the surface, but considering that people actually trust their lives to the telephone system, it is not unappropriate.

On the Internet, there is a standard protocol for network adminstration: the *Standard Network Management Protocol* (SNMP). There is a *Management Information Base* (MIB) for Web servers, but it is not used in many systems. There is no MIB for application servers in location-based systems, and the requirement is higher to integrate with the management system of the network operator in case the application server sits in their data center.

There are no standards for management systems in the telecommunications world, however. Each manufacturer has its own, and delivering it is a way of achieving a lock on the customer—because once you have a system running thousands of systems, you do not change easily.

One way is to create an API that maps to the internal interfaces of the management system. The application server connects to the application server, and any event that causes a management trigger is sent to the management system through the API.

The operators of mobile networks see the location-based services as a potential revenue-spinner. They would like to offer their information to as many different providers as possible to maximize their revenues. But at the same time, they do not want to expose and share their most sensitive systems and information. For this reason, it is probable that you have to use an application server or even an application service provider that has an exclusive agreement with the mobile network operator.

This situation is precisely the use case for application servers functioning as a middleware between the different systems providing the information and integrating them into a presentation that can be sent to the user. Some operators are likely to be even more careful (like NTT DoCoMo of Japan) and only allow *application service providers* (ASPs) access to their network (that DoCoMo started their own ASP does not, of course, make their business case worse). For the application developer, however, it is a fact of life that the position data in its raw form is inaccessible, and if you want to get it you have to use an application server (which, of course, also brings a number of advantages, such as access to a number of other services).

Looking back at the conceptual architecture I presented in Chapter 1, the application server can sit in the service owner's data center, or it can be outsourced. It can be used to provide one single service, or it can be shared by several users (if provided by an ASP). The application providing the service addresses the functions provided by the application server through an API.

Application servers will also have to connect to content distribution networks, which are now emerging. Syndicated information is typically only news and stock prices, but the same mechanisms used to personalize that information could be used to adapt it to the location. Then, the data flow would come from an external source and the application server would only contain the template and profile management. The platform would enable each content database to support multiple information categories.

Content Distribution Network and Web Service Interfaces

Most servers use XML as the data interchange format. With XML, the content provider can maintain and store its database by using its own convenient format and structure. Once connected to the platform via its XML channel, the content becomes available to all platform applications and users. Once a request for information is received, the application server must determine the appropriate content provider and direct a request to that provider. If multiple pieces of information are required from different providers (for example, simultaneous requests for restaurants and traffic information), the original request is segmented and sent to the relevant providers.

The heuristics for how this process is done can be built into the application server, but it requires an intimate knowledge of the databases and their content. Either an application service provider can take this role and create the XML templates or business rules that manage the databases, or this role might be fulfilled by the content provider. Which model will prevail is yet uncertain because this industry has not started yet (even on the Internet).

The XML-based information channel isolates the platform from the various database structures. It enables the platform to interchange database and geographic objects between different providers and applications via a unified language (actually, a metalanguage). These channels can be designed to handle database information (with a single request to the database), or they can be designed to support real-time information that is interchanged between the platform and content provider.

There are a number of proprietary formats for information channels (one per application service provider and content provider, as things currently stand), but there is standardization work ongoing in both the IETF and other places, which will most likely create a standardized format for information syndication. One example is RSS, an open-source format based on RDF which lets you describe each channel at the provider's end.

Another issue is event information provisioning. It might not be appropriate to provide the information in the same way as for news, for instance. I discuss some formats for event information in Chapter 7, "Data Formats for Geography-Related Information," but the event information management is still not something either standards bodies or application service providers have addressed. If event information is handled at all, the assumption is typically that it resides in a database, and that is how it is delivered to the users.

Key functionalities of the information channel management system that are emerging include the following:

- Self-identification of the content (information) provider
- Adding and deletion of information on the platform objects
- Editing information fields of an information object

The metainformation about the information channel is managed in the application server, but the information provider will most likely want to handle the information himself or herself and have it on his or her own location. In other words, data will have to be requested when it is required. A problem with this approach is that it compounds the latency of delivering information to the user. Database retrieval from a local database is fast, but if the retrieval has to be done over the Internet, an unpredictable delay has to be included into the application. That can be very irritating to the user, especially if he or she is on a mobile connection that also introduces added latency compared to the fixed environment. Most likely, methods will emerge for syndicators to put their databases in the application services. You still need to think about the updating of the database, however. Waiting for a platform to get batched content updates from information providers is almost as bad as getting information slowly.

Another problem is when the information is not in a traditional database management system; for instance, when a small restaurant puts a coupon on its home page that can only be accessed by users who use a mobile phone (this procedure is actually very easy to do, as I will show you in Chapter 10). But easy as it is to do, it is not easy to manage. No standards exist, but several application server providers have their own ideas about how it should be done. For instance, the LocatioNet platform enables content owners to update their content (for example, my restaurant now has another location, and our happy hour is longer) through an easy XML interface over the Internet. The content provider maintains full control and monitoring of every use of its content, unlike when it has to be exported to another external database.

There are several standards groups that develop content networks. As yet, XML over HTTP using proprietary formats are the only thing that seems to be used. If the management of data can be delegated to a proxy that is integrated with the application server, however, the system will have much less latency while maintaining the advantage of keeping the data on the Internet. One way is using SyncML, but that only works for entire databases, not for applications that change part of the data. Several groups in the *Internet Engineering Task Force* (IETF) are addressing this issue.

Location application servers, however, are on their way to being integrated not only in content distribution networks, but also in Web service networks. Web services are currently just a discussion item in the *World Wide Web Consortium* (W3C) but are using SOAP to embed an XML document to describe the data in the same way as the OMG IDL describes a data structure and a method. The difference is that there will not be one method per function, but rather only the methods defined by HTTP that will use the Web server to forward the data to the actual application. The Web server insulates the application from the Internet, as we discussed in Chapter 4, "APIs and Protocols"

Billing System Interfaces

The final thing you have to consider is billing mechanisms. As I talked about in Chapter 1, "Developing Location-Dependent Services," there is nothing that says that information has to be gratis on the Internet (even if it is free).

There are four different stakeholder segments that are involved in the development of a location-based application: the information providers, the application developers, the location information providers, and the end users. In Chapter 8, "The User Interface to Location Information Services," we will look at the user interface aspects of the development, and in Chapter 6, "Providing Databases and Doing Searches," we look at how you get the data and handle it in a database (which is no different from the management of another database, really). The network operators, though, take the position that they will own the location information and sell it to the providers of location-based services. This scenario will probably be true because they have a good claim to owning the network, and they are able to insert terms into the subscription agreements that give them the right to sell the information. Whether they will sell it in detail or en gros, though, is another matter. And whichever applies, it is very likely that the user will understand that he or she has to pay for the added value.

Many operators have looked at the model of NTT DoCoMo in Japan, which has built a system where the most popular Web sites are registered in its gateway and the operator handles the subscription administration. The charge comes on the telephone bill, and NTT DoCoMo charges 9 percent of the transaction sum for the service. They also take the risk, of course. Because the sums involved are usually much lower than the $5 to $7 which typically are the lowest limit for credit card transactions, and because users frequently are younger than 18 years old and therefore do not have their own credit cards or even bank accounts, this proposition is attractive to many content providers (who can also calculate with the fee in their business plans).

In other words, both the operator and the service provider (if they are not the same) have an interest in providing and using a billing system to be able to charge the user. Logging events that should be charged for is not very difficult; the problem is instead getting it into a system where a bill can be sent to the user. Different LBS applications tend to require different billing procedures, further complicating the overall management challenge for the carrier.

Operators are often tied up into their billing systems, which are large applications that cannot be shared easily with other applications. Most are old applications that have been developed since the operator companies bought computers the first time. Different location-based applications tend to require different management procedures, and this situation of course complicates the connection with the operator, which has an old and inflexible system that cannot be modified for new business models.

The billing in most systems is tied to the logging. Transactions are typically recorded by creating a record, the *Customer Data Record* (CDR), in a database. This database can be the billing database, or it can be a separate database that reports into the billing system—either in real time or in batch mode (for instance, once a month). In real time, the data can either be pushed to, or fetched by, the billing system. For each transaction, successful or unsuccessful, a CDR is created. A transaction in that sense is a complete round of service request and response between the application server and the request initiator. For example, an application request for the user location, mapping/address, and relevant content will result is a single CDR detailing this transaction with its full process and attributes.

The method also allows for the handling of prepaid accounts (but typically, the batch method is not used in that case). If the user has a prepaid account, the transaction is registered directly and checked against the account to see that there is coverage for the transaction; if not, the transaction will be terminated. The policy for how much to charge for each transaction can either reside in the billing system or the application server, but the interfaces between them have to be secure because the system is otherwise open for hacking and fraud (which in reality means that the service provider registers the amount to be charged with the network operator by some other, out-of-band means, and the only thing that is sent over the wire is an event).

Depending on the architecture, there are various ways of connecting to a billing system. Mobile operators often have very large billing systems, which in some cases dates back to the early '50s when computers became available for administrative purposes. Interfacing with billing systems is no trivial task, and in most cases there should be an intermediary who connects the application server to a billing system (providing the necessary interfaces and maybe also doing the data collection, because many billing systems are batch-oriented).

In the LocatioNet system, the billing gateway links with a billing client. The billing client, once connected to the gateway, can submit requests for billing information captured by the billing gateway—in essence, all transactions that are executed within the platform. The billing gateway submits a response with the required data. The billing client then becomes the owner of the data that it queried. It can then manipulate the billing data in a way that is consistent with the business models adapted by the wireless carrier. The billing client can also act as an adapter between the platform XML-based API and the API used by the billing system of the carrier and its billing databases.

How the interface works depends on the architecture of the application server. In principle, each transaction executed by the application server should be able to create a record, with all the relevant transaction details, in the billing CDR database. A single CDR is recorded for every transaction in the platform, successful or unsuccessful, for every and all triggering sources, such as service (positioning) request by the user, by application, or by any GMLC (typically for roaming or push services). A transaction in that sense is a complete round of service request and response between the platform and the request initiator. For example, an application request for the user location, mapping/address and relevant content will result is a single CDR detailing this transaction with its full process and attributes.

All CDRs are stored in the billing database, and transferred to the required billing and/or prepaid system either off line or in real time, one by one or in batch mode. Interface specifics, such as time interval, CDR format, and which

Customer Data Records in a Billing System

In a telephony system, each CDR includes some or all of the following attributes:

- **Status (successful or unsuccessful)**
- **MSISDN/UserID request**
- **MSISDN/UserID response**
- **Application ID**
- **Content Provider ID**
- **Transaction type (information, map, location or any combination)**
- **Time stamp**
- **Transaction duration**
- **Transaction size (the amount of data transferred)**

interface and protocol were used, can be customized. There is also a connection to the provisioning system, which determines which services the user should be able to access.

The definition of users and corporate authorization, which attributes that can access and modify in the database of course becomes crucial. For instance, prohibiting users from subscribing to new applications in case such subscription involves charging is one such policy that can be set. Alternatively, operators can allow subscribers to add any new application to their portfolio, as part of their marketing policy to encourage users to experience location-based services. It is usually possible to set different rules depending on the policy (for instance, users with a corporate billing plan gets unlimited access, whereas users with prepaid cards get no access at all).

Running the application server is no different from running other servers, however. When you have designed the application that goes into the application server, you need to worry about the databases and especially the data that goes into them. That is the topic of the next chapter.

CHAPTER 6

Providing Databases and Doing Searches

When a request for information comes in to the application server, it needs to retrieve the information from some location. That location is a database, and it is usually handled by a *database management system* (DBMS). But databases do not emerge by themselves. Someone has to create them, and someone has to maintain them. Another database problem is when the information is not in a traditional database management system; for instance, when a small restaurant puts a coupon on its home page that can only be accessed by users who use a mobile phone.

As a content provider, you will probably not want to miss any location-dependent service that comes your way if it means increasing your user base or even a new source of revenue. If you can influence it, however, you want to make sure that the process of registering your content is as simple as possible. The more integration work there is, the less your revenue—unless, of course, the decrease in integration work means a decrease in flexibility. You do not want to sacrifice the user's experience for ease of providing content. That is a bad equation.

The service provider will have to build a brand that the customers—the users—are likely to come back to. That is more difficult than it sounds, and it is actually not easier on the Internet. (It was easier when there were only a couple of million people using it, but those days are long past.) The mobile system is likely to start with several millions of users, and if they have Web phones, they can all get access to your services at the same time. When DoCoMo released its iArea service, for instance, all users of iMode (some 30 million by then) got access to it at the same time.

There are also copyright aspects to consider. In the United States, databases are not protected by copyright. In Europe, they are. In other words, in the United States, it is perfectly legal to scan the telephone book into an *optical character recognition* (OCR) system, and because it represents a structured data set, you can feed it into a database management system and convert the addresses to geocodes and presto—you have a database for the position-dependent application.

Not so in Europe (and the rest of the world). Here, you have to pay the content provider (which probably is cheaper than paying someone to correct all the errors of the OCR program, anyway). For the information provider, the core competence is to collect data (restaurant reviews, store locations, real-time traffic reports, local sports content, and so on) and maintain it up to date. Their challenge is to add position information to it and make it available to as many different paying customers as possible.

On the Internet, this situation has meant setting up a site and selling advertisements. But there are other options, such as publishing guidebooks (which you can sell), syndicating information to magazines or newspapers, and many others. I discussed business models briefly in Chapter 1, "Developing Location-Dependent Services," but do not forget that someone has to do the job of getting the information you want to present on your site together and that they are very unlikely to do it for free (although that can be an option for certain information types, such as a church services directory).

Position-Dependent Databases

As you can understand from Chapter 1, the database system is a crucial component of any position-dependent information system. Database systems, however, consist of two different pieces: the database itself and the database management system. Often, these are mixed up and the DBMS is taken to mean the database, which is reasonable in a way because the internal format that DBMSs use to store the database is proprietary and the DBMS cannot be changed without a major conversion effort. (Or at least, that is how it used to be.)

With XML, this situation has changed, however—and the entire Internet became a database. Some database functions already existed, such as the search function, before the event of HTML. Applying free-text search to HTML was relatively easy; the major problem that faced the early search engines (and which faces search engines today) is how to gather the information.

In principle, the DBMS has two parts: the part that manages the data in terms of files and the part containing the logic that programs enable the DBMS to apply in order to create answers. With XML, the file storage part (necessary in a day when disk space was expensive) becomes redundant. XML can be seen as a way of creating DBMS-independent tables of relational data, which the DBMS can then manipulate. Think of an XML document like a serialized database (what you get if you take a line from a table and print it out with the labels from the column heads attached), and you have a good idea of how this system works.

In practice, the DBMS will not go away, however. There are too many advantages; for instance, when it comes to updating information, having a system that handles everything in one neat system is preferable instead of handling it as an enormous collection of text files. But it does mean that the database can be spread out, for instance, in a city where each Web server that belongs to a business has its own set of XML data and maintains it itself. The database is the collection of all the different metadata files. They will probably be aggregated and managed in a central DBMS anyway, because it would be inefficient to have to query the Web servers of each business each time someone wants to see what is available at his or her position, but that gives you both the advantages of distributed updates and the advantage of having a central point of access for the data.

Most databases are stored as relational databases, which means that they store data in tables and the tables are used to express relations about the data. Data that belongs together is stored in one table and linked to other tables by using common values called keys. The same thinking can be applied to an XML document. If you create the XML documents, thinking about them as rows in a database table, they can be handled as a database. Using key values can link the tables, or document rows, together. Forget the allegory between documents and printed pages; they are database rows, and regarding them that way makes much more sense than handling them as virtual paper pages.

Say, for instance, that you have a database table that looks like Table 6.1.

That table would look like this in an XML serialization (with line breaks inserted to make it a little clearer):

```
<UserID>54621</UserID>
<Name>Johan Hjelm</Name>
<Home> Yamate House, 11 Suwa-Cho, Naka-ku, Yokohama</Home>
<Phone>045-624-5253</Phone>
<Department>Research</Department>
```

Most relational databases use a standardized query language to manipulate data and make queries on the data sets. In fact, *Standardized Query Language* (SQL) is an international (ANSI/ISO) standard for DBMSs. It has all the func-

Table 6.1 A Sample Database Table

USERID	NAME	HOME	PHONE	DEPARTMENT
54621	Johan Hjelm	Yamate House, 11 Suwa-Cho, Naka-ku, Yokohama	045-624-5253	Research

tions needed to handle the data tables; the only problem being that the major manufacturers have produced proprietary extensions that are used in such a way that if your data is stored by an Oracle DBMS, you can not query it other than in the most basic way from an IBM DB2 DBMS. The W3C, the people who invented XML, are working on a query language for XML that would let you do basically the same thing for an XML dataset, but it is not ready yet and it is not clear how it would work in relation to an application server either, so we will not attempt to use it in this book.

Data Modeling for Location-Dependent Information

There are two database-related aspects to developing a location-dependent system that you probably have not encountered before, though, especially if you are a database engineer. The first is that you might have to work with legacy databases, which have to be integrated in the system. While it would be convenient to develop everything from scratch, there is simply too much information in existing systems to just throw it away. This situation applies to pseudodatabases, such as the Web, as well as systems handled by DBMSs, such as customer information databases. The conversion to a position-related database is relatively easy. Many of the companies providing maps can also provide geocoded and reverse-geocoded information, which can be used to link addresses to positions. If your database is structured correctly, this feature will mean that you can integrate the position information into the data and start operating on it.

The second aspect is that the development methodology might be different. In traditional computer systems development, you started with the data and developed the software to handle it, and in the end you put a user interface on it. If you were following a development method that allowed for it, you might integrate use cases. Developing the database starts with creating a data model, which essentially is a way of describing the tables that will be used to store data and how they are connected (for relational databases, that is, which consist of tables. For some geographical applications, object data models are more suited).

The data themselves are not so different, although they might appear to be at first glance. A map is a database expressed on paper. In Chapter 7, "Data Formats for Geography-Related Information," we will look at markup for geographical information, which shows how you can serialize a map (just like you can serialize a relational table). In Chapter 9, "Maps as User Interfaces," we will look at how you can convert an XML serialization of geographical features to an image in a vector graphics format, which could be a map.

The traditional development process starts with developing the data model, but there are plenty of data models for geographical information in specialized areas, however. The *Public Petroleum Data Model* (PPDM), for instance, was developed by an association that represents more than 100 oil and gas companies, vendors, and regulatory agencies worldwide. Their mandate is to deliver "... a vendor-independent standard petroleum data model that serves as the industry foundation for managing information as an essential asset in the global business of oil and gas exploration and production." Although PPDM is both broad and deep, it is focused on the needs of the oil and gas industry. It is being developed to include stratigraphy (the layering of rocks) and being spatially enabled. But although there are plenty of standardization efforts in this area, it is still far more usual that organizations use their own proprietary data models for location-dependent data. Often, they use their own data because they want to include some parameter that is not part of the data model, and they do not see any need to use a standard data model (which the petroleum industry does, because trading rights to exploration often means trading databases, as well). If you use the existing LBS data formats like GML, you will have to use the data models of those formats as well. But in XML, it is easy to create your own data model by re-using elements which are already defined, using the XML namespaces mechanism.

But if you look at the most frequent applications of location information that is foreseen by industry pundits, it is not business-to-business applications but rather tourist information. At least seven different research projects have been performed with funding from the European Union, and there are several cities that have started to look at providing tourist information based on position (in Florence, you can already get tourist information by dialing a special telephone number, and as you move along the square in front of the cathedral, the guide voice shifts to the cell ID which the user happens to be in).

Developing for the large class of the consumer services that are based on position information, especially those that are geared toward creating a user experience, is very different from developing for users who are professionals. The expectations on the service will be very different, as will the level of support that can be offered. The data might not be so different as you might think (although there are not enough examples yet for us to really make a definite statement), but the way the data is presented is entirely different. The presentation can make or break the service, especially if you intend to charge the user for using it. And that makes the development methodology different, because your business model is based on creating an attractive presentation for the user instead of maintaining as complete a data set as possible.

Instead of starting with the data, you start with the user interface. Often, the result is the same anyway, but frequently you find yourself having to add data elements that you would not have thought about if you had developed the application in the way with which you were familiar. Working from the user interface toward the core databases instead of the other way around might feel different, but when it comes to the development of the database module, it will not be so different.

Perhaps you are working with, or even are, a Web developer who thinks that all this formal stuff is not necessary. After all, the Web is just pages with links, right? And, in reality, that part about method is all well and good if you sit in a classroom, but what actually happens is that everything takes place simultaneously and chaotically as usual.

Well, you are wrong. What we are talking about here is not simple pages with links. We are talking about an application that changes with the situation to become something different. Of course, you can start developing and see what happens. But if you do not try to create a model of what will happen before you start developing, you risk creating something that will not work and definitely will not fulfill the stakeholders' requirements.

There are a number of advantages to using a structured design method. It forces you to standardize the logic throughout the system you are analyzing, and it also gives you a number of other benefits that you will find map well to the development of location-based services, where you might start with one subsystem and develop further subsystems later. The benefits of structured analysis are as follows:

- Large systems can be partitioned into component subsystems or subfunctions for further analysis.
- Specifications for individual components are easier, faster, and more accurate to define than the total system.
- The interaction between the parts can be planned, designed, evaluated, and implemented to reflect improved information flows and controls.
- More than one person can work with the same system simultaneously.
- Standardized format and grammar enhance and simplify communication and maintenance.

The last point is important in developing for a location-dependent system, because you will need to communicate the function of the system not only to your own staff but also to other stakeholder groups, such as the people who will be creating the data sets and those who will maintain it. This statement is especially true if there are many different groups that will be developing the databases, such as Web site owners.

If you have been doing any database work before, you are probably familiar with some data modeling method and know a thing or two about structured design. You can apply the method that is most familiar to you to the development of databases for location-dependent services, but there are some methods that work better than others. But if you have not, let me give you a quick introduction to structured analysis.

When you are developing a database system, and when you are restructuring and maintaining it, you need to be able to keep track of how the data in it is related. This task is normally done by using a technique called entity relationship modeling or data modeling, which uses diagrams to illustrate the tables in the relational database. Lines represent relations, and depending on the exact method, there are a few different ways of drawing the tables themselves.

The important thing is that the purpose of this technique is to graphically demonstrate how entities are related to one another. An entity represents a real or abstract thing that is important to an enterprise about which data needs to be stored. For example, an entity could be Customer, Product, Inventory, Supplier, Sale, Purchase Order, or some other label generally in the form of a singular noun. An entity would typically correspond to a table in a relational database.

The *entity relationship diagram* (ERD) is a way to graphically show the data model. It provides a clear and concise method for describing data by using entity symbols that are interconnected by relationship lines. Relationships between entities consist of specific associations that are described in terms of their cardinality and are generally labeled by using action verbs. Cardinality refers to the numerical scope of associations between entities, such as a one-to-one association (one sale is associated with one customer); a one-to-many association (one supplier supplies many products); or a many-to-many association (many salesmen sell many products). The terms "one-to-one," "one-to-many," and "many-to-many" are common statements used to describe the cardinality of a relationship. There are specific ERD symbols used to signify cardinality, the terminators on relationship lines.

Depending on what type of system you are designing and what method you are using, there are different starting points for your design. Some feel that entity relationship modeling should be the starting point for designing a system because it is necessary to know the nature of the data in order to determine the processing done upon it. Others feel that the process model is the best starting place because the processing of the data is the system and the data and its storage can be designed to fit the necessary processing.

There is nothing that prohibits you from mixing and matching different design methods, and one that is particularly useful together with the entity-relationship diagram is structured design. Structured design is the partitioning of a system into a hierarchy of modules that performs the activities internal to your system. It is a technique used for representing the internal structure of a program or system and its components.

Structured design is complementary to structured analysis (or process modeling) and implements another stage in the software life cycle. If data flow diagramming is the "what" of your system, structured design is the "how."

The modeling technique used in structured design is the structure chart. A tree or hierarchical diagram defines the overall architecture of a program or system by showing the program modules and their interrelationships. The structural information contained in the system model is used to generate code in automated systems and to create the precise infrastructure of your system. This information includes the passing of control and parameters between program modules as well as the specific order in which the modules are arranged in your code.

A module represents a collection of program statements with four basic attributes: input and output, function, mechanics, and internal data. It could also be referred to as a program, a procedure, a function, a subroutine, or any other similar concept. A structure chart shows the interrelationships of the modules in a system by arranging the modules in hierarchical levels and connecting the levels by invocation arrows designating the flow of control. Data couples and control couples, designated by arrows, show the passing of data or control flags from one module to the next. This process is equivalent to passing parameters between functions or procedures in an actual program.

Process modeling, otherwise known as structured analysis, is one of the techniques for graphically depicting a system. Process modeling describes a system by focusing on the transformations of data inputs and outputs by processes. Whether examining an existing system or designing a new one (or both), this step is essential toward fully understanding the system. The diagrams you draw enable you to show, at levels of increasing detail, how data flows through your system and what is being done to it along the way. Using process modeling is a way to identify the data flowing into a process, the business rules or logic used to transform the data, and the resulting output data flow.

Process models are documented by using *data flow diagrams* (DFDs). The DFD consists of data flows, processes, data stores, and external entities. A data flow is data that is in motion in your system. An arrow that indicates the direction of the flow of data represents it. A data flow is labeled as a noun, indicating the particular data that is being transferred. A process is a procedural component, a transformation agent, in the system. It transforms inputs to outputs. A process is indicated by an action verb describing the sort of transformation that occurs. A data store, also called a file, represents a logical file, a database, or even a filing cabinet. In a system, it is data at rest within the scope of the system. An external entity, also called a source/sink, provides data to the system from outside the scope of the system or receives data from the system. External entities are outside the system, so they are beyond the scope of analysis. A data store, a source, and a sink are all generally labeled as nouns.

Process models are generally useful for systems where the interactions between users and data will change the data in some way. If the data is static, an entity-relationship diagram might describe it better. This diagram, however, is as much a matter of taste, education, and tradition in your company as anything else, and it is perfectly possible to model a location-dependent system by using a process model.

The entities in a data model and the data stores in a process model are very similar, if not the same—although just how, and if, they correspond to each other is not generally agreed upon in the structured modeling community. But if you are going to be using both models for different applications (say, the entity-relationship model for currency exchange rate databases, but the process model for a bank system, and a location related component to automatically reconfigure the bank balance of the user to the local currency), you might find that specifying such a correspondence helps you to ensure that all data is accounted for between your models. You can specify that every entity must correspond to a data store with the same composition. Another link between your process and data models is the ability to create a view of the portion of your data model that is affected by a particular process.

Object-oriented modeling is extremely useful for systems that have limited data sets but where the functions can be repeated. The telephone system was one of the first use cases for object-oriented modeling. But while object-oriented modeling works fine for modeling databases, it is a longer step from the object model to the database table than from the entitiy-relationship diagram. In this book, we will not use object-oriented modeling (although we will use object-oriented concepts). But then, we will not be developing a large system. Instead, we will be applying a mix of structured design methodologies. There are a number of different methodologies, and there are plenty of books documenting them, so there is no need for us to go into them here. We should say one thing though. You might think that position-dependent services would be easy to model as state transitions, but it is not the information object that transitions through the system; it is the user.

Depending on how you develop, you might take the use cases and transform them to *functional decomposition diagrams* (FDDs). They give you the capability to perform high-level planning of business functions and their hierarchical relationships while concurrently populating the repository. You can enter business functions that you define onto diagrams and break them down into successively finer gradations. At some point, one that is entirely up to you, you can break down functions into processes. These processes are equivalent to the processes that appear on data flow diagrams. The processes can themselves be broken down into smaller parts (still lower-level processes) on the FDDs.

Depending on your development process, you might want to involve the user community in this process, or you might want to start with the use case descriptions and then check back with them when you have developed the FDDs. The diagrams describe organization responsibilities, which translates into functions. Once they begin describing what they do, it indicates the transfer to processes. This point is probably where you want to start showing finer gradations of functionality as processes.

All of the processes of a branch of your functional decomposition diagram can be laid out in your modeling tool, together with their hierarchical relationships. Most modeling tools can automatically spawn DFDs from an FDD function. You can then work on the DFDs and add data flows, files, external entities, and so on. The functional decomposition diagram is very different from a process decomposition diagram. The former is a full diagramming methodology for doing business planning. The latter is simply an unstructured diagram laying out the hierarchy of processes that are descendants of an indicated process.

Regardless of which development method you use, there are tools that will enable you to create the diagrams in a graphical editor and generate the code for the database system automatically. Developers with much experience will recognize that there are performance gains to do in the manual optimization of databases and programs. The SQL programs can usually be optimized, and the structure of the tables usually can, too. Performance-wise, the programs running the logic can usually be optimized over the result of development tools.

Application servers for location-dependent applications, as we saw in Chapter 5, "The Application Server," are typically connected into databases. The language they use is normally Java, and the method of connecting to the database is JDBC or ODBC. It is usually easier to model the databases first and do the application development later, however.

Which XML language you select also can determine your data model. The Geographic Markup Language (GML) is built on a hierarchical vector model, with feature collections comprising features that can comprise more features (and feature collections). Combining data sets also becomes very easy, because you can just combine them in a feature collection. You can also create entirely new data models by combining elements from different XML namespaces, if you do not want to use the GML data model.

This means that a GML database in itself can be distributed, using Xlink and Xpointer to connect the features to the collections (unfortunately, how to do this was not sorted out at the time of this book's publication). Another advantage to using GML is that the data validation functions of XML can be used, which enables a higher data quality, especially in distributed databases.

A disadvantage to storage in XML is if you want to use the native processing functions of the DBMS, since no major database management system currently uses XML as its base format.

Like any XML data, it is not necessary to store the data itself as XML; that can easily be generated when the data is output. If the data is to be used only as GML, it makes sense to store it as GML, however, and likewise if the database is already distributed (that is, one file at each location). This makes distributed updates possible, which may correspond to the actual organization of the data creation. As of this writing, GML servers are available from CubeWerx Ltd, Ionic Software, ESRI, Oracle Corporation, and Laser Scan. More may have appeared when you read it.

Data Quality in Geographical Databases

The number of features of the real world, and the different ways they can be approached and used, are infinite. In other words, any attempt to design a total system for an area will be hopeless. You simply cannot cover everything. You can create reasonably comprehensive databases of certain aspects of it, but it is not even possible to cover exactly everything (quick: When was the pavement outside your home relaid last time?). So, if you are compiling an entirely new database, you will have to concentrate on one aspect.

It is, of course, different if you are using existing databases, especially if they are from the City Office of Paving, for instance. In that case, you will probably know perfectly well when the pavement was laid, who was responsible, how it was financed, what materials were used, and so on.

You would still have to contend with all the normal problems of database compilation, though. The data might be comprehensive for that certain aspect, but the supervisor who signed off the work might have filled in the wrong coordinates, and it might look as if the pavement was covering the entire street.

As you will see in Chapters 8, "The User Interface to Location Information Services," and 9, "Maps as User Interfaces," designing Web pages is very different from designing location-dependent applications. That goes for designing the data elements that go on the display, too. When you think about database design for position-dependent applications, you should consider that the elements are much smaller than what would normally be used in a Web page, for instance. First, the user has a much smaller screen that you can work with, and second, you are composing a page of different elements that relate to the user's current position. Displaying *Crime and Punishment* on the screen might be nice, but the passages that relate to each location in St. Petersburg are very long and would be unreadable.

Also, remember that you probably would like to compose the presentation out of several separate elements rather than having to fill in the blanks in one large element. This feature gives you a much bigger flexibility, and the reason why you want flexibility is very simple: You know what you are doing now, but databases survive much longer than applications, and you never know what you will be doing with your data tomorrow. Future-proofing it now is relatively simple if you are building a database from scratch or modifying an existing database. It is very difficult if you are using a database, which was originally intended for something completely different. In those cases, it might be better to use it as a data source, dump it out to the new database, and set an update link so that any updates to the old database are also done in the new one.

Spatial Processing in Database Systems

That positioning systems have become available is one of the factors that have made location-dependent information services possible. But there is another, at least as important, technology that has become available during the last few years: spatial database processing.

The way we perceive space is as a set of objects that have specific relationships, such as distance, size, and closeness. Depending on which relationships we are interested in, different models of space such as topological space, network space, and metric space can be used. Topological space uses the basic notion of a neighborhood and points to formalize relationships, which are invariant under elastic deformation. Topological relationships include closed, within, connected, and overlap. Network space deals with relationships such as shortest paths and connectivity. Metric spaces formalize the distance relationships by using positive, symmetric functions that obey the triangle inequality. The different spatial relations that we can perceive are as follows:

- Absolute and relative location
- Distance between features
- Proximity of features
- Features in the neigborhood of other features
- Direction and movement from place to place
- Boolean relationships of and, or, inside, outside, intersecting, non-intersecting, and so on

Spatial referencing, geopositioning, or fixing the location of an object in space and time, is a central aspect of geographic data modeling. Entities and phenomena are not meaningful in a geoprocessing context unless they are expressed in terms of a model that fixes them in time and that places them relative to the surface of the Earth. Spatial reference systems define how coordinates are interpreted within the geometry of a feature. In a raster system, the symbol representing a feature is a grid cell location in a matrix. In a vector system, the locational symbol might be a one-dimensional point; a two-dimensional line, curve, boundary, or vector; or a three-dimensional area, region, or polygon.

Spatial relationships between objects are, at the most abstract level, a relation between a location symbol and its meaning. But it is also a question about linking other data to the object by using the coordinates as the key (for instance, the symbol and the location; the spatial data and any non-spatial data that there might be; and the geographical features and their attributes). There are actually three different levels of relationships involved: between the object and its representation; between the geographical features of the object and the object; and between the non-geographical features and the object. Non-geographical features can be things such as opening and closing times, which are socially determined (not determined by anything that is linked to the feature).

The different types of relationships include functional relationships between and among geographic features and their attributes, which means information about how features are connected and interact in real life. An example could be the status of different roads; for instance, from a two-track rut in a forest to an eight-lane highway. The rut leads to the highway (or the other way around), but it is very unlikely that they are directly linked. Zoning in cities is another type of functional relationship where different properties have been assigned to different zones (industrial, housing, and so on). Another type of relationship is a logical relationship between the features and their attributes, such as a property being a habitat for an endangered species and therefore being zoned as a conservation area, and a road being put on the list for reinforcing because it is on a hillside that is at risk for mudslides.

Functional and logical relationships are assigned by human intervention, because there is nothing in a natural feature set that says that a property should be in the habitat of an endangered species (and interestingly enough, we often tend to forget that and reverse the relationship). Topology is a branch of mathematics that enables the calculation on points and lines (called nodes and arcs, and strictly theoretically speaking not points and lines at all but rather curves—and, as you saw in Chapter 3, "Position Technologies and Coordinate Systems," curves can be used to express most shapes). Nodes are endpoints of arcs, which really are curves. The order in which the nodes are connected defines the shape of the arc or polygon. This information is stored in a spatial database, and it also contains the functions for calculations on the data that are available for the developer to call and use in his or her programs. Associating attributes with the nodes (essentially, creating a table that holds all the relevant information and linking it to the node by using the coordinates of the node) can then enable you to express information about the node itself (its properties).

One application of topological theory is network analysis, which uses topological modeling to determine shortest paths and alternate routes. This information is used in route calculation applications to determine the shortest route between two points, for instance. Of course, this process also needs to take into account functional attributes, such as roads being one-way or too narrow to hold the current vehicle (which is very relevant if you are driving an 18-wheel truck).

Figure 6.1 shows a set of nodes on an imaginary map with arcs between them, enclosing polygons. We can assign the nodes coordinates to identify them.

The nodes define arcs, which have the nodes as endpoints, but also directions (which makes them vectors, mathematically speaking). As you note, there can be arcs that have the same starting point and endpoint, enclosing an ellipse. Arcs define polygons. By tracing arcs around an area, you can create any polygon because the ellipse arcs can be small enough. The neighboring arcs and polygons are also recorded. The edge of the area studied (in the figure, the rectangle) is the boundary of the universe of discourse of the current database.

Once we have this information, we can ask ourselves questions about the areas, arcs, nodes, and polygons. For instance, "Which polygons are neighbors to polygon D?" "What is the quickest route from node 1 to node 4 (which you can calculate using various algorithms, such as selecting all the arcs that lead from 1 to 4, and measuring the distance by measuring their length)?"

The comparison between geographical features is very different from the comparison between data in insurance accounts. The mathematics are different, so you need to create different functions to compare the data. The data types are not numerical, so you cannot apply arithmetic and need to create new functions for calculation. Spatial data have different sets of properties, so you need

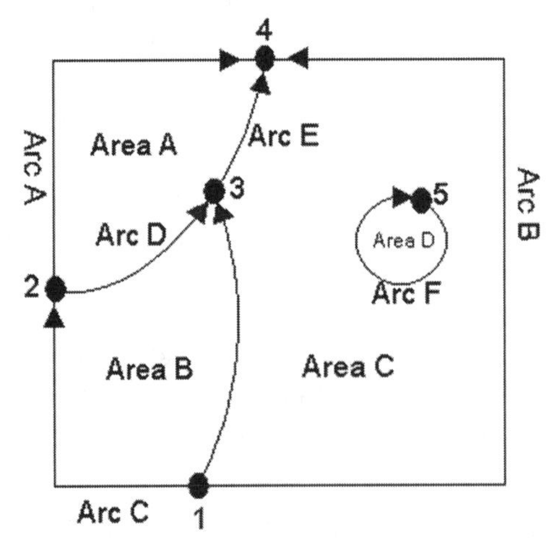

Figure 6.1 Nodes, arcs, and polygons.

the DBMS to manage the data differently. But the *Geographic Information Systems* (GIS) industry has been an important market for several years, and the major DBMS managers have all developed spatial processing systems.

The programming logic of early database systems allowed little more than primitive arithmetic. But sophistication grew rapidly, and relational database management systems as a rule have very sophisticated systems built in to process different data types, separately or in combination. Some of those rules are reflected in SQL, the Structured Query Language; others are proprietary extensions by the DBMS manufacturers. Some might not even be in SQL but in the programming language you use to address the logic.

To handle spatial data, you have to create a mechanism to represent it in the database, and you need a way to index and search for them in the database. Oracle Spatial, to take one example (Sybase, IBM, and other leading DBMS vendors have similar functions), addresses these issues by providing an object data type (SDO_GEOMETRY) for spatial data, an indexing capability, and functions/operators on SDO_GEOMETRY.

Complicating the matter is the fact that the collections of features that are used in the geographical databases have both non-spatial and spatial attributes. Examples of non-spatial attributes are customer name, address, orders, and part_number. The spatial attribute, such as a geocoded address or sales_region, is a coordinate geometry (or vector-based) representation of the shape of the feature. Also, the geometry is an ordered sequence of one or more vertices. The structure and semantics of the geometry are determined by its type, which might be one of point, line, or polygon.

There are a number of standardization groups that work on standardization of different aspects of spatial database management; for instance, Working Group 3 in ISO/TC204, which works on standards for geographic databases for ITS applications. It deals with a number of different work items:

1. *Geographic Data File* (GDF): A standard for the definition and exchange of geographic information for ITS applications
2. *Physical Storage Format* (PSF): A standard for the application format of map data in on-board navigation systems
3. *Application Programming Interface* (API): A standard for the programming interface between the application software and the data access software resident in a navigation system
4. *Location Referencing* (LR): A standard for unique and unambiguous referencing of geographic locations

As the standardization work points out, spatial processing has also seen some changes in how DBMSs handle data. Using an object relational system (which stores data in objects in relational tables), you can handle spatial information as a spatial object data type and access or manipulate it by using spatial index methods and functions. Because spatial is now just another attribute represented in the database, users can use it as another qualifier or criteria when searching or browsing the database.

Object relational databases provide a higher abstraction of spatial data by incorporating concepts closer to humans' perception of space. This task is accomplished by incorporating the object-oriented concept of a user-defined *abstract data type* (ADT) and associated functions. The data types that the spatial databases contain are more complex than traditional data types, however. They include complex data types such as points, lines, and polygons. Also, the operations on these data types are more complex when compared to the operations on simple data types. There are four main properties of the spatial data that make it different from traditional relational data: geometry, distribution in space, temporal data changes, and large volumes. A satellite survey of even a small area of the Earth can contain data in the order of several terabytes. But we will not consider it in this book.

Geometry is a main property in any kind of spatial data. The geometry deals with the mathematical properties of an object. That includes its measurements (the metric of the geometry), the relationships of points, lines, angles, surfaces, and solids (topology), and their order. A simple geometry is usually constructed from geometric primitives such as points, lines, curves, and areas. Complex geometries are constructed from collections of simple geometries.

For instance, the Oracle Spatial DBMSs support three geometric primitive types and geometries composed of collections of these types. The three primitive types are Point, Line String, and N-point polygon (where all these primitive types are in two dimensions). A 2D point is an element composed of two coordinates, X and Y. Line strings are composed of an ordered sequence of two or more points that define line segments. Line strings can be composed of straight-line segments, arc segments, or a mixture of both. Polygons are composed of connected line strings that form a closed ring, and the interior of the polygon is implied. Because polygons are composed of line strings, this situation implies that polygons can have some edges as straight lines and some edges as circular arcs. The spatial data model is a hierarchical structure consisting of elements, geometries, and layers. Spatial layers are composed of geometries that are in turn composed of elements.

In addition, there are a number of geometric relationships between geometries that are important in handling spatial data. On a road map, a connectivity relation describes how one intersection is connected to another intersection (describing how two geometries interact). Metric relationships deal with the distances between two geometries, such as how objects are related through a distance ("find all gasoline stations within 10 miles"). Another factor that has to be considered, however, is the irregular distribution of objects in space. There are very few mountains in the central plains of the United States and very few cities in the Sahara. Objects are also very different in size (and how the size is defined also depends on the definition of the object). For instance, the Hekla volcano in Iceland is much smaller in area (and much lower) than Japan's Mt. Fuji or Mt. Etna in Sicily. Cities can be defined as large or small, depending on whether you count the administrative boundaries or the population boundaries.

Spatial objects also tend to change over time. Cities grow, roads are built, and on a different time scale, mountains erode and disappear. Other properties also change over time: The road is under construction, or it can be opened for traffic. Or, it might even be in the planning stage and only exist as a line on the map. Which property applies will be very important for anyone who wants to use it to calculate a route from A to B.

Searching and indexing the data, however, requires that the spatial types and operations on them should be part of the standard query language that is used to access and manipulate non-spatial data in the system. In relational database systems, this situation means that SQL should be extended to support spatial data types and operations. It must also be able to index spatial data, so spatial queries (range and join queries) can be made in the same way as for nonspatial data (so the database can be addressed by using standard programming languages such as SQL or APIs such as ODBC and JDBC). Traditional SQL is, however, not enough to express typical spatial queries. Several standards bodies are working on specifications that extend SQL with spatial functionality.

Spatial queries are often processed by using filter-and-refine techniques. In the first filter step, an approximate representation of a spatial object is used to determine a set of candidate objects that are likely to satisfy the given spatial query. The approximations are chosen so that if the approximations of objects A and B do satisfy a relationship, then the objects A and B are likely to have that spatial relationship. For example, if the approximations are disjoint, then the objects A and B will be disjoint. If the approximations are non-disjoint, however, objects A and B might still be disjoint.

Using methods such as these to compute searches and object retrieval speeds up the searches, because the approximations can be made less complex than the actual objects as they are stored in the database. Spatial queries fall into two categories: Window queries and Join queries.

Window queries take a spatial object (called the window object) and locate spatial objects in a table that satisfy a binary relation with the window object. For example, the query, "Find all the roads in the City of Minneapolis which overlap the park Minnetonka" is a window query. Here, the object representing the park Minnetonka is the window object. A join query seeks all the object pairs from two tables that satisfy a given relation. For example, the query, "Find all the roads and parks which overlap the city of Minneapolis" is a join query.

Indexing a database speeds up searches by enabling searches to take place in the index instead of the actual data. Most DBMSs provide indexing mechanisms that work with scalar data but might not be suitable for spatial data. A spatial index would search not only the subset of objects embedded in the defined space to retrieve the query answer set. Apart from spatial selection, spatial indexing also supports other operations such as spatial joins, finding the object closest to a query value, and so on.

When you index spatial data, you can build a separate spatial index, and store it in the server as binary objects. The second method is space partitioning, where the space is divided into cells in a hierarchical structure; the cells are then assigned numbers, and the spatial objects assigned numbers according to the cells they are in. But space is not linear, and many spatial features are irregular, which complicates queries.

R-tree is an indexing mechanism used in most spatial databases (actually, all databases which have spatial data types). It splits the area to be indexed in hierarchically nested boxes, which are used to create an index, as in Figure 6.2.

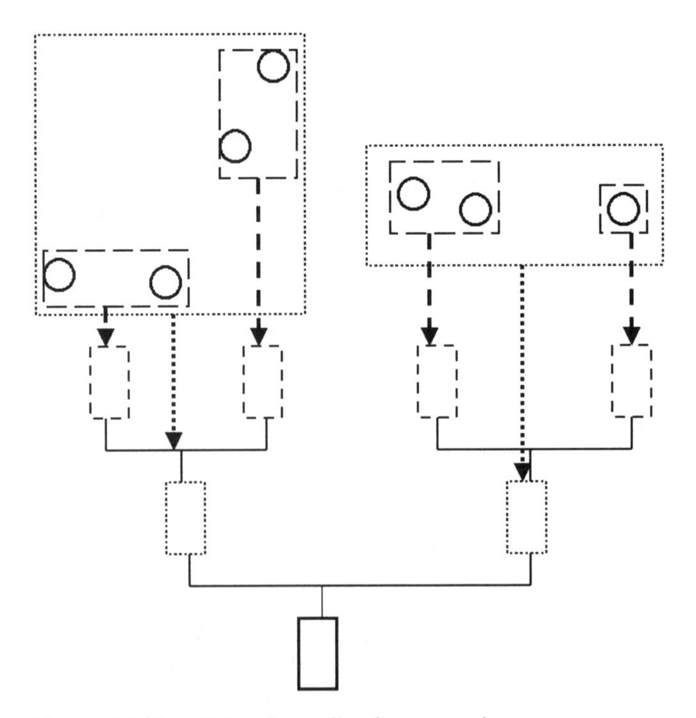

Figure 6.2 How R-tree bounding boxes work.

The boxes surrounding each object are determined by the shape of the object. It works extremely well for data that is static, but objects that are moving will change the index, which causes problems. It is, however, used in exactly the same way as other indexing mechanisms. It uses the minimum bounding rectangle (MBR) as the abstracted representation of such spatial objects. R-tree does not divide space into a full coverage of non-overlapping, adjacent cells. These bounding boxes may overlap and do not need to cover the entire space; there is no need to know the spatial extent of the data in advance.

Catalog Interfaces (LDAP)

The Catalog Services Specification of the Open GIS Forum will require the specified metadata operations to handle all common data types and values for metadata elements, including any subset of the metadata currently defined in the ISO TC211 15046-15 draft document on metadata and extended (user-defined) metadata adhering to that ISO working document. Proposals must also address how to exploit or connect with existing catalog and geospatial technologies, including the Z39.50 (Z39.50 Maintenance Agency) and LDAP protocols.

Lightweight Data Access Protocol (LDAP) is an IETF standard for accessing X.500 structured data catalogs. X.500 is one of the surviving *Open System Interconnect* (OSI) data formats, and it structures data in object hierarchies for use for catalogs. The X.400 mail system is probably the most widespread example of a system using an X.500 catalog.

X.500 catalogs have rather restricted use because it turned out that most of the foreseen functions of X.500 can be handled as easily by a DBMS, and the data access can be done by using HTTP. The Open GIS Forum is defining a standard for handling data access (especially metadata) using LDAP, however.

- Query (or selection) operations will support discovery and access of metadata elements having spatial, temporal, text, and/or numeric values. These operations will:
  - Enable users to specify metadata element names and values (for example, spatial, temporal, numeric, keyword, and other types) in the query predicate
  - Enable the use of Boolean operations (for example, and, or, and not) between search terms in the query predicate
  - Allow the use of spatial query operations such as intersect, contains, contained by, within, and beyond as search terms in the query predicate

OpenGIS Catalog services will include the following:

- Operations for creating and maintaining catalogs and catalog entries
- Operations for creating and maintaining collections of geospatial datasets, including associated metadata sets and metadata entities related to each

stored dataset. An analogy would be as follows: Catalog services give a librarian the ability to put new books on the shelf and new cards in the card catalog.

- Operations for recording and reporting the accessibility and retrieval status of geospatial datasets
- Operations for discovering the content and structure of geospatial resources described in the metadata for those resources

There are at least two kinds of books on the shelves: information sources and information processing. The OpenGIS Catalog Services Specification must accommodate both of these as well as hybrids.

Geocoding of Information

A location is much more than an area delimited by a set of coordinates. It is a place and usually has a name. There are several types of place names, and several of them have been standardized. Country names (although the definition of a country can be fuzzy at times) have been defined in ISO-3166-1 for use on the Internet so that Australia is abbreviated AU, Austria is AT, USA is US, and so on. There are also standard names for states and similar administrative regions, defined in ISO-3166-2. For instance, ca is for California, ma is for Massachusetts, vi is for Victoria, and so on. This system is used in the unfortunately almost-forgotten U.S. top domain on the Internet, where the states have names like ma.us and Greenville, Massachusetts does not have to fight with Greenville, California over the name Greenville (greenville.ma.us and greenville.ca.us). Not all countries are federal republics, however, nor do administrative divisions (however important they might be) conform to the same logic (for instance, the Swedish region of Västra Götaland is not internationally standardized).

Depending on which culture you are applying the names to, you will have different levels of subdivisions. In Europe, it would be natural to follow the county region with a town or a city; in Japan, the prefecture would be followed by the city, which would be followed by the ward. There are a number of databases (gazetteers) for names, but cities sometimes have different definitions depending on context (the historic city of Stockholm is something entirely different in area and boundaries from the administrative region of Stockholm). Another problem is caused by cities that have names in multiple languages, whether official or unofficial (something that also applies to areas). Romanization of non-Latin forms of writing also causes a problem; there are, for instance, two official ways to Romanize Chinese place names in Hong Kong.

Another way of naming areas is to look at some format that uses them for distribution purposes. The one that is most frequently considered in this context is the postal system, which varies considerably from country to country, however, and which does not at all follow the same logic everywhere. For instance, postal codes are instructions to the postman in Britain, giving a very high precision of the address whereas they are assigned at random in other places. In some areas, they are numbers; in others, they are strings of letters. Granularity varies with where the letters are to be received and sometimes might not have anything to do with location at all (such as when a post box is used, in which case it actually is probable that it does not represent the user's home address).

Telephone numbers used to be encoded geographically, but most numbering systems are now divorced from the actual location of the telephone, especially in countries with deregulated telephone systems. (In Sweden, when there was a telephone monopoly, there were maps that showed the exact location of each telephone, and the telephones were numbered consecutively in the order they were connected to the exchanges, which were hierarchically numbered. In the United States, AT&T had a similar system.) With a mobile phone, however, the relation between the telephone number and the location has of course disappeared altogether, and in countries with deregulated telephone systems, there is no relationship between the numbers and the telephone systems anymore.

Probably very few customer databases are without street addresses. Street addresses are usually perceived to give some geographic precision, but that assumes that the streets themselves have been mapped to coordinates with the same level of precision. It also assumes that they are encoded in a standardized way. This situation is actually rarely the case. As for the encoding of a street address with precision, this situation is not true in the case of many European cities. In some cases, numbers might be missing or be illogically ordered (often for historical reasons). And, in some countries, such as Japan, street names do not exist. Houses are numbered in blocks, and the numbering is done in the order that they were first built. In rural areas, there are no house numbers either, and the nearest mail stop might be several kilometers away from the house (or several tens of kilometers, in the example of the Australian outback). In other words, there cannot be, except in some exceptions such as American cities, a rule for how addresses should be mapped to coordinates. It has to be done on a case-by-case basis.

Acquiring this data and getting it into the database is a major operation. There are a large number of sources for the data, and if the information frequently changes, it will require frequent updates. Using the same mechanism as Internet search engines, as demonstrated by the Kokono example, is a relatively simple and straightforward way to acquire the data from distributed locations on the Internet.

A street address is a simple example of spatial (or locational). Geocoding that address (converting it to a latitude/longitude coordinate pair) results in a location that can then be used to determine the spatial relationships among street addresses or between a street address and some linear spatial feature, such as a road, or an area feature, such as a census block. Reverse geocoding means taking the coordinates and applying street addresses to them.

Most of the location-dependent systems we have looked at in this book assume that the geocoded information is stored in a database, and the processing takes place through a DBMS. The WAP Forum Location Framework does actually have a geocode data type, though, so you can make requests for positions by using their addresses.

The element is used empty in request messages to request specific geocodes (for instance, postal code) and in answer and report messages to deliver a geocode value. Internationalization of the value is supported through the encoding element. A geocode can be viewed as a combination of a type (for example, "postal-code") and a value, such as "BT1 2FJ." The WAP location specification only specifies a small, basic, interoperable set of geocode types (picked from the ITU-T X.520 specification). The current version of the WAP location specification does not specify any syntax for the geo-code values; these are treated as string values.

The type attribute MUST be used in both invocations and responses to indicate the geocode type. Table 6.2 lists values that are specified by the WAP location specification.

In a response, the actual value MUST be included as a string value.

Example: Request for postal code in an invocation.

```
<geo-code type="postal-code"/>
```

Example: Delivery of postal code in a response.

```
<geo-code type="postal-code">"BT1 2FJ"</geo-code>
```

Table 6.2 Types of Geocodes as Listed in the WAP Location Framework

TYPE VALUE	EXPLANATION
country	Country name, e.g., "United Kingdom"
locality	A named location, e.g., "Edinburgh"
state-or-province	Name of state or province, e.g., "Kent" or "Ohio"
street-address	Specifies a site for the local distribution and physical delivery in a postal address, i.e., the street name, place, avenue, and the house number. For example "101-111 Donegal Street" or "Siechenmarschstrasse 11A"
postal-code	Postal code or zip code, e.g., "BT1 2FJ" or "CA 94063"

Semistructured Database Searches: Kokono and Other Search Engines

Kokono is an experimental system developed by NTT Labs in Japan that indexes the Japanese Internet according to position information. It uses the addresses to derive the position information and relates it to other databases with the position information as the key.

Kokono (which means "here" in Japanese) is a search engine, an indexing system, and a portal for location-dependent information (in Japanese). It can be used from a PC, a PDC phone, or a PHS phone, and it can use the positioning methods of those systems to sense where you are. In the PC version, you have to fill in your own position, which can be selected from your street address or from the nearest railway station (especially around Tokyo, there is hardly anyplace that is farther than a kilometer from a railway station).

The search engine indexes information from the Internet based on addresses. Only pages with Japanese addresses are stored; others are thrown away. Content that is gathered is timetables, shop opening times, local television listings, and so on. NTT has managed to get maps from a number of map providers into the service. Users search by filling in a form and selecting the appropriate categories, as shown in Figure 6.3. The queries are reverse-geocoded in case they are given as coordinates (or if they come from a mobile client), and objects that appear close to the target are retrieved.

Using the addresses as the key and applying geocoding to them, they are linked to the maps. Data from the information providers are metacoded with position data, and that data is matched to the user's position when data are searched. The user can choose to have a map displayed with the information requested or can choose a listing. Recommendations to the user are based on spatial closeness to the address the user requests.

Because addresses in Japan are always structured in the same way, the search engine can retrieve the addresses from the pages, mark them up, and store them in a common format. About half of the documents retrieved contained an address, but for smaller subdivisions than cities, the figures started to drop. Only 7 percent of all documents contained the block number (there are no street addresses in Japan; instead, all cities are divided in wards, which are divided in blocks, which are numbered). Pages were fairly evenly distributed over Japan (correlating with the population) with a slight overweight for Tokyo.

The service has been going since 1997 and is still run as a research project. Up to now, maps are the dominant search term, with 38 percent of users searching

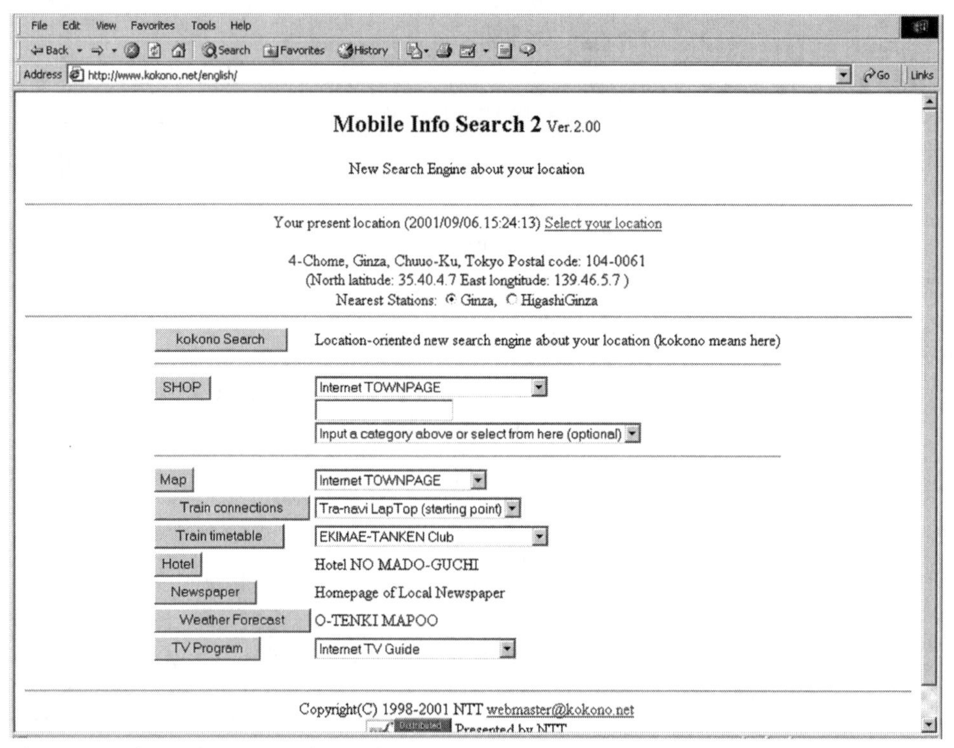

Figure 6.3 The Kokono search window (in English).

for maps. Fourteen percent searched for shops, and 14 percent searched for transportation.

In the United Kingdom, you can use the Somewherenear search engine from your WAP phone. It does not yet handle automated position input, but it does give you a list of the restaurants closest to you—with reviews from people who have visited them. The listings are all user-compiled.

The system is based on geocodes derived from the address (post codes in the U.K. cover small, well-defined areas). Users can add new entries themselves, as well as review entries. The group behind it adds new categories, as well. The original base for it was a pub list compiled by the developers.

The system came about based on the perceived needs of the developers, or to quote Mike Banahan, its chief architect, "Our goal was to be able to go to a strange town, press the 'nearest pub' button on whatever the device was, and get a map showing establishments nearby, together with visitors' ratings. Having gone there, you should be able to add your own vote on the quality of the beer, the food, the surroundings, and so on and then ask for the nearest good curry house and cashpoint to pay for the food. (Non-U.K. readers may not be aware of the mystic significance of curry and beer, but believe me, it is extremely important to the U.K. software industry. Curry houses are by far the most common food outlets in Britain. Even Macdonalds had to introduce Chicken McTikka to be able to compete; it's just as appealing as their other offerings.)"

The most interesting thing about it, however, is that it is entirely built on an open standards base, leveraging CGI-scripts and PHP3 for the configuration. The entire platform is built on Open Source software, such as Apache and MySQL.

PHP3 is used as the template language, but the templates deal only with presentation—the CGI scripts do all data management. This has turned out to be very effective, allowing them to change the presentation at will—creating a WAP frontend in less than a day, for example.

Other search engines take a different approach to geocoding. (Both Kokono and Somewherenear are really metadata search engines, the metadata being the address and geocodes.) The Geotags search engine requires special geotags in the HTML of the site. Because it is not standardized and not widely publicized, very few pages have these tags. Why, when addresses will serve the same purpose, and using GML will give you full GIS functionality?

The search engine clearly illustrates that it is possible, and perhaps even relatively easy, to build searches using geographically defined areas.

The third example of a geographical search engine is American. Go2 will present its content on a variety of wireless devices, from Web-enabled phones to PDAs. The base for the system is its database of what it claims to be more than 16 million locations (this is, of course, impossible to check).

The locations are referenced using a proprietary technology and not geocodes at all. The technology uses an abbreviation method that is similar to the way the SOUNDEX operator in SQL works (by looking at the consonants of the word, you can determine if two words are likely to sound similarly). The first part of the Go2 address identifies the country, state, and city for the location. The last part of the Go2 Address is the unique identifier for that real world location, in the specified city. For example, a merchant might choose to have "BURGERKING" or "BK" as the unique identifier for their locations. This would enable real world searches for the closest Burger King using "US.CA.IRV.BK".

A second version encodes the geocodes in the same type of address. This address takes the form of "US.CA.IRV.XX.XX.XX.XX". The first part of the address identifies the country, state, and city for the location. The remainder of the address uses several sets of number pairs to pinpoint the real world location with increasing levels of accuracy. The first number pair in the sequence denotes a 100-meter square area in which the location can be found. The entire five-pair combination denotes a 10-meter square area in which the location can be found. For example, US.CA.IRV.55.52.13.70.58 identifies the location of a B. Dalton bookstore in Irvine, California. Note that this method does not work for countries with different administrative structures—that is, for almost all places outside the US.

The system does have a problem with the advertising-based model, however. It is almost impossible to know whether a location for a chain of restaurants, which is a competitor of their current sponsors, is not listed because it was not known or because it is a competitor. All directory services suffer from these network effects (it is not until you become so large that your margin for error is negligible that you become credible), especially during a starting phase. But, the user-driven approach of Somewherenear creates a community, which even though it will be biased will be more credible, because you can identify the bias.

In countries that have a system of addressing that is tied to the position of the address, such as the United Kingdom or Japan, it is feasible to build this type of application. It does, however, also require that you can collect the information in a simple way—and here, the Japanese have been helped by their language and the paucity of the links from the Japanese Internet. Building such a search engine in Sweden would be extremely difficult, because the collection of pages, which can occur in the .se, .com, .nu, or any of a number of other domains, becomes unwieldy.

If you want to build a search engine, it is an advantage if you can convince users to mark up their information in a way that is automatically retrievable. And that is what we will look at in Chapter 7.

CHAPTER 7

Data Formats for Geography-Related Information

Building a new application is relatively easy, and gathering and organizing the data is not so hard, as we saw in the previous chapter. However, if you want to tie it into data about the rest of the world, you need to use a common data model and one that is navigable at that. As we discussed in Chapters 5 and 6, there are two basic ways of expressing this position-relevant information. One is to put it all into a database; in other words, you are constrained by having everything in one single database, but it does have the advantage that if you are using a DBMS that can handle relations between spatial objects, such as Oracle 8 and 9i, you can build programs that use those functions to relate the position of the user to the position of the object in which the user is interested. It is also relatively easy to build stored procedures that filter the information based on other predicates, such as whether the user is riding in a car or in an 18-wheel truck.

The other way is to put the information in XML documents, which can be metadata documents that relate to other XML or HTML documents on the World Wide Web (or they might stand alone). As a matter of fact, this method does not preclude the use of a DBMS, because most DBMSs nowadays can import and export XML. What does become important, however, is that the information is in a structured, predefined format so that a program in the DBMS can import or export it. After all, no human is going to read it like an HTML document, but it will be imported into the process used to personalize the information for the user.

The information about an object at a position further breaks down into two different areas: the geographically related information (for instance, the position, the depth, or altitude of the object, its shape, whether it is an object of certain predefined types like a building, a church, a temple, or something else, and so forth) and the interest-related information, such as whether this place is a gas station, its hours of operation, and so on. There might be a third type, which is a subset of this category, namely the event-related information. This information would, for instance, determine whether the object was a rock concert. Then, it would be interesting to know the times and locations where tickets could be bought, which doors of the arena holders of tickets in different sections should enter, how to get there through public transport or where to park, and so forth. Traffic information is a subset of this category or can be classified as time-related information in parallel with the weather, which can vary for a location (or not matter) but does not change the location itself. It only changes the interest at the particular time you are looking for information (a beach is not much fun in the rain, for instance).

XML is the foundation of most modern, interchangeable, structural data formats. While there are alternatives (such as SGML and ASN.1), XML is totally dominant when it comes to defining new data formats. This situation is due to the relative ease by which data formats can be defined but also to the possibility to extend a document by using other XML formats through the namespace mechanism.

XML is based on the use of Unicode for the character encoding and the use of URIs (which are based on the domain name system, DNS, of the Internet). It reserves certain characters for system use, and the format has to be declared in a special document (a Data Type Description, or DTD, or a schema in the XML Schema language).

A longer description of XML and how it works can be found in Appendix A, "Who Does What in Location-Dependent Standards?"

GIS Concepts and XML Formats

There are two types of information about a geographical feature that you might want to provide in a location-dependent service (and, as we will see in Chapter 9, maps are just visualizations of databases that encode the information in a different way). The first is the information, which can be visualized as a map. Those features (the selection of which are largely directed by custom) are static, because maps had to depend on them for a long time. Maps had to be printed and stored for a long time. They represented the world for explorers who saw the locations for the first time, several years after the mapmaker got his information. This situation meant that they could not be based on features that were likely to change, and definitely not on cultural phenomena (because the people making the maps did not share the culture of the people whose phenomena they would have illustrated).In other words, maps use objects like mountains, rivers, and other landmarks to determine the location of routes.

A peculiar thing about a map, compared to other image representations of information, is that the entire area of the image carries meaning. In a diagram the relevant thing is the slope of the curve and the points in terms of X and Y values. The diagram can be stretched in one dimension as long as the relation remains the same. The white space around the curve does not convey any meaning in itself, either.

A map is different. If you want to expand the scale, you have to expand both the X and Y axes proportionally. Otherwise, the representation of reality of the map will be skewed. Except for violent earthquakes and the land rising in the Baltic, the Earth does not stretch in one direction (and certainly not suddenly).

Because all points on a map have meaning and their relation to each other also has meaning, describing the points and areas becomes a very tricky job. Mapping agencies use vector graphics systems, where several object planes are overlaid to represent the different aspects of a terrain (for instance, the surface plane representing the altitudes of objects, the plane of rivers, lakes, and other topographical features, and the plane of roads and buildings). Also, you could easily layer an additional plane, an information plane, which describes the features of the objects.

This process is the key to the mapping between databases and descriptions. We looked more at it in Chapter 6, but the descriptions can be used as a schema language to create tables that can be filled with property values describing the different features (the mountain table could be filled with coordinates and altitudes, for instance). Once you have a schema, it is relatively easy to translate the information formatted with that schema to other formats, however.

Data in one XML data format can be combined with any other XML data. This means that you could add a POIX document into a GML document or a GML document to an XHTML document. You can use the XML-base format to combine a database of demographic data with a GML document, creating a description of who lives where. The *geographic information systems* (GIS) information industry has been working on creating a standardized feature set of descriptions of geographical objects for many years. There is an ISO standard group, ISO TC 211, that works on creating standardized descriptions of geographical features.

They, like maps, relate to a static representation of the world. With dynamic displays, you have new possibilities (to be used wisely, as we will discuss in Chapter 8). But meanwhile, there are two types of geography-related information: that which is related to objects that are static, like information about mountains and rivers, and information about objects that are dynamic, like a rock concert. First, we will look at the first type.

Open GIS GML

The Open GIS Forum has developed an XML-based language to describe geographical features. It is part of an entire framework, which includes the means for digitally representing the Earth and Earth phenomena, mathematically and conceptually; a common model for implementing services for access, management, manipulation, representation, and sharing of geodata between communities; and a framework for using the Open Geodata Model and the OpenGIS Services Model. The intention is to solve not just the technical problem but also to define a vocabulary and a way to extend it, which provides a common ground for information exchange between different information communities (sets of users with different meanings, semantics, and syntax for geodata and spatial processing).

There are, of course, already a number of encoding standards for geographic information, including COGIF, MDIFF, SAIF, DLG, and SDTS. And those are just a few examples of a vast number of specialized data encodings used in different communities. GML is based on a common model of geography (OGC Abstract Specification), however, that has been developed and agreed to by the vast majority of all GIS vendors in the world. It uses a hierarchically organized vector data model, which is well suited to the representation of geographic data, because objects are often contained in other objects, or themselves contain objects (countries contain states that contain cities, for example). It is also based on XML, and why that is important I think you have already understood from reading Chapter 6, "Providing Databases and Doing Searches."

GML does not contain any styling information. How a feature is represented on the screen is up to the application designer. Maps can be created using SVG or Java or generated as raster output. Other ways of viewing the information than maps are of course also possible.

The OpenGIS GML Specification defines a standard set of primitive data types and operations on those data types. It also describes how to provide a service; how to request a service; and how to determine whether a request is a request for data, a request for an operation on data, or both. It uses the W3C XLink and Xpointer specifications to express relationships between geospatial entities. In other words, such relationships can be expressed between features in the same database or between features across the Internet. You can also construct relationships between GML feature elements in different databases without requiring any modification of the participating databases. No more than read access is required to establish a relationship between data elements.

GML is not perfect, however. It lacks formats for storing topology, which means that conversions from map formats using vector graphics may be problematic. There is no way of expressing coordinates in higher than three dimensions (you can have coordinates in four dimensions, if you have a point with elevation and time, for example). You can store elevation and other aspects as attributes, but it becomes problematic if you want to use them as the main key. GML also can handle only lines, not curves, which means conversion from CAD formats can be difficult. The OpenGIS GML Specification is tiered, with three conceptual levels (the model can be extended to accommodate more tiers):

1. The Essential (Abstract) Model—The "facts" consisting of "real" objects (entities, attributes, and relations) and instantaneous events. This structure is the codeable structure of the real world as the specification writers perceive it.
2. The Specification Model—A generic model of the software, what states it can be in, and the way it responds to stimuli (events or messages) by changing state and generating responses (also events). That is, the model consists of "ideal" software objects and "ideal" events.
3. The Implementation Models—Models of the software objects in specific executing software environments and how the software objects communicate in those software environments. The models are models of actual software objects, which have types, states, and properties and communicate by sending messages. The Open GIS Forum wants there to be one implementation model for each computing environment in which the specification is implemented.

In this book, we will not discuss the last two models, because the emerging location-dependent information industry does not use them but instead uses other implicit data models (as we saw in Chapter 6).

GML also is organized in three sets of profiles of different complexity. The first profile is defined in three DTDs: The Feature DTD, the Geometry DTD, and the SRS DTD. The Feature DTD defines the structure of features and feature collections. The Geometry DTD defines the structure of the geometric properties and geometric elements. The last DTD is the SRS DTD, which defines the Spatial Reference Systems (SRSs).

The second profile also consists of three DTDs: the Geometry DTD, the SRS DTD, and an application-specific DTD that specifies structures of features, so that features can store their properties in other elements than "Property" and "GeometricProperty". Depending on the application, this might make sense, for instance if you want to highlight specific features in your data model.

In the third profile, an RDF schema is defined for each application.

The OpenGIS Specification is based on a lexicon of common geodata types defined in terms of primitive data types (such as those available in all programming languages). This model is the Open Geodata Model. It is limited to those vocabulary elements that are needed to communicate geospatial information.

The model is an object model, but it is not a pure object model defining the interfaces of the objects. It also defines an API for the manipulation of the objects. This API is abstract, not mapped to the individual systems—that is something to be done for each specific system, such as CORBA or COM, or indeed the World Wide Web. The object-oriented model provides the same capability but is usually presented to the application developer as an object API (in other words, a class library).

Every geoprocessing system has a geodata model that serves as a guide for digitally representing Earth features and phenomena. The Open Geodata Model is a universal geodata model that enables interoperability interfaces to be specified, referring to those parts of the Open GIS Essential Model that focus on data: geometry, spatial reference systems, transformations, shapes, locational geometry structures, topology, the well-known structures from which feature geometries are constructed, coverage extents, schema range functions, coverage generation functions, and so on.Simple features are features whose geometric properties are restricted to simple geometries for which coordinates are defined in two dimensions and the delineation of a curve is subject to linear interpolation. A geographic feature is "an abstraction of a real world phenomenon; it is a geographic feature if it is associated with a location relative to the Earth."

In other words, a digital representation of the real world can be thought of as a set of features. The state of a feature is defined by a set of properties, where each property can be thought of as a {name, type, value} triple. The number of properties a feature can have, together with their names and types, are determined by its type definition. Geographic features are those with properties that can be geometry-valued. A feature collection is a collection of features that can be regarded as a feature; as a consequence, a feature collection has a feature type and thus might have distinct properties of its own in addition to the features that it contains.

There are two fundamental geographic types recognized in the OpenGIS Specification: features and coverages. Both features and coverages can be used to map real-world entities or phenomena.

- A feature, as described previously, is a representation of a real-world entity or an abstraction of the real world. It has a spatial domain, a temporal domain, or a spatial/temporal domain as an attribute. Examples of features include almost anything that can be placed in time and space, including desks, buildings, cities, trees, forest stands, ecosystems, delivery vehicles, snow removal routes, oil wells, oil pipelines, oil spills, and so on. Features are usually managed in groups as feature collections. A GIS thematic map layer for a city that shows only roads, for example, is a collection of features—each of which is a feature of type road. Features usually represent entities.

- A coverage is an association of points within a spatial/temporal domain to a value (of a defined data type, possibly a complex type). That is, in a coverage, each point has a particular simple or complex value. A coverage is a function from a spatial/temporal domain to an attribute domain. That is, a coverage in the OpenGIS Specification is simply a function that can return its value at a geometric point. Scalar fields (such as temperature distribution), terrain models, population distributions, satellite images and digital aerial photographs, bathymetry data, gravitometric surveys, and soil maps can all be regarded as coverages. Coverages usually represent phenomena.

Coverages have all of the characteristics of features, so they are a subtype of a feature. Therefore, features and feature collections are the central OpenGIS Geodata Model elements. A coverage has a data value associated with every location. For instance, a city defined as a feature does not return a value for each point. At a given point, it might contain another feature or it might contain a coverage, but by itself it does not return a value. A city defined as a coverage returns a value for each point, such as an elevation or an air quality index value. A coverage can be derived from a collection of features. A collection of features can be used as the starting point, and one or more attributes of these features can be used to define a coverage—the value of the coverage at a point being the value of the attribute of the feature located at that point.

On a map, there are certain features that you cannot help agree on being part of the representation of the Earth. A mountain is undeniably there, and its height can be unambiguously measured. But when you involve other information types, the composition of features will be a social phenomenon. Someone has to suggest a feature, and others have to accept it. The population density of a city will be meaningful to some users and irrelevant to others. The more a feature is grounded in society and less in the purely physical reality, the more it will be subject to interpretation and therefore not unambiguous. The Open GIS Consortium maintains a record of consensus about features, coverages, OpenGIS services, and so on in a number of Bookshelves containing Topic volumes that together form the OpenGIS Abstract Specification.

A collection of features in GML is comprised of features, the basic unit of digital geospatial information. Features can be defined recursively, so there can be considerable variation in feature granularity. For example, depending on the application or interests of the information gatherer, any of the following items could be a feature:

- A segment of a road between consecutive intersections
- A numbered highway consisting of many road segments
- A georeferenced satellite image
- A single pixel from a georeferenced satellite image

- A temperature overlay on a weather map
- A triangulated irregular network
- Those segments of a dynamically segmented road that fall between two other roads
- A drainage network
- A single seismic event magnitude contour

Features can be recursive (for example, features can contain many subfeatures that contain, in turn, subfeatures) and can contain a collection of subfeatures or coverages (the other type of element in the OpenGIS specifications) that form a logically consistent grouping in terms of resolution, accuracy, content, and context. Because the design enables recursion, it might also lead to the creation of logically inconsistent collections. Such inconsistencies will be a serious problem for anyone integrating data from different sources. XML validation is one way to get rid of part of the problem, but the main task lies with the database designer, who has to make sure that the collections are consistent. Although the spatial reference system has to be declared in all data, there is no way of declaring other attributes that might affect consistency, such as accuracy. This may become a problem when using GML data from different sources in applications, which demand knowledge about accuracy, such as construction work applications. It is, of course, possible to define an attribute in the application-specific data model for storing the accuracy. That attribute, however, cannot be expected to be present in all GML data, and this will not solve the problem.

In the Feature DTD, every feature element has a type-name attribute describing the object type of the feature itself. A feature also can contain an arbitrary number of property and geometric property elements. Property elements have the attributes typeName and type. The value of the typeName attribute describes what the relation the property has to the feature, and the value of the type attribute specifies the type of the contained data.

The DTD itself can constrain the data type of only the contained data to be character data, which is why information about data type must be specified in an attribute. The value of the type attribute must be boolean, integer, real, or string.

Features consist of three basic elements:

- ***Geometry* with an associated spatial and/or temporal reference system, including a statement of the resolution and accuracy of the geometric model.** They can be defined by using simple, primitive geometric shapes defined as instances of well-known types of objects in the OpenGIS geometry, such as polygons, line strings, polyhedrons, and other OpenGIS shapes. The rules for representing feature types with well-known types must be explicit. (For example, a rule can specify that a brick house is seen as a polyhedron). Given a description of the objects in a feature collection, this situation will mean that they can be translated to a graphic representation. Instances of well-known types are called well-known structures, and the model of the well-known structures must carry sufficient information to enable the reconstruction of the extents of the features to which they contribute. That is, the geometry components must know how they contribute to complex geometries. (For example, the highway segment geometries must know the sequence in which they concatenate to become an entire highway feature.)

A geometric property element has the attribute typeName and contains a geometry element that belongs to the entity GeometryClasses. The geometry elements belonging to this entity are:

"Point"

"LineString"

"Polygon"

"Multipoint"

"MultiLineString"

"MultiPolygon"

"GeometryCollection"The OpenGIS Abstract Specification manages geometry independently of the representation used by the specific geoprocessing applications. All geometries must be able to use a coordinate geometry representation (for example, x, y). Just like other properties, geometric properties must be named. So, the River feature type might have a geometric property called centerLineOf and the Road feature type might have a geometric property called linearGeometry. It is possible to be more precise about the type of geometry that can be used as a property value. In the previous examples, the geometric property could be specialized to be a line string property. Just as it is common to have multiple simple properties defined on a single feature type, so too can a feature type have multiple geometric properties.

- ***Semantic properties*, or the definition of the entity or phenomenon.** Unlike geometries, which are the same between information communities, semantics can vary from group to group in the same way that the conceptual view does. Definitions and meanings might vary slightly or radically, as in the different ways that farmers and civil engineers might define roads.

- ***Metadata* is other information that might be needed to position the phenomenon in the context of the application environment or user community.** Metadata content requirements and standards are often defined through professional societies and are used to trace lineage and provide a measure of quality assurance to the user. Metadata is a subset of the properties of a feature (or, more typically, of a feature collection), but it is data that describes the data (or the instance), not data that contributes to the presentation or modeling (in the current application—what is metadata to one application might be metadata to another). For example, a property of a coverage representing an aerial photograph might contain simply a name, such as date flown and a value from a date type. The complexity of the metadata can be adjusted to meet the demands of the application. A metadata dictionary (schema or data mapping) listing metadata categories and their data type might look like the following:

```
element name
type
acquisition date
date
percent cloud cover
float
```

A feature is made up of geometry, semantics, and metadata. But there is no rigorous requirement that mandates a value for each of these three elements. For example, a developer or an application can create a feature that has no geometry and therefore no spatial/temporal reference system (in effect, no location). Because the GML specification is designed for use in geographic applications, most features will have a location. There are also better ways of representing features that do not have a location.

Feature instances are identified in operational software by an object identity (often known as an object ID, or OID). The OID is, ideally, unique for each feature in all data sets everywhere through time. What the OID should be is implementation dependent. Because OID is a proper type (typically implemented as a formatted string or a long integer), it can be used as a value of a property. The use of OIDs corresponds to pointers in object systems and reference values in SQL3-based relational systems.

Geographic elements fall into two broad categories: entities and phenomena.

- Entities are recognizable, discreet objects that have relatively well-defined boundaries or spatial extent. Examples include trucks, buildings, streams, certain landforms, and measurement stations.
- Phenomena vary over space and have no specific extent. Examples include temperature, soil composition, and topography. A value or description of a phenomenon is only meaningful at a particular point in space (and possibly time). The phenomenon called temperature, for example, only takes on specific value at defined locations, whether measured or interpolated from other locations. Tourist information is another example of a phenomenon.

These are not mutually exclusive sets of information. In fact, there are many components of the landscape that are part entity and part phenomena, making their ultimate classification subjective and open to interpretation. For tourist information, the distinction might not be so difficult, but a highway can be thought of as a feature or a collection of observations measuring accidents, shoulder quality, asphalt composition, or other structural status items.

The basic unit of the geospatial information interchange between applications is the feature collection. Feature collections can be any size. They can contain only a single point, grid post, pixel, road segment, or several terabytes of data depending on the context in which the feature collection will be used (the context of the interchange transaction in the language of the Open GIS specifications). A feature collection, for instance, might take the form of an online database with a front end to support on-the-fly map generation and retrieval. A database of geospatial information can be made available online for users to access and request tailored subsets of the feature collection (which might, in itself, become a feature collection).

GML provides information about which SRS is used in the srs-name attribute. This attribute can be specified for individual geographic elements. In a hierarchical structure, the top-level element contains the SRS declaration, and this can be overrridden by other SRS declarations further down in the structure, which is standard XML. When combining data, the srs-name attribute can be used to set up the proper coordinate transformations so that all data in a document references the same SRS. Another approach is to keep the original coordinate values and carry out the transformations when the data is displayed. If the reference systems do not match, one of the data sets can be sent to an automated coordinate transformation service. This service uses the information about the reference system to set up the appropriate transformation parameters.The measuring stick is to be capable of encoding all of the reference systems, which can be found at the *European Petroleum Survey Group* (EPSG) Web site, and the specifiers' claim. In addition, the encoding scheme allows for user defined units and reference system parameters.

Feature collections can employ a number of representation methods. They can contain vector, gridded, and raster data or any combination of these. They are assumed to be owned by information communities, which can be a group of people who share a trade or who are concerned with a feature from different aspects (for instance, a highway might concern ecologists, engineers, and politicians but from different perspectives).

The community groups its collection of features into a catalog, which is the way it shares the collection and its semantics with the world. Translating between different catalogs is done by using a set of XSLT transformation sheets, mapping the semantics of one catalog into the semantics of the other (a semantic translator; for more about how XSLT works, see Appendix A).

It would be nice to say that all these problems disappeared if you used XML, but that is unfortunately not true. XML only provides the mechanics; it does not define meanings. To understand why data sharing is complex, consider the analogy between human language and spatial/temporal computing. When we share a context (such as a culture, or even at a lower level such as a workplace or an association), we use a common language to describe that context and to set up a similar frame of reference in regard to it. The members of the culture see the world through the same eyes and characterize it by using shared descriptors. Standardization of meanings facilitates unimpeded, accurate communication. But although the semantic intent of a feature might be consistent across two feature class definition schemes, the content of the supporting schemes might diverge from one another. Even when the attribute sets for two comparable feature classes in different schemes match to a large degree (but not completely), there are still opportunities for the loss of attribute information.

There are a number of cases in which information can be lost when communicating between different language groups, and by analogy, between information communities. Defining a vocabulary in a formal language avoids this problem, because there is a common reference. The Open GIS Forum gives a hydrological example of how misunderstandings can occur and what results they can have.

Hydrography, according to the Department of Defense Glossary of Mapping, Charting, and Geodetic Terms, is "the science which deals with the measurements and description of the physical features of the oceans, seas, and lakes, and their adjoining coastal areas, with particular reference to their use for navigational purposes." Hydrographic data has an important geospatial component. Two well-known systems of feature class definitions used to describe features within the hydrographic discipline are the S-57 Object Catalogue and the *Feature and Attribute Coding Catalog* (FACC). The S-57 Object Catalogue is part of the IHO Transfer Standard For Digital Hydrographic Data developed by the International Hydrographic Organization. The FACC is part of the *Digital Geographic Exchange Standard* (DIGEST) developed through an international cooperative effort by the member nations of the *Digital Geographic Information Working Group* (DGIWG).

Both of these schemes have a robust ontology of hydrographic features and attributes to support the geospatial use of these data, although each is used for a slightly different purpose. The S-57 Object Catalogue is primarily intended to support the visual display component of electronic charts on board commercial seagoing ships and is used in the United States by the Department of Commerce in producing digital hydrographic charts for the *Electronic Chart*

Display Information System (ECDIS). The FACC, as part of DIGEST, was developed to support broadly applicable geospatial analysis requirements and is used in the United States primarily by the Department of Defense in the generation of the Defense Mapping Agency's *Vector Product Format* (VPF) products, including the *Digital Nautical Chart* (DNC). Semantic translation between the two schemes might be necessary to support the exchange of hydrographic data under international exchange agreements aimed at updating and improving the hydrographic charts and safety of navigation information produced and maintained by both communities.

Therefore, there are two different feature class definition schemes for the same objects. When translating between two feature class definition schemes, there are at least six different results that can occur. It is possible to create an exact match of the meanings of the definitions between the two feature classes with no loss of information in the translation. For instance, an AQ070 Ferry Crossing in the FACC is the same as a FERYRT Ferry route in the S-57 Object Catalogue.

If there is no direct semantic match between definitions in the two feature class definition schemes, it is possible to do an exact translation by using information in the attributes of one or both feature definitions. For example, a LITFLT Light float in the S-57 Object Catalogue is not an exact match for the FACC feature class BC00 Light, but through the use of the FACC attribute BTC Beacon / Buoy Type Category, the information can be recovered without loss when an attribute value of BTC006 Light Float is used.

To translate between feature classes, you need to aggregate features. Essentially, you need to map the classes in one schema to the classes in another. For example, the FACC feature classes "BJ040 Ice Cliff," "BJ065 Ice Shelf," "BJ070 Pack Ice," "BJ080 Polar Ice," and "BJ100 Snow Field/Ice Field" can be aggregated to the feature class "ICEARE Ice area" in the S-57 Object Catalogue. If you aggregate features, however, you cannot determine whether "ICEARE Ice area" was originally an ice shelf, an area of pack ice, or something else. The feature content is lost in the translation, but you need to have a description somewhere of how the aggregate is created and what translation process was used (for instance, an XML Schema and XSLT transformation sheet used to create the collection).

But you do not only need to create new aggregates, you also need to deconstruct them to translate between feature classes. Using the feature classes from the previous example, a translation of an "ICEARE Ice area" feature from a dataset based on the S-57 Object Catalogue to one based on the FACC is not possible. If you want to make such a translation, you would have to modify the FACC schema required to create a new Ice Area feature class. If it is not feasible to modify the schema, you need to verify the source data (or perform an image analysis or ground truth reconnaissance process) to classify the Ice Area as one of the existing snow-and-ice-related feature classes supported in the FACC. That might take weeks or even months. The only feasible option is to classify “ICEARE Ice area” in the FACC, incorrectly and misleadingly, as one of the snow-and-ice-related feature classes. In this case, it would be necessary to include a caveat with the feature to capture the discontinuities in the translation process and to maintain an accurate lineage of the feature.

Matches between the meanings of two comparable features in the schemas of different information communities is possible, but you need further clarification on the definition of the supporting feature classes before the match can be verified or the translation might be dependent on the representation of the feature in its respective information communities. For instance, a “BA020 Foreshore” in the FACC might be the same as an “ITDARE Intertidal area” in the S-57 Object Catalogue. Both definitions specifically refer to measurement of the shoreline, but in respect to different datums. “BA020 Freshore” references Mean Low Water, while “ITDARE Intertidal area” references Mean High Water. You need to define a conversion that takes these differences in the reference datum into account so that the information community you are targeting might use the same instance of a shoreline differently. But you need to document the nature of the conversion, and the translation process, to maintain an accurate lineage of the feature.

Matching two feature classes from different information communities might not be possible without losing significant information. An example would be the FACC feature class “GB040 Launch Pad,” which has no counterpart in the S-57 Object Catalogue.

Geographic feature definitions become more specialized the more they are focused on narrow applications. A road might seem like a fairly simple object, but take four different GIS information communities and they will define four very different phenomena that have different sets of definition information. If other communities are involved, the definitions of objects can be defined in even more divergent terms.

Depending on what you want to represent in the real world, there are different ways you can use GML. If you want to represent a city for tourist purposes, you would represent its buildings, roads, and rivers as features and create a feature collection describing the city. If you wanted to represent the city as a climactic feature, you would probably represent it as a coverage of measurements.

```
<os:Road>
 <gml:description>Georgia Street</gml:description>
 <os:numberLanes>4</os:numberLanes>
 <gml:centerLineOf>
  <gml:LineString srsName="EPSG:4326">
    <gml:coordinates>0.0,100.0 100.0,0.0</gml:coordinates>
```

```
    </gml:LineString>
  </gml:centerLineOf>
  </os:Road>
```

The geometry elements in GML consist of coordinate lists, which consist of coordinate tuples (value pairs). When working with simple features, you have only two dimensions, and, therefore, every tuple has two coordinates. In the structure of a GML document, every geometry element has to include a coordinate element. The coordinate element of a point element has one coordinate tuple. The corresponding coordinate element for a line string has more than one coordinate tuple, and a linear ring, used for defining the extent of polygons, has at least four tuples, in which the last tuple duplicates the first. The reason for this is that both the starting point and the endpoint have to be declared, even if they are the same. The following is the GML for the linear ring in Figure 7.1.

```
<LinearRing>
        <coordinates>
      0.0,0.0
      100.0,0.0
      50.0,100.0
      0.0,0.0
        </coordinates>
</LinearRing>
```

When defining a polygon in GML, you define a linear ring containing the outer boundary and then (if necessary) a linear ring—or more linear rings—defining the inner boundary. If a polygon declares the SRS attribute, the spatial reference system will apply to everything that is inside the polygon.

The linear ring is the base for polygons. Other possibilities exist, based on the homogenous aggregates of geometry elements: "MultiPoint", "MultiLineString",

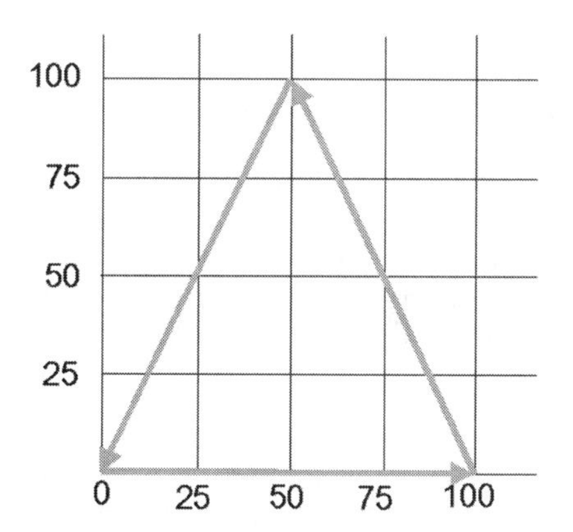

Figure 7.1 A linear ring in GML.

and "MultiPolygon". If different kinds of elements are to be aggregated, the element for heterogeneous aggregation, called "GeometryCollection" is used.

Here is another GML example:

```
<MiddleSchool ID ="145l">
<extentOf
<Polygon srsName="epsg:27354">
                            <outerBoundaryIs>
                                <LinearRing>

                <coordinates>491888.999999459,5458045.99963358
491904.999999458,5458044.99963358
                                    491908.999999462,5458064.99963358
491924.999999461,5458064.99963358
                                    491925.999999462,5458079.99963359
491977.999999466,5458120.9996336
                                    491953.999999466,5458017.99963357
</coordinates>
</LinearRing >
                            </outerBoundaryIs>
                        </Polygon>
                    </extentOf>
</MiddleSchool >
```

Note that this coding has no properties (other than the geometry). If we add properties for the building, we can get something that looks like the following example:

```
<MiddleSchool ID ="145l">
               <description>Balmoral Middle School</description>
               <NumStudents>987</NumStudents>
               <NumFloors>3</NumFloors>
               <extentOf>
                    <Polygon srsName="epsg:27354">
                         <outerBoundaryIs>
                              <LinearRing>

                <coordinates>491888.999999459,5458045.99963358
491904.999999458,5458044.99963358
491908.999999462,5458064.99963358 491924.999999461,5458064.99963358
491925.999999462,5458079.99963359 491977.999999466,5458120.9996336
491953.999999466,5458017.99963357</coordinates>
                              </LinearRing >
                        </outerBoundaryIs>
                   </Polygon>
               </extentOf>
</MiddleSchool >
```

If you are working with a large number of geometry elements or a number of geometry elements close together that have common properties (representing only one feature or feature collection), you can specify a grouping element to give the application a hint of which elements are present within the area. This element, a bounding box, is a box with the edges parallel to the axes, containing all elements in a defined set of elements. A feature collection must always contain an element called "bounded by", which in turn contains a box element. It is also possible for features to have this property, but it is not required.

Bounding boxes are defined in GML using the geometry element "Box". A box element is defined by only two coordinate tuples, which both are depending on the coordinates of the elements in the set of geometry elements. The first represents the corner having the lowest coordinate values on all axes, and the second represents the corner having the highest coordinate values. When the box defines a bounding box for a set of elements, the highest coordinate values are the same as the highest coordinate value on each axis, and the lowest coordinate values are the lowest value on each axis, of the elements contained in the box.

Data Formats for Dynamic Objects

The map, as well as GML, describes the geographical features of an area. It does not, except for some conventional symbols, say anything about what properties those features might have. That is a different plane of information, if you will, and belongs with the navigation plane. To describe these in a standardized way, several navigation languages (and no doubt a vast number of database encodings) have been created. Here, we will look at the XML versions of three of those.

It is often the case that when you want to describe a specific information domain, there is no data structure that covers what you want to say precisely. The objects become either too specific or too generic. In such cases, application designers often feel that they need to design a totally new description language.

It turns out that this idea is more often bad than not. Using XML, you can reuse elements from an existing encoding of the information. The risk in inventing your own is that you miss something, either in terms of data description or in terms of geographical reach (for instance, the zip code element in NVML cannot be used for zip codes in British addresses, because it was defined in Japan where zip codes are always numeric). It is far better to reuse elements that describe what you want to say; for instance, a zip code element from an international standard (which, in an XML application, you can do by using namespaces).

There is a tradeoff between system efficiency and standard availability that must be taken into account here. If you have a specialized domain of information, which you are certain will not interact with other location-dependent data sets, you can then create your own encoding of the data. If your data set is going to be related to other data that is dependent on location, however, you should consider using a standardized markup even if that might not fit your data precisely.

Looking at several of the different formats available, you can see that there are two things a user needs from his or her geographical application: the coordinates for a point and the properties of that point. The properties, of course, will vary greatly with the current interests of the user as well as the context of the user. If you are driving a car and your gasoline is running out, you will be interested in the nearest filling station. If you are driving an 18-wheel truck, you will be constrained in which filling stations you can access. And, if you are walking and looking for the nearest subway station, you will watch the cars passing with indifference. The navigational data are the data that helps the user relate the properties of the point to his or her own context.

If you are using several different markup systems—for instance, GML to mark up the geographic features and POIX to mark up the interesting features of the point—you need to relate the two descriptions. This problem is the same as any database system using foreign keys, and it means that you should think about which element is the key element that binds the descriptions together. The element that will always be unique is the coordinates, but if your description is related to a set of features where all the objects have unique names, you could use the names of the objects as keys.

Of course, when using XML it is always possible to mix data elements from different name spaces and use a mix of different markup languages. This situation assumes, however, that your application can handle that. You can specify a subset of the different markup languages that you intend to use, but that is not necessarily the right way to do things. If you use elements from different namespaces, your application should at least in principle be able to handle those different namespaces in their entirety. In a location-dependent application, it is better to put the processing in the application server, which means that there will be enough capacity to handle generic XML processing.

There are several types of languages that can be used to mark up dynamic features. Here, we will look at three languages for the markup of dynamic features: POIX for point-of-interest markup, NVML for the markup of routes, and SKiCAL for the markup of event information. We will also look at a metaformat, the Dublin Core format, which contains a coverage element that you can use to embed geographical descriptions in your markup.

The Point-of-Interest Exchange Language (POIX)

A group in the Mobile Information Standard Technical Committee, a mostly Japanese organization, designed the Point of Interest exchange language. It was submitted to the W3C in 1999 as a proposal for future work, although the W3C later decided not to work on specific XML applications.

POIX is an XML-based language intended to describe a position and information related to that position only. Other information, such as opening times, cannot be contained in a POIX document but has to be represented elsewhere. The position, however, can be moving as well as fixed—the intent is that you should be able to describe your car as a point of interest as well as the restaurant along the road.

It is intended to describe the information in a navigation system—more specifically, a car navigation system—although it is general enough to work for any navigational system.The intention brings some restrictions to the format, however—most specifically that it tends to describe objects as strings of points (loci) and not as areas, for instance. While this feature makes it possible to describe a route in terms of sample points, it does not lend itself to describing that the surface of Highway 101 is slippery in the southbound lane because there was an oil spill south of Burlingame. There are tradeoffs in any description format, and you have to select what is relevant.Because you can mix different description formats as long as they are in XML, however, this problem will tend to become smaller once XML formats become widespread.

Point-of-interest descriptions are intended to be monolithic, but it is possible to link several points together; for instance, representing the entrances and exits to a service station. The intent, however, is that only one information item (about a point) should be exchanged at the time.

A document containing a point-of-interest description starts with a POIX element, which is the root of the description. It in turn has a child element that describes the format of the geographic information. The children of the format element describe the geodetic datum (in other words, the type of coordinate system used), the units in that coordinate system, the target type (whether it is mobile or not), information creator, and information creation date/time. They are listed, with their attributes and data types, in Table 7.1.

If the format information goes in the head of the document, the POI element corresponds to the body (although, of course, both of these can be inside some other element of some other format, such as NVML, and that in turn could well be inside something completely different—like the Dublin Core Coverage element). Note that there are several inherent assumptions about the format of the data (for example, the POI element is described in terms of latitude and longitude, but not all coordinate formats need or use those). The POI child elements are listed in Table 7.2.

The point in Table 7.2 gives the position, but it does not do so directly. Instead, it uses a subelement (which, it should be noted, is not an attribute but a child

Table 7.1 The Format Elements of POIX

ELEMENT	ATTRIBUTE	VALUES	FUNCTION	COMMENT
datum			Defines the geodetic system used	Text format. Three values are given in the current specification: WGS-84, IRTF, and Japanese Geodetic System.
unit			Defines the angular unit system used	Text format. Delineates systems like degree, radians, etc. In the specification, only degrees and the geodetic degrees/minutes/seconds (dms) formats are given. You need this if you are to describe a two-dimensional vector, which you need for a moving object.
type	object	fix or mobile	Describes whether the object is moving in a short time or not	"short time" is a tricky context. It is rather about the intention of the object - is it built to move or be fixed?
author	xml:lang		Language identifiers per RFC 1766	Describes the name of the author of the information. The purpose of the xml:lang element is to describe the language use for the name of the author (which may be relevant e.g. in China, where people sometimes have both Western and Chinese names).
time			Indicates when the information object was created.	This element can have the ISO standard designators for time: YYYY = four-digit year (1998), MM = two-digit month (a value 01 to 12), DD = two-digit day (a value from 01 to 31), hh = two-from digit hour (a value from 00 to 23 in 24-hour notation), mm = two-digit minute (a value from 00 to 59), ss = two-digit second (a value from 00 to 59); and a time zone value, which is described in the form of +hh:mm or -hh:mm according to the time zone. For UTC, Z is designated.

Table 7.2 The POI Child Elements

ELEMENT	ATTRIBUTE	VALUES	FUNCTION	COMMENT
point	Child elements, see table 7.3	See table 7.3	To give the coordinates for the point	The data structure contains assumptions about the coordinate system that enforces the use of WGS-84 or something similar.
move	Child elements, see table 7.4	See table 7.4	Expresses a vector and movement means	It is not clear from the specification why this is interesting, but if the movement means is rail, you can conclude that the route will be fixed in a different way from it being foot, for instance.
name	style	formal or popular	Describes the name of the point of interest. The actual name is expressed in the sub elements.	Does not consider multilingual names.
access	Child elements, see table 7.6	See table 7.6	Expresses the route for how to get to the target (POI)	Essentially designates a route
contact	xml:lang, xml:link, href	URI	Designates the contact for the point	Contact can be a person, company, etc.
note			Free text information	Not defined in the specification, funnily enough
mate	xml:lang, xml:link, href	URI	Indicates another point of interest relevant at the target	The relevance has to be expressed as content of the element.

Table 7.3 The Children of the Point Element

ELEMENT	ATTRIBUTE	VALUES	FUNCTION	COMMENT
lat			Designates the latitude of the point	Value according to the datum element
lon			Designates the longitude of the point	Value according to the datum element
herror		Integers in meters	The horizontal error	Estimated error in determining the horizontal position
alt		Integers in meters	The altitude, positive if higher than the datum, negative if lower	Can be determined by some positioning methods, e.g. GPS
verror		Integers in meters	The vertical error	Estimated error in determining the vertical position (altitude)

element to the point element—although this construction is very confusing). The children of the point element are described in Table 7.3.

Table 7.4 contains the children of the move element, which describes the modality of movement (in other words, how the point of interest is transporting itself or being transported).

The locus element describes a route or area by using the lat and lon subelements, as follows: <pos><lat>35.7700</lat><lon>139.8800</lon></pos>. A sequence of these points will comprise the movement history of the mobile entity. Points are indicated in reverse historical order starting from the current point.

The name of the object is given by two subelements of the name element that betray the Japanese roots of the system (they are not meaningful outside Japan) and that do not contain facilities for the expression of multilingual names, something that would be required in a place like Quebec, for instance. The name elements are described in Table 7.5.

Since the creation of this specification, there has been a separate specification in the W3C which describes Ruby markup, and that should be used instead of this clunky method.

How to get to the target is described in the subelements of the access element, which are described in Table 7.6. Essentially, it describes a starting point, a route, and the properties of the target.

Table 7.4 The Children of the move Element

ELEMENT	ATTRIBUTE	VALUES	FUNCTION	COMMENT
method		car, motorcycle, railway, onfoot, bus, airplane, ship, others	Describes how the movement is accomplished	Only the values designated are possible
speed		Kilometers per hour, no decimals	Current moving velocity of the mobile entity	Other measurement means would have to be recalculated
dir		Positive integer in degrees, counted clockwise from north as zero	Current direction of the mobile entity	Curiously, does not use the angular units described in the unit element
locus	Pos child element		Designates the route as pos child elements	Essentially describes a route

The ipoint, the introductory point, and the tpoint, the terminal point, have a number of subelements that describe the point's position and class (which has values designated in the specification). The tclass element can have a limited

Table 7.5 The Children of the name Element

ELEMENT	ATTRIBUTE	VALUES	FUNCTION	COMMENT
nb			Presents the formal name, i.e. the name in Kanji characters in the original specification.	In most places, the distinction is not meaningful, but it is often the case that an object has both formal and informal names, Studenternas Gymnastikinstitution in Uppsala being known popularly as Svettis, for instance.
rt			Presents the "ruby" phonetical transcription of the formal name.	In Japanese text, the phonetic Hiragana syllabary is often used to state how a Kanji character should be pronounced.

Table 7.6 The Access Subelements

ELEMENT	ATTRIBUTE	VALUES	FUNCTION	COMMENT
method		car, motorcycle, railway, onfoot, bus, airplane, ship, others	Describes how the movement is accomplished	Only the values designated are possible
ipoint	Child elements: iclass,pos,name	The starting point of the access locus		
tpoint	Child elements: tclass,pos,name	The end point of the access locus		
route	Child element: pol	The route from the ipoint to the tpoint		pol is short for "polyline", i.e. a line of coordinates
note		Free text information		Not defined in the specification, funnily enough

number of values: crossing, street, station, stop, airport, harbor, and others. The iclass element can have the values parking, entrance, and others. This situation would seem to imply that you would always end up in a parking lot.

The following is a description of the subway entrance to the Mitsukoshi department store in Ginza, a very fashionable part of Tokyo:

```
<?xml version="1.0" encoding="Shift_JIS" ?>
<!DOCTYPE poix PUBLIC "-//MOSTEC//POIX V2.0//EN" "poix.dtd">
<poix version="2.0">
<format>
<datum>wgs84</datum>
<unit>degree</unit>
</format>
<poi>
<point>
<pos>
<lat>35.6680</lat>
<lon>139.76887</lon>
</pos>
</point>
<name><nb>Mitsukoshi Ginza Store</nb></name>
<access>
<method>onfoot</method>
<ipoint>
<iclass>station</iclass>
<pos>
```

```
<lat>35.66805</lat>
<lon>139.76833</lon>
</pos>
<name><nb>Ginza station of Ginza subway line</nb></name>
</ipoint>
<tpoint>
<tclass>entrance</tclass>
<pos>
<lat>35.667778</lat>
<lon>139.7686</lon>
</pos>
<name><nb>Subway entrance</nb></name>
</tpoint>
<note>
You may enter the store from entrance A7 of Ginza station of Ginza
subway line.
</note>
</access>
<contact href="tel:81-3-3562-1111" />
<note>Not closed on Monday</note>
<mate href="http://www.toyota.co.jp/0223.poi">Annex</mate>
</poi>
</poix>
```

Navigation Markup Language (NvML)

The *Navigation Markup Language* (NVML) is an XML format to express locations as points along a route. Developed by Fujitsu and submitted to the W3C, it is intended as a data exchange format between navigation systems, such as car navigation systems, and to enable route data to be used in other contexts (for instance, for tourist information). It does not attempt to define the route calculation mechanisms (which can be rather complicated), but it does define routes that can, for instance, be walking tours of cities for tourists or truck routes for cargo transport.

NVML defines points, routes, information elements, and child elements under those that can hold information of various kinds. The points are intended to be points along a route, which can be defined by an external agency (such as a tourist agency). One possible use case for NVML is a tourist guide system that triggers at different positions along the route and gives the user information.

Because NVML is an XML format, it is possible to include other document types in an NVML document and relatively easy to transform from other formats into NVML. That would mean that given the appropriate heuristics (for instance, a program), NVML route descriptions could be generated dynamically and on-the-fly—adapting the information to the path the user is actually taking, instead of a route that has been predetermined by the tourist agency (something that is relevant in cities such as Kyoto, which has thousands of points of interest). The route could also be pre-filtered depending on the user's interest; for example, to present only Buddhist temples or sites related to Genji Monogatari or architectural masterpieces. This feature, of course, requires that the language describing the points of interest can contain all these information planes.

The concept of routes is as difficult to define as that of areas. In the context of NVML, it implies a vector with duration in time. Route calculation and definition is a difficult concept at best, however, as we saw in Chapter 6.

A NVML document is divided in two parts: the head part, which contains metainformation about the route, and the body, which contains the route. Both elements are children of the root element, the nvml element. The child elements of the head element must be given in the same order as in Table 7.7.

Note that if there are several routes (possible or otherwise), they can be compared by using the duration and distance elements. That does require that they use a data type that is shared by the application, though.

The body element contains two elements, the navi and guide elements, which describe the route. The navi element contains details about the navigation of the route; the guide element specifies the details of a specific point of interest. They share several child elements with each other and with the head element, but for clarity, I have repeated them on the tables.

The navi element can have two child elements: point and route. The point element specifies the point of interest, and the route element shows the details of the route. In Table 7.8, you will find the child elements of the point element. The point element itself can have an attribute, the area attribute, which specifies the area of output as a circle with the radius specified in kilometers or meters. It is only allowed when the point element is a child of the guide element.

The point element can also be a child of the guide element, in which case the same applies for the navi element.

If the latitude element and the longitude element are specified in the point element, the NVML system can use this information to determine the point location. If they are not specified, the system might use information contained in the other child elements. If the information is not enough to determine the point location, the system might try to deduce it as well as it can, but if it is not possible, it should produce an error. The error messages are not specified in the NVML document.

The route element gives additional details about the route in the same way as the point element defines information about the point. It can have a number of daughter elements: name, category, number, means, duration, distance, expense, note, and info.

Table 7.7 The NVML Children of the head Element

HEAD CHILD ELEMENTS	SPECIFIES	DATA FORMAT	COMMENT
title	The title of the route	Text	
category	Type of information (e.g. tourist)	Text	Arbitrary categories. Can also apply to the point and route elements.
transport	Mode of transport for the route	Text	Arbitrary categories. Only one per route.
duration	Time required for the route (given the mode of transportation)	Text	Arbitrary. Assignment can be dependent on situation (e.g. if there is a traffic jam, the duration will be longer than if the road is clear). Only one per route.
distance	How long the route is	Text	Arbitrary units. No data type specified.
expense	Describes the direct expense of taking the route (bus fare, road toll, etc).	Text	No data type. No monetary unit specified. Can be used to specify congestion, according to the specification, so it could also be used to specify environmental cost (e.g. it is environmentally more expensive to fly than to take the train).
geodetic-system	Defines the geodetic system of the route description.	Text	Default value is irtf, which means the International Terrestrial Reference Frame (see chapter 3). The other specified value, wgs84, means the World Geodetic System 1984.

continues

Table 7.7 The NVML Children of the head Element (continued)

HEAD CHILD ELEMENTS	SPECIFIES	DATA FORMAT	COMMENT
note	The author and the content of the route, using the author and content attributes.	Author and content attributes, no other content	Specifies the content of the description and who wrote it.
info	Output of the information.	Delay and duration attributes. Delay specifies the beginning time (before or after a given time) that the information should be displayed this time is dependent on the system. Duration is the display time, and can be set to "full time", which is application dependent.	In other elements, this has different functions. In the head element, it specifies the duration during which the information should be presented.

Table 7.8 The Children of the point and route Elements

POINT CHILD ELEMENTS	SPECIFIES	DATA FORMAT	COMMENT
name	name of the point or the route	Text	Arbitrary names, not tied to a gazeteer. Can be repeated only once.
category	Category of the parent elements information	Text	Arbitrary categories. Can be repeated many times.
latitude	The latitude of the point of i nterest	Depends on the geodetic system specified in the head element	Can only be specified once per point element. Requires the longditude element.
longitude	The longitude of the point of interest	Depends on the geodetic system specified in the head element	Can only be specified once per point. Requires the latitute element.
address	The address of the point of interest.	Text	Can only be specified once per point.
zip-code	The postal code of the point of interest	Numeric and hyphen	Can only be specified once per point. Will not work in countries which do not use numeric postal codes
phone	The phone number related to the point	Numbers, hyphen, and plus characters	Can only be specified once per point.
fax	The fax number related to the point	Numbers, hyphen, and plus characters	Can only be specified once per point.
email	The email (MIME/SMTP) address that is related to the point	letters (small and large caps), numbers, at sign, and dot	Address defined as in RFC 822, which means that the specification should give a larger character set

continues

Table 7.8 The Children of the point and route Elements (continued)

POINT CHILD ELEMENTS	SPECIFIES	DATA FORMAT	COMMENT
url	The URI of a feature located at the point	Letters, numbers, dot, slash, colon	According to RFC 1738, which means the NVML DVD specifies a too small character set. Also, it only allows for one URI per point, which may not be true.expense See head element See head element See head element.
note	See head element	See head element	See head element.
number	The number of the route	Text	Can only apply to the route.
means	The way the route is to be traversed	Any of the following: foot, bicycle, motorcycle, car, bus, train, plane, ship, others	There is no way of adding means. Applies only to the route element.
duration	See head element	See head element	See head element. Does not apply to the point.
distance	See head element	See head element	Does not apply to the point.

The guide element can have another child element, the info element, which describes at which point or area the information should be displayed. Where the element occurs determines the way it is displayed. If it occurs in the head element, the application will have to determine the output; if it occurs in a navi or guide element, the point element in the navi element will determine where the information will be displayed. It will also determine the display format (such as text, voice, image, and so on). There are three different display formats specified (text, voice, and image) and a note format, as shown in Table 7.9.

Note that the presentation mechanism only enables very simple presentations, such as text without formatting and images without additional data planes.

In addition, the info element can have three attributes: delay, duration, and times. Here is where it gets complex, because the combinations of area, delay, and duration must be handled by the system in a way that ensures the display of the information at or around the point that the author has declared. The delay value can be a number of hours, minutes, and seconds. It can be negative, so the display of the information can begin before the user passes into the area. The duration can also be set by using the same values, plus a value called full-time, which means that the information is displayed either until a new navi element starts to be valid (if it is part of the navi element) or for as long as the user is in the area specified in the point element. The times attribute on the info element will determine the number of times the information will be displayed, if applicable.

A NVML document first declares a point or a guide, then an info element; following is the next pair, so that each information item has a duration during which it will be displayed.

Table 7.9 The Presentation Elements and Their Attributes

ELEMENT	ATTRIBUTES	FUNCTION	DATA TYPE	COMMENT
text	none	contains the text to be shown on the screen	text	no formatting possible
voice	src	specifies the address of a voice (sound) file containing the information	text	format is entirely dependent of application
image	src	specifies the file name of the file to be displayed as an image	text	depends on application

An example NVML route could look as follows. It defines a (fictive) tour of Tokyo, the Rainbow Town Tour:

```
        <nvml version="0.60">
    <head>
            <title>           Rainbow Town Tour </title>
            <category>        sightseeing       </category>
            <transport>       car, foot         </transport>
            <duration>        3 hours           </duration>
            <distance>        9.6 km            </distance>
            <expense>         700 yen           </expense>
            <geodetic-system> itrf              </geodetic-system>
            <note name="author" content="XYZ Tour"/>
            <info>
                <text>
                    This is the Rainbow Town Tour.
                </text>
                <voice>
                    Welcome to the Rainbow Town Tour!
                </voice>
                <image src="image/rainbow-town-tour.jpg"/>
                <note name="copyright" content="XYZ Tour"/>
            </info>
        </head>
    <body>

        <navi>
            <point>
                <name>      Tokyo Station      </name>
                <category>  Station            </category>
                <latitude>  N35.40.39.0        </latitude>
                <longitude> E139.46.18.1       </longitude>
                <address>   Chiyoda-ku, Tokyo  </address>
                <zip-code>  123-4567           </zip-code>
                <phone>     012-345-6789       </phone>
                <fax>       012-345-6789       </fax>
                <e-mail>    xxx@yyy.zzz        </e-mail>
                <url>       http://xxx.yyy.zzz </url>
            </point>
            <info duration="3min">
                <text>
                    Tokyo Station
                </text>
                <voice>
                    This is Tokyo Station.
                </voice>
                <image src="image/tokyo-station.jpg"/>
                <note name="copyright" content="XYZ Tour"/>
```

```
            </info>
        </navi>
<navi>
            <route>
                <name>     Toumei Highway </name>
                <category> highway        </category>
                <means>    car            </means>
                <duration> 10 minutes     </duration>
                <distance> 20 km          </distance>
                <expense>  500 yen        </expense>
            </route>
            <info duration="1min">
                <text>
                    Toumei Highway
                </text>
                <voice>
                    This is Toumei Highway.
                </voice>
                <image src="image/tomei-highway.jpg"/>
                <note name="copyright" content="XYZ Tour"/>
            </info>
        </navi>
<guide>
            <point area="1.0km">
                <name>      Tokyo Station      </name>
                <category>  Station            </category>
                <latitude>  N35.40.39.0        </latitude>
                <longitude> E139.46.18.1       </longitude>
                <address>   Chiyoda-ku, Tokyo  </address>
                <zip-code>  123-4567           </zip-code>
                <phone>     012-345-6789       </phone>
                <fax>       012-345-6789       </fax>
                <e-mail>    xxx@yyy.zzz        </e-mail>
                <url>       http://xxx.yyy.zzz </url>
            </point>
            <info duration="3min">
                <text>
                    Tokyo Station
                </text>
                <voice>
                    This is Tokyo Station.
                </voice>
                <image src="image/tokyo-station.jpg"/>
                <note name="copyright" content="XYZ Tour"/>
            </info>
        </guide>
        </body>
</nvml>
```

SKiCAL and iCAL

When you plan your weekend, it is no different from scheduling your workday. The weekend is probably more fun, but the process is the same. This common denominator is why a SKi-object is a special case of the iCalendar VEVENT. The abbreviation was originally Svenska Kalenderinitiativet, because it was intended to be used with calendars for tourism information and is now being deployed by local government tourist offices and newspapers all around Sweden.

The SKi standard proposal describes the parameters that a SKi-object can have, including those defined in the iCalendar standard. The description follows the model that Greg FitzPatrick, the inventor of the format, usually uses when he describes his proposal. What, when, which, and where are not a perfect fit, but it does help you understand how the different vocabulary entities (because that is what they are) belong together.

A SKi-object can have all the fields in a VEVENT but also a number of special fields with the prefix X-SKI-, which describes information that is specific to events. Together with the fields in the iCalendar specification, the special fields make for a very rich description of an event. Now, we will look at fields that are intended to describe events that are closely tied to calendars. I will not cover fields that are not related to position-dependent applications or event descriptions here.

The idea behind the iCalendar specification is to create a format that can be used as a vendor-independent exchange format for calendaring and scheduling between applications and systems. It enables users in different organizations by using different calendar software to book meetings and schedule events. Because it is a MIME content type, it can be transported with any protocol that can handle MIME, such as SMTP and HTTP.

SKi takes you one step further than iCAL. Not just intended for the exchange of information between calendaring applications for display to the users, it also aims to provide a machine-understandable format for metainformation about events. The specification contains a vocabulary to describe the location, time, and other relevant information about any type of event imaginable. And machine-understandability is the key to creating contextual applications with adaptive user interfaces. If your computer does not understand what you do not want, how can it exclude it?

If several organizations that arrange events publish their information in a standardized way, it will become possible for anyone who creates a calendar of events to use a search engine to gather that information and relate it to the views of the users. To make sure the search engine does not have to search the entire Internet, it can either look up a specified list of servers or search a domain that is limited in some other way.

iCAL Components

Most associates of any business use a calendaring system to keep track of their meetings. Most users also use their organizers after work. Planning is a feature of our personal lives as well as our professional existence. Monday night football is not available on Thursdays, and if you want to see the Olympics, you can do it only once every four years. Foliage season in New England is only in autumn, not in spring. If you want to experience them, you have to plan ahead. And you have to know when and where the event occurs. If you had gone to Sydney in summer for the 2000 Olympics, you would not have much funbecause they were held in winter, because Australia is in the southern hemisphere.

The SKi format is based on VEVENT, defined in the IETF standard RFC-2445, Internet Calendaring and Scheduling Core Object Specification (iCalendar). VEVENT was originally intended to be used for the exchange of information between desktop calendars, but using SKi, it can be extended to contain all the information needed to describe an event—much more than what you need to describe a meeting. Besides VEVENT, there are a number of other events in the iCalendar format, such as VTODO, VALARM, and so on. They do not relate to descriptions of events, however.

The iCal format is based on a series of objects. The Calendaring and Scheduling Core Object is a collection of calendaring and scheduling information. Today, iCalendar is a text format, but it is not very difficult to express in RDF. Once expressed in RDF, the calendar information can be matched with the information that users provide about themselves and their terminals. Such matching can be done in other formats, of course, but it is easier to do in RDF and it ties directly into intelligent agent systems.

An iCalendar object is organized into individual lines of text called content lines. Long lines have to be split into multiple lines. The object can have a set of properties, which can have parameters and attributes. Where properties and parameters allow a list of values, a comma character must separate them. There is no significance to the order of values in a list. Some property values, which are defined as multiple parts, must be separated by a semicolon. A property can have attributes associated with it. These property parameters contain metainformation about the property or the property value; for example, such information as the location of an alternate text representation for a property value, the language of a text property value, the data type of the property value, and other attributes. The general rule for encoding multivalued items is to simply create a new content line for each value, including the property name. Binary content information in an iCalendar object should be referenced by using a URI within a property value (and indeed, when it is inline, it must be).

Besides the fields (component properties) that are described in the iCalendar specification, the iCalendar files can contain information fields that are extensions to it, which is what SKi is. That the SKi objects are an extension of iCalendar also means that it uses the same content type (text/calendar) as iCalendar. The file extension of ics is used to designate a file containing calendaring and scheduling information consistent with this MIME content type, although you might want to use ski as your file type if you are publishing it as metadata or rdf if you use the RDF version.

Typically, the information in an iCalendar file will consist of a single iCalendar object. The calendar properties are attributes that apply to the calendar as a whole, and calendar components are collections of properties that express a particular calendar semantic in a parameter-specific format. For example, the calendar component can specify an event, a to-do list, a journal entry, time zone information, free/busy time information, or an alarm. The first and last line of the iCalendar object must contain a pair of iCalendar object delimiter strings.

The following is a simple example of an iCalendar object that contains a VEVENT—in this case, a celebration of Bastille Day (July 14):

```
BEGIN:VCALENDAR
VERSION:2.0
PRODID:-//hacksw/handcal//NONSGML v1.0//EN
BEGIN:VEVENT
DTSTART:19970714T170000Z
DTEND:19970715T035959Z
SUMMARY:Bastille Day Party
END:VEVENT
END:VCALENDAR
```

To make sure that you as well as I understand it, let's go through it line by line:

```
BEGIN:VCALENDAR
```

The BEGIN:VCALENDAR field shows that this document is an iCalendar document and nothing else. Because the standard does not use XML, it has to specify what is meaningful text to the system and what is not. Per definition, everything that is between BEGIN and END is meaningful. Calendar Properties, attributes that apply to the iCalendar object as a whole, are also specified after the BEGIN:VCALENDAR property and before any calendar component. These properties do not appear within a calendar component.

```
VERSION:2.0
```

Version 2.0 tells you which version of the iCalendar standard is being used (the highest version number or the minimum and maximum range of the iCalendar specification that is required to interpret the iCalendar object). The current version is 2.0.

```
PRODID:-//hacksw/handcal//NONSGML v1.0//EN
```

The *Product Identifier Property* (PRODID) specifies the identifier for the product that created the iCalendar object. It must be specified once in an iCalendar object. The vendor of the implementation is supposed to have ensured that this identifier is globally unique. This declaration is similar to the doctype declaration in XML-based applications, and as you can see it states that this application is not SGML.

```
BEGIN:VEVENT
```

BEGIN:VEVENT is where the VEVENT description actually begins. A number of properties can appear within calendar components, as specified by each component property definition. I will go through those relevant for the VEVENT and SKi formats later in this chapter.

```
DTSTART:19970714T170000Z
```

The DTSTART property defines when the activity begins. The T marks the beginning of the time, and the Z is a way of stating that this zone is a UTC (Universal Time Zone, the global time zone that used to be known as Greenwich Mean Time) time, which means that when July 14 (Bastille Day) begins according to this description, it will actually be 01.00 in France. The time is hours-minutes-seconds; no split seconds are allowed by the specification, and it uses a 24-hour clock. Time zones are a particularly tricky problem because there is no standard for the definition of a time zone. Not all countries in the world use time zones in the same way (Saudi Arabia is on solar time, for instance, and Nova Scotia has a half-hour offset from the time zone of the U.S. east coast). Nor is there a standardized description of time zones, and there is no guarantee that they include Daylight Savings Time (reversed in the northern and southern hemispheres).

```
DTEND:19970715T035959Z
```

DTEND describes when the activity ends. This event is quite some party, going on for 28 hours (ending at one second to four in the morning the day after Bastille Day).

```
SUMMARY:Bastille Day Party
```

The SUMMARY property is a description of what the event is. The SKi properties are also of the same kind and could indeed go in here to make this description more complete.

```
END:VEVENT
```

END marks the end of the VEVENT, which means that other iCalendar objects could follow this one. Then, there would have been a BEGIN: and then the properties. You cannot have two VEVENTs in the same iCalendar file, however.

```
END:VCALENDAR
```

Here, the iCalendar description ends.

```
VEVENT
```

The VEVENT is the calendar component used to specify an anniversary or daily reminder within a calendar. A VEVENT has DTSTART and DTEND properties, which are given as dates instead of the default date-time. If a start time is given but no end, the event is assumed to begin and end on the same day. In the preceding example, there is nothing that says where the party happens (not even whether it is in France), so we have no way of knowing where to go. iCalendar defines a GEO tag, using the geocoordinate format that is the least common denominator for all position information.

What

The What parameters in the SKi format describe what an event is about. This description includes things such as who the performer is, if it is part of a larger event, and other description items. The elements describing it are listed in Table 7.10.

A number of properties belong in the What section, but they have to do with the integrity of the information rather than a description of it, as shown in Table 7.11. They include for instance properties to calculate a checksum (which can be used to verify that the message has not been changed on the way), timestamps, and the timestamp for creation.

One problem with any information published on the Internet is that there is no way of ensuring the identity of the publisher. Anyone can publish event information and claim that someone else publishes it. Today, the only reliable solution is to encrypt the information end to end and publish the public key so anyone who has the private key can decrypt it. It is also possible to use digital signatures to sign the information, which gets rid of one of the major problems of the end-to-end encryption: that you are unable to cache information. Of course, using only information from a reliable source is the easiest way, but if you want to ensure that the information has not been tampered with in transit, you must use encryption. Another possibility is to manually check all files. The SKi specification has a parameter, X-SKI-CHECKSUM, which is intended to verify the integrity of a file. It does nothing but verify that the information was not corrupted since the transmission started, however, so for the transmission to

Table 7.10 The What Parameters of SKi

TITLE	PURPOSE	EXAMPLE	COMMENT
X-SKI-TITLE	To describe the event title.	X-SKI-TITLE;LANGUAGE=en: "One night with Victor Borge"	Note the language parameter on the title. It is possible to have titles in several languages.
X-SKI-PERFORMER	Lists actors in the event, for example, artist, sports team, or guide.	X-SKI-PERFORMER:The Rolling Stones	If this uses well-defined performer names or refers to a list of performers, it can be machine-understandable.
X-SKI-CREATOR	Describes the creator of the work being performed. Creator can be the author, composer, and so on.	X-SKI-CREATOR:Beethoven	Same as X-SKI-PERFORMER.
X-SKI-PARTOF	Describes this event as being part of another event, for example, a festival.	X-SKI-PART OF;CN= Vattenfestivalen: <19990401T080045Z-F192713 @stoinfo.se>	The format describing the "container event" is the same as the one describing the object.
X-SKI-EVENT-LANGUAGE	The original event language. If it is being translated, the RESOURCE field should be used.	X-SKI-EVENT-LANGUAGE:en	Text identifying a language using a two-letter code, as defined in [RFC 1766].
ATTACH	Associates a document object with a calendar component through a URI.	ATTACH:http://www.w3.org/ people#hjelm	URIs can refer to many other things besides Web pages, for example, email messages. Can be used in any of the different sections.

continues

Table 7.10 The What Parameters of SKi (continued)

TITLE	PURPOSE	EXAMPLE	COMMENT
SUMMARY	Defines a short summary or subject for the event.	SUMMARY:Department Party	Intended to be really short. An implementation can truncate a "SUMMARY" property value to 255 characters. If you need a longer description, use DESCRIPTION or reference a description with a URI.
DESCRIPTION	More complete description than the SUMMARY property.	DESCRIPTION:Last draft of the new novel is to be completed for the editor's proof today	Intended for "lengthy textual descriptions" of the event. The recommendation is that lines longer than 75 byte should be folded.
X-SKI-EVENT-LANGUAGE	The original event language. If it is being translated, the RESOURCE field should be used.	X-SKI-EVENT-LANGUAGE:en	Text identifying a language using a two-letter code, as defined in [RFC 1766].

Table 7.11 SKi What Properties Related to Information Integrity

TITLE	PURPOSE	EXAMPLE	COMMENT
X-SKI-CHECKSUM	The checksum that is calculated on the entire calendar object. The intent is to protect the information from manipulation while in transit.	X-SKI-EVENT-CHECKSUM:948	Checksum is a weak protection, and it needs to be combined with digital signatures to establish the provenance of the information.
DTSTAMP	Date/Time stamp property.	DTSTAMP:19971210T080000Z	Indicates when the object was created. Must be specified in UTC (the "Z" time zone).
CREATED	Specifies the date and time that the calendar information was created by the calendar user agent in the calendar store.	CREATED:19960329T133000Z	The creation date is analogous to the creation date and time for a file in the file system, and it should be generated automatically.
LAST-MODIFIED	Describes when the object was last modified.	LAST-MODIFIED:19960817T133000Z	Must be specified in UTC. Analogous to the file creation date in the file system.
SEQUENCE	Defines the revision sequence number, within a sequence of revisions.	SEQUENCE:2	If the object has been modified several times, this can be useful to know.

be secure, you have to use other methods such as digital signatures or even encryption.

When

The iCalendar format defines time very thoroughly. The SKi additions are mainly a format to describe when something can take place (open times being a good example), and they can be used to give a number of possible dates for an event (for example, a weather-dependent event that is scheduled in advance). See Table 7.12.

The positive and negative time raises questions about the values of the iCalendar properties. iCalendar properties have to have a value that corresponds to the data type specified for that value. These types can be BINARY, which identifies properties that contain inline binary data; BOOLEAN, which is used to identify properties that contain either a TRUE or FALSE Boolean value; CAL-ADDRESS, which identifies properties that contain a calendar user address; DATE, which identifies values that contain a calendar date; DATE-TIME, which identifies values that specify a precise calendar date and time of day (in ISO 8601 format); DURATION, which defines a duration of time (which can be any time interval that can be specified); and PERIOD, which is used to identify values that contain a precise period of time.

Its start and its end identify a period of time. This format is expressed as the complete representation (from-to) in ISO 8601 format. A period can also be defined by a start and a positive duration of time (that is, how long an event lasts). TIME is used to identify values that contain a time of day. The format is based on ISO 8601 (an international format that Europeans use when they talk about time—A.M. and P.M. are out) and consists of a two-digit hour of the day (that is, values 0-24), a two-digit minute in the hour (that is, values 00 to 60), and two-digit seconds in the minute (that is, values 00 to 60). Fractions of a second are not supported by this format. The form of time with UTC offset, for example, 230000-0800 (which would be read "UTC 23.00.00 minus eight hours" and which is the same as "2 P.M. San Francisco time"), is not valid unless you specify the time zone as UTC-OFFSET. Then, it is calculated from the UTC difference, and you do not give the actual time. If seconds of the minute are not supported by an implementation, then a value of 00 should be specified for the seconds component in a time value.

Time values of this type are said to be floating and are not bound to any time zone in particular. They are used to represent the same hour, minute, and second value regardless of the time zone in which the event occurs. This floating characteristic is why you should use UTC (identified by adding a Z to the time, 0700Z, or giving the time zone ID). UTC-OFFSET enables you to give the offset from UTC to local time. The UTC offset for New York standard time (five hours

Table 7.12 The When Parameters of SKi

NAME	DESCRIPTION	EXAMPLE	COMMENTS
X-SKI-OPENINGTIMES	Opening times.		The OPENINGTIMES property is under discussion in the CalSched working group.
X-SKI-SCHEDULINGTIMES	Preliminary scheduled times.		See X-SKI-OPENINGTIMES.
DTEND	Specifies when the event ends.	DTEND:19960401T235959Z	Default is DATE-TIME format. Time must be given in UTC time. Can also be set to DATE, in which case you do not have to give the time.
DTSTART	Specifies when the event begins.	DTSTART:19980118T073000Z	See DTEND. When the "DTSTART" and "DTEND" have the same value data type (for example, DATE-TIME), they should specify values in the same time format (for example, UTC time format, "Z").
DURATION	Specifies a (positive) duration of time.	DURATION:PT15M	The PT signifies Positive Time, the M minutes. If you have an event lasting 1 hour, it is written as 1H0M0S (0 minutes, 0 seconds). Instead of specifying DTEND, you can specify DTSTART and a DURATION. While positive time is possible to imagine, negative time is harder. It might occur when you are delayed, for instance.

behind UTC) is -0500, and for Geneva (one hour ahead of UTC) it is +0100. VTIMEZONE provides a grouping of component properties that defines a time zone (for example, by UTC-OFFSET). VTIMEZONE is necessary because there are no formal names for time zones, so if you use one, you must define it by using TZID. Remember that the offsets vary with the season (Daylight Savings Time is used in many countries in the world; the British time zone specified by GMT has Daylight Savings Time, but UTC does not).

A calendar entry with a DTSTART property but no DTEND property is considered not to take up any time. It is intended to represent an event that is associated with a given calendar date and time of day, such as an anniversary.

iCalendar has a very comprehensive system for specifying recurrence of events, too. RECUR is used to identify properties that contain a recurrence rule specification. The value must have a list defining the recurrence rules (for example, INTERVAL, BYYEARDAY, and BYMINUTE). It is also possible to set exceptions to recurrences using the EXDATE property (with the value DATE-TIME). The exception dates, if specified, are used in computing the recurrence set. The RRULE property defines a rule or repeating pattern for recurring events. The list of recurrence rules is quite long and complicated, and if you want it all, you will have to look it up in the iCalendar specification.

Which

Knowing the whats of an event is only the start, however. You must also know what type of event it is to see whether it is interesting to you. Table 7.13 describes the Which properties of the SKi format, which do this exactly.

The RELATED-TO property is used to represent a relationship or a reference between one calendar component and another. It consists of the persistent, globally unique identifier of another calendar component represented in a calendar component by the UID property (which is set by the system according to a non-defined method). You can actually define a limited set of object-oriented inheritances. Changes to a calendar component related to another can have an impact on the related calendar component, but the property provides only information.

Where

Having determined whether the event interests you, finding out whether it takes place in a location and time that makes it available to you is the next step. The parameters describing the location of an event are listed in Table 7.14.

The GEO value specifies latitude and longitude in that order (that is, LAT LON ordering). The longitude represents the location east or west of the prime meridian as a positive or negative real number, respectively. Latitudes (the values parallel to the equator) north of the equator are specified by a plus sign (+)

Table 7.13 The Which Parameters of the SKi Specification

NAME	DESCRIPTION	EXAMPLE	COMMENTS
X-SKI-VENUE	Describes where an event takes place (to make it possible to filter out all non-interesting ones).	X-SKI-VENUE="Internet"	Must be one of the following: Internet, Radio, TV, Outdoors, Indoors, Travel-transit. Whether to describe a broadcast football game as two or one events is a problem, but the recommendation in the specification is that it be given as two separate events: the game and the broadcast.
X-SKI-ORIENTATION	Describes the intended target audience.	X-SKI-ORIENTATION;Political:Liberal	The intention is that there should be a keyword list to which this parameter could refer.
CATEGORIES	Describes into which categories the event falls. There are no standard descriptions for the categories.	CATEGORIES:SHOW,MUSIC,DANCE,CROONER	The only restriction is that the ategories should be given in ASCII and separated by commas. As it is, you can invent your own.

Table 7.14 The Where Parameters of the SKi Specification

NAME	DESCRIPTION	EXAMPLE	COMMENTS
X-SKI-DIRECTIONS	Gives directions in a human-readable format.	X-SKI-DIRECTIONS: TRAVELBY=walk:Turn left after the waterfall, follow yellow signs	You have to specify the method of transportation that you intend people to use to get to the event. There may be multiple descriptions. By referring to a syllabus of place and venue names, the parameter can become machine-readable, but a better solution is to give the venue position in geocoordinates.
GEO	The global position of the activity specified by the calendar component.	GEO:37.386013;-122.082932	See below for more comments. Easiest to determine with a GPS receiver.
LOCATION	Defines the intended venue for the activity.	LOCATION:Conference Room-F123, Bldg. 002 LOCATION;ALTREP= "http://xyzcorp.com/conf-rooms/ f123.vcf": Conference Room-F123, Bldg. 002	Specific venues such as conference or meeting rooms may be explicitly named using this property. An alternate representation, for example, a URI, can be included. In principle identical to the X-SKI-VENUE, except that it is intended for conference rooms and not concert halls.

or by the absence of a minus sign (−) (which makes them positive anyhow), which precedes the digits designating the degrees. Latitudes south of the equator are shown by a minus sign (−) preceding the digits that show the number of degrees; the equator itself is considered to be in the northern hemisphere. Longitudes east of the prime meridian (which passes through the city of Greenwich, outside London) are specified by a plus sign (+) or by the fact that they are not preceded by a minus sign (−) before the digits that show the number of degrees. Longitudes west of the meridian are shown by a minus sign (−) before the number that shows degrees. A point on the prime meridian (such as London's Millennium Dome) is considered to be in the Eastern hemisphere. A point on the 180th meridian (which does not pass through any land because it is in the middle of the Pacific Ocean) is considered to be in the Western hemisphere. In addition, there are also exceptions for the North and South Poles and for describing a band around the Earth.

The longitude and latitude values can be specified with an accuracy of up to six decimal places, which will allow for accuracy to within one meter of geographical position. Receiving applications can truncate values of greater precision (that is, values that are longer). Values for latitude and longitude are expressed as decimal fractions of degrees. A two-digit decimal number ranging from 0 through 90 represents whole degrees of latitude. A decimal number ranging from 0 through 180 represents whole degrees of longitude. When a decimal fraction of a degree is specified, it must be separated from the whole number of degrees by a decimal point. This way of specifying location is probably familiar to anyone who has used a GPS receiver.

There is a simple formula for converting the traditional degrees-minutes-seconds format into decimal degrees: decimal = degrees + minutes/60 + seconds/3600.

Who/How

There are, however, some additional qualifications that you need to take into account before going to an event. For instance, can you get tickets? Can you afford them? The parameters describing this information are listed in Table 7.15.

Why

Public events must be marketed differently from business meetings. iCalendar has methods for agreeing on an event. With SKi, a number of properties can be used (see Table 7.16).

The Who/How parameters are intended to describe the capabilities of the event location (if there are restrooms for physically challenged persons and so on). The Who parameters describe the individuals (or organizations) involved in the event. I have divided them in this way for increased clarity.

Table 7.15 The Who and How Parameters of the SKi Specification

NAME	DESCRIPTION	EXAMPLE	COMMENTS
X-SKI-QUALIFICATION	Gives the necessary or recommended qualifications for participating in the event.	X-SKI-QUALIFICATION;Age: >7 with adult accompaniment-REQUIRED	Keywords can be retrieved from a list and given with parameters, such as REQUIRED or RECOMMENDED.
X-SKI-PRICE	Gives the price for the event.	X-SKI-PRICE;Admission:+ 135SEK	Uses the same structure as that defined in W3C micropayments specification, and uses currency code as defined in ISO 4217. Charging for the event is determined according to a list of keywords, such as Admission, Breakfast, Room, Happy hour, Monthly fee. There is no standard for the keywords.
X-SKI-TICKETS-AT	Tells you where to get tickets.	X-SKI-TICKETS-AT;DTSTART= 19980313T141711Z:Ticketron kiosks	X-SKI-TICKETSAT not only gives the place where tickets can be bought, but also tells when (start and end dates in the same way as the duration of an event).
X-SKI-RESERVATIONS	Tells where you can register your interest for the event.	X-SKI-RESERVATIONS;DTEND= 19990514:"http://interested .event.nu"	Can have a start- and end date and time. The format should be given as a URI.
X-SKI-AVAILABLE	Tells you if there are seats left or other restrictions for the admission.	X-SKI-AVAILABLE:55 seats	If a time is given, it is the number of seats at that time; otherwise, it is the total number of seats. A problem is numbered seats; you essentially have to express how these are described.

continues

Table 7.15 The Who and How Parameters of the SKi Specification (continued)

NAME	DESCRIPTION	EXAMPLE	COMMENTS
X-SKI-HANDICAP-FACILITIES	Whether the event is accessible for people with special needs.	X-SKI-HANDICAP-FACILITIES; wheelchair-ramps:TRUE	Keywords should be picked from a list. No standard for lists exists, but the data type should be Boolean.
X-SKI-PAYMENT-METHOD	Which payment methods can be used for the event.		Still being developed. There are a couple of formats for this, but the problem is covering all possible forms of payment (which also vary between countries).
CLASS	Describes who should have access to the calendar information. Default is PUBLIC.	CLASS:PUBLIC	Again, you are free to define your own classes. The intention is that this should be integrated with the semantics of the calendaring systems.
RESOURCES	Defines which equipment or resources are needed for an activity.	RESOURCES:EASEL, PROJECTOR,VCR	There are no standards for naming resources.

Table 7.16 The Why Parameters of the SKi Specification

NAME	DESCRIPTION	EXAMPLE	COMMENTS
X-SKI-MOTIVATION	MOTIVATION is intended to be a subjective description by the arranger to market the event.	X-SKI-MOTIVATION;LANGUAGE=en:Yo Bro! This is going to be a great party. Everybody will be there. If you are not—you will regret it.	Note that this can be a URI, but there will are also further fields.
X-SKI-ADVERT	Gives commercial information about the event.	X-SKI-ADVERT:http://www.stadsteatern.se/affischer/othello2000.gif	Can be a Web page, newspaper advertisement, etc.
X-SKI-REVIEW	Independent review of the event.	X-SKI-REVIEW;LANGUAGE=en:http://www.alltomstockholm.se/reviews/12345/	Of course, this is subjective because the arranger selects which reviews to present.

Who

Most events take place without problems. But if there are, whom do you sue? Finding out that and other information that puts you in contact with the arranger is what the who parameters of SKi are about, as listed in Table 7.17.

Other iCalendar Parameters

These are not all the parameters that an iCalendar object can have (you have to look up the specification to find all of those), but those are merely relevant to describe it as an event. You have seen the CN parameter used in several places, for instance. Additional iCalendar parameters are listed in Table 7.18.

It is also possible for a calendar user to have a number of users delegate his or her participation; for instance, football fans deciding to delegate the decision about their attendance at a game to a fan club. I will not look into that issue further here, but it does give you an idea of possible future applications.

Metafiles for Events: Some Examples

Now that you know the principles for declaring the event information, let's create a few metafiles. The idea is that the Web owner creates a file containing the descriptive metainformation; this information is then harvested by a spider or other similar piece of software, and this information is compared to the sphere of interest that the user has set up for different times of day and different event types (when it starts to get really close to lunch and I ask for a lunch restaurant, I will want a restaurant that is within 150 meters or a 10-minutes walk; when I want a restaurant in the evening, itshould be fancier and distance does not matter as much). As you understand, it is a lot easier for the matching system if the information is presented to it in RDF. At the end of the chapter, I describe how this system could work.

Metafile for the Networld Conference

Meetings and other arrangements probably fill your calendar as well as mine. An example of what this situation might look like is as follows:

```
BEGIN:VCALENDAR
PRODID:-//xyz Corp//NONSGML PDA Calendar Version  1.0//EN
VERSION:2.0
BEGIN:VEVENT
DTSTAMP:19960704T120000Z
UID:uid1@host.com
ORGANIZER:MAILTO:jsmith@host.com
DTSTART:19960918T143000Z
DTEND:19960920T220000Z
STATUS:CONFIRMED
CATEGORIES:CONFERENCE
```

Table 7.17 The Who Parameters of the SKi Specification

NAME	DESCRIPTION	EXAMPLE	COMMENTS
X-SKI-RESPONSIBLE	The person or organization that is formally in charge of the event.	X-SKI-RESPONSIBLE;CN=Nobelkommitten:MAILTO:info@nobel.se	Can be a URI, an email address, or any other way to reach the individual or organization in question. In the event that there is a legal party responsible, this ought to be it (for example, publisher for newspapers).
X-SKI-OTHERAGENTS	Other arranging agents.	X-SKI-OTHERAGENTS:MAILTO:info@ematelstar.se	Could refer to the group responsible for logistics, e.g. the caterer for a wedding.
X-SKI-ASSOCIATION	The organization to which the arranger belongs.	X-SKI-ASSOCIATION;Rotary international	If a register of organizations existed, this could refer to it.
CONTACT	The arranger or a reference to the arranger.	CONTACT:Jim Dolittle\, ABC Industries\, +1-919-555-1234	You can also use ALTREP to include a URI. It will not be automatically resolved (however. it would if this was XML).
ORGANIZER	The arranger.	ORGANIZER;CN=John Smith:MAILTO:jsmith@host1.com	Overlaps the other parameters to some degree.

Table 7.18 Other Relevant iCalendar Parameters

NAME	DESCRIPTION	EXAMPLE	COMMENTS
CN	Common name parameter that is associated with the user.	ORGANIZER;CN="John Smith": MAILTO:jsmith@host.com	The common name is intended to give a real name for a real person.
DIR	Specifies reference to a directory entry associated with the calendar user specified by the property.	ORGANIZER;DIR="ldap://host.com: 6666/o=eDABC%20Industries, c=3DUS?? (cn=3DBJim%20Dolittle)": MAILTO:jimdo@host1.com	DIR can be used with X-SKI-RESPONSIBLE, X-SKI-OTHERAGENTS, and X-SKI-ASSOCIATION, for instance, but also with the parameters relating to performers, composers, and venues. It can be an entry in the yellow pages. It is always given as a URI.
ALTREP	Specifies an alternate text representation for a property (in the same way as the ALT-tag in HTML).	DESCRIPTION;ALTREP="CID: <part3.msg.970415T083000 @host.com>"	Must also include a reference to the default representation of the text value.
LANGUAGE	Specifies language values for the property.	LOCATION;LANGUAGE=en:Germany LOCATION;LANGUAGE=se:Tyskland	Text identifying a language (two-letter code), as defined in RFC 1766. Note that it applies to the parameter.
RANGE	The parameter specifies the effective range of recurrence instances that is specified by a property that has a recurrence parameter.	RECURRENCE-ID;RANGE= THISANDPRIOR:19980401T133000Z	The parameter value can be "THISANDPRIOR" to indicate a range defined by the recurrence identified value of the property and all prior instances. The parameter value can also be "THISANDFUTURE" to indicate a range defined by the recurrence identifier and all subsequent instances.
RSVP	To specify whether there is an expectation of a favor of a reply from the calendar user specified by the	ATTENDEE;RSVP=TRUE:MAILTO: jsmith@host.com	The RSVP parameter is used by the "Organizer" to request a participation status reply from an "Attendee" of a group scheduled event or to-do. If not specified on a property that allows this

```
SUMMARY:Networld+Interop Conference
DESCRIPTION:Networld+Interop Conference and Exhibition\n
Atlanta World Congress Center\n
Atlanta, Georgia
END:VEVENT
END:VCALENDAR
```

So the Interop+Networld conference and exhibition opened in Atlanta, Georgia at 2:00 P.M. September 18, 1996, and it closed at 10:00 P.M. two days later. Well, that was a short conference.

Metafile for Cross Brothers Concert

Let's look at a rock concert instead. Following is what an event description would look like for the Christian rock band Cross Brothers giving a concert in the Masonic Hall of Decatur, Illinois as part of their Universal Salvation Tour on February 29, 2000 from 7 P.M. to 11 P.M. during Decatur for Christ week. There is a $5 admission fee, and you buy your tickets at the church secretary's office from 9 A.M. on February 1. If you want to know more, contact Dean Wittman at the Decatur Church of the Apostles:

```
BEGIN:VCALENDAR
PRODID:-//xyz Corp//NONSGML PDA Calendar Version  1.0//EN
VERSION:2.0
BEGIN:VEVENT
DTSTAMP:200002012T120000Z
UID:uid1@decaturapostlechurch.com
ORGANIZER:CN: Dean Walt Wittman MAILTO:wittman@decaturapostlechurch.com
X-SKI-OTHERAGENTS:Christian Concerts of America, Inc.
X-SKI-ASSOCIATION:Catholic Church of North America, Diocese of Chicago
X-SKI-TITLE:Universal Salvation Tour
X-SKI-PERFORMER:Cross Brothers
X-SKI-PART OF:Decatur For Christ Week
ATTACH:http://crossbrothers.org/salvationtour/
SUMMARY:Cross Brothers plays the hottest hits from heaven - and saves
Decatur in the bargain!
DTSTART:20000229T010000Z
DTEND:20000229T050000Z
X-SKI-VENUE:"Indoors"
X-SKI-ORIENTATION:Christian
CATEGORIES:CONCERT,ROCK
X-SKI-DIRECTIONS:Parking behind the hall
LOCATION:Masonic Hall, 10 Main St, Decatur, Ill.
X-SKI-PRICE;Admission:5USD
X-SKI-TICKETS-AT;DTSTART=20000201T150000Z:Church Secretary Office, 14
Main St. Decatur, Ill.
X-SKI-RESERVATIONS;DTEND=20000229T1500Z
X-SKI-AVAILABLE;DTSTART=20000228T100000:10 seats
X-SKI-HANDICAPFACILITIES;assistance:TRUE
```

```
X-SKI-PAYMENTMETHOD:Check,credit card,cash
CLASS:PUBLIC
X-SKI-MOTIVATION:Saving the youth of this poor city for Christ
X-SKI-
ADVERT:http://www.decaturapostlechurch.com/Christ_week_2000/crossbrother
s/
X-SKI-
REVIEW:http://www.Americanchristian.com/events/reviews/crossbrothers/tou
r99.html
END:VEVENT
END:VCALENDAR
Let's go through that one line by line, too:
BEGIN:VCALENDAR
PRODID:-//xyz Corp//NONSGML PDA Calendar Version  1.0//EN
VERSION:2.0
```

The preceding code is the header of the declaration. In an RDF file, this header would be the XML and namespace declarations.

```
BEGIN:VEVENT
```

The event description begins:

```
DTSTAMP:200002012T120000Z
UID:uid1@decaturapostlechurch.com
ORGANIZER:CN: Dean Walt Wittman MAILTO:wittman@decaturapostlechurch.com
```

It was created February 12, 2000. The ID for the description is of the e-mail address type; as you can see, it is the first description that is created in the church this year. The organizer's *common name* (CN) is Dean Walt Wittman, and his email address is wittman@decaturapostlechurch.com.

```
X-SKI-OTHERAGENTS:Christian Concerts of America, Inc.
X-SKI-ASSOCIATION:Catholic Church of North America, Diocese of Chicago
```

The concert is organized by Christian Concerts of America, Inc., and the Church of the Apostles belongs to the Catholic Church of North America.

```
X-SKI-TITLE: Universal Salvation Tour
X-SKI-PERFORMER: Cross Brothers
X-SKI-PART OF: Decatur For Christ Week
ATTACH: http://crossbrothers.org/salvationtour/
SUMMARY: Cross Brothers plays the hottest hits from heaven - and saves
Decatur in the bargain!
```

The title of the event is Universal Salvation Tour (strictly speaking, this event is a part of that tour, too). The performers are Cross Brothers, and this event is part of the Decatur for Christ Week. There is more information at the Cross Brothers Web site, and there is a summary of the concert information there.

```
DTSTART: 20000229T010000Z
DTEND: 20000229T050000Z
```

The concert starts at 7 P.M. (1 P.M. UTC minus 6 hours) and ends at 11 P.M. (5 P.M. UTC minus 6 hours) on February 29, 2000.

```
X-SKI-VENUE: "Indoors"
X-SKI-ORIENTATION: Christian
CATEGORIES:CONCERT,ROCK
X-SKI-DIRECTIONS: Parking behind the hall
LOCATION: Masonic Hall, 10 Main St, Decatur, Ill.
```

The concert venue is indoors (remember, "venue" is used to describe the logical location of the event, such as on the Internet, broadcast, and so on). It has a Christian orientation (no surprise there), and it is in the concert and rock categories. You can park behind the hall (the European orientation of the SKi format shows in that there is no X-SKI-PARKING field). And the concert takes place in the Masonic Hall of Decatur.

```
X-SKI-PRICE;Admission: 5USD
X-SKI-TICKETS-AT;DTSTART=20000201T150000Z:Church Secretary Office, 14
Main St. Decatur, Ill.
X-SKI-RESERVATIONS;DTEND=20000229T1500Z:tel://+1-800-CRO-BROS
X-SKI-AVAILABLE;DTSTART=20000228T180000Z:10 seats
X-SKI-PAYMENTMETHOD;Check,credit card,cash
```

The price is $5, and it is an admission fee. You can buy your tickets at the church secretary's office starting February 1 at 9 A.M., and you can make reservations by calling the 800 number of the Cross Brothers (1-800-CRO-BROS) until 9 A.M. the same day. There were 10 seats available on February 28 at 12 A.M., and you can pay by using check, credit card, or cash.

```
X-SKI-HANDICAPFACILITIES;assistance: TRUE
CLASS:PUBLIC
X-SKI-MOTIVATION:Saving the youth of this poor city for Christ
X-SKI-
ADVERT:http://www.decaturapostlechurch.com/Christ_week_2000/crossbrother
s/
X-SKI-
REVIEW:http://www.Americanchristian.com/events/reviews/crossbrothers/tou
r99.html
```

There is assistance available for handicapped people, and the event is public. It is intended to save Decatur's youth for Christ. There is an advertisement at the Web site of the church, and *American Christian* magazine has reviewed the tour.

```
END:VEVENT
END:VCALENDAR
```

The description ends.

The intention is, of course, not that you should have to either read or write all of this information. The computer is intended to take care of it for you, using some kind of editing software. But remember, Tim Berners-Lee believed that this process was how people would edit the Web, too. Instead, Notepad has become one of the most popular editors, which does nothing to decrease the frequency of errors in the editing of the Web.

The Concert Description as RDF

Now, let's dress up this example as the XML-format RDF, or Resource Description Framework, which enables us to relate the description to the object. That might seem like an unnecessary exercise; it is already structured, isn't it? But using RDF and XML gives you a lot for free. It becomes possible to compare the description with other descriptions, although you do not know what the tags mean. Otherwise, you have to create a separate parser for each description format, which will mean that you cannot do ad-hoc comparisons and that you have to program the parser for the format with the specific rule set for the format. It also becomes possible to mix elements from different namespaces in the same document, which means that you can use the elements that fit the data best instead of having to squeeze them into an element type that might not be the best descriptor of what you want to achieve.

In the following example, I have created a namespace for both iCal and SKi. Strictly speaking, SKi is a subset of iCal, so I did not really have to do that. It will illustrate the use of multiple name spaces better. One problem, though, is the collapsed form of RDF. To be able to compare it to other RDF files, you would have to express it as full XML, so this description is in the full XML format. In other words, among other things, the parameter names must be in lower-case. In some cases, such as the facilities for physically challenged persons, we also have to switch things around. The attribute TRUE applies to any handicapped facilities available, and we have to give them as the parameter value while the fact that they exist means (TRUE) becomes an attribute value:

```
<?xml version='1.0'?>
<rdf
xmlns:rdf="http://www.w3.org/TR/WD-rdf-syntax#"
xmlns:ical="http://www.w3.org/TR/WD-ical-RDF#"
xmlns:ski="http://www.w3.org/TR/WD-skical-RDF#">
<rdf:Bag>
      <rdf:Description about="VCALENDAR">
      <ical:vevent>
<ical:dtstamp>200002012T120000Z</ical:dtstamp>
<ical:uid>uid1@decaturapostlechurch.com</ical:uid>
```

```
<ical:organizer_cn>Dean Walt Wittman</ical:organizer_cn>
<ical:organizer_mailto>wittman@decaturapostlechurch.com
</ical:organizer_mailto>
<ski:otheragents>Christian Concerts of America Inc. </ski:otheragents>
<ski:association>Catholic Church of North America, Diocese of
Chicago</ski:association>
<ski:title>Universal Salvation Tour</ski:title>
<ski:performer>Cross Brothers</ski:performer>
<ski:part_of>Decatur For Christ Week</ski:part_of>
<ical:attach>http://crossbrothers.org/salvationtour/</ical:attach>
<ical:summary>Cross Brothers plays the hottest hits from heaven - and
saves Decatur in the bargain! </ical:summary>
<ical:dtstart>20000229T010000Z</ical:dtstart>
<ical:dtend>20000229T050000Z</ical:dtend>
<ski:venue>Indoors</ski:venue>
<ski:orientation>Christian</ski:orientation>
<ical:categories>CONCERT,ROCK</ical:categories>
<ski:directions>Parking behind the hall</ski:directions>
<ical:location>Masonic Hall, 10 Main St, Decatur, Ill.</ical:location>
<ski:price admission="true">5USD</ski:price>
<ski:tickets-at dtstart="20000201T150000Z">Church Secretary Office, 14
Main St. Decatur, Ill.</ski:tickets-at>
<ski:reservations dtend="20000229T1500Z">tel://+1-800-CRO-
BROS</ski:reservations>
<ski:available dtstart="20000228T100000">10</ski:available>
<ski:handicapfacilities
facilities="TRUE">assistance</ski:handicapfacilities>
<ski:paymentmethod>Check,credit card,cash</ski:paymentmethod>
<ical:class>PUBLIC</ical:class>
<ski:motivation>Saving the youth of this poor city for
Christ</ski:motivation>
<ski:advert>http://www.decaturapostlechurch.com/Christ_week_2000/crossbr
others/</ski:advert>
<ski:review>http://www.Americanchristian.com/events/reviews/crossbrother
s/tour99.html</ski:review>
</ical:vevent>
</rdf:Description>
</rdf:Bag>
</rdf>
```

Geographic Markup in Metadata

In HTML, there was a META element that was intended for users to put in information about the data in the document, such as who wrote it, when, who was responsible, and so forth. Metadata can be vastly more sophisticated, being used (for instance) in database schemas, in schemas for RDF and XML documents (specifying what the elements used in the document mean), and so on.

When the Web was new, it lent itself to the traditional publishing model of creating documents and making them available through publication. That is still true, but as you could see in Chapter 5, there is also an emerging model for publishing database data, which can be dynamic and constantly changing.

As long as your data is in the shape of documents, however, you can take advantage of the extended metadata mechanisms to publish information about your information in standardized data formats.

If you have a Web site that consists only of a few documents, you might still want to make it available to people who are interested in your neighborhood. The URI of the pages—the address—expresses only one semantic (and it might be as bad as www.famousisp.com/users/pages/homepages/pagelocations/~yourusername/homepage.html, which does not actually say anything about the pages except that they have a location). That is relevant in some contexts but not if you want to make it available for people who are interested in, say, restaurants on Silom Road in Bangkok.

In theory, your page could have any number of URIs, but it would become unmanageable because you would have to register it with the maintainers of the URIs. It is bad enough to try to get it into Google and Yahoo!. Instead, you can publish metadata on the page that describes it to search engines, and they can then generate an automatic index and take care of setting up and presenting the relation.

If there are several search engines that want to make use of your data, you need to use a standardized format to mark it up. GML, POIX, NVML, and SKiCAL are such formats. But to make it even more accessible to search engines, you can use a standardized format in which to embed the metadata. That format is Dublin Core, which has been developed by the library community (actually, the Online Computing Library Centre, or OCLC, in Dublin, Ohio). It represents the most important data about a static document that you need to put into it to describe its relations to others.

Dublin Core contains data elements for the description of author, publisher, and other publication-related information, but it also has an element, Coverage, which is especially intended for geographic and time-related metadata. The Coverage element can contain an identifier for a place. If a name or geocode is used, then the scheme from which that is selected determines valid values. There is also a qualifying element, the DCMI Point, which is an identifier that specifies the coordinates of the point location of a place. Rectangular regions can be encoded by using the Box qualifier.

The coverage as defined in the Coverage element can be anything to which the document relates. The novel *Brideshead Revisited* by Evelyn Waugh relates to Morocco, London, Brideshead, England, and the beginning of the 20th century (it also covers fin-de-siecle, homosexuality, British upper class, and a lot of other things, of course). Coverage does not have to be along a single dimension, but a document can cover many things (and be related to others, such as the birthplace of the author).

The Dublin Core Point is defined in terms of Cartesian coordinates in a defined coordinate system. The elements are east and north for geocoordinates elevation; a separate value for units (and zunits, which defines how the units are defined if they are not defined in standard values), the projection (in other words, coordinate system), and the name, which is the place name. An example of how it could be encoded in XML is:

```
<Point projection="v6" name="v7">
  <east units="v4a">v1</east>
  <north units="v4b">v2</north>
  <elevation zunits="v5">v3</elevation>
</Point>
```

For Perth in west Australia, this coding would look like:

```
<Point name="Perth, W.A.">
        <east>115.85717</east>
        <north>-31.95301</north>
</Point>
```

The Dublin Core metadata format also contains a method to encode a box. The box identifies a region of space using its geographic limits. Components of the value correspond to the bounding coordinates in north, south, east and west directions, plus optionally up and down, and enable the coordinate system and units to be specified and a name to be added if desired (and appropriate). The elements introduced for the two-dimensional box delimitation are northlimit, eastlimit, southlimit, westlimit, and for the three-dimensional delimitations uplimit and downlimit. The units and zunits define the units, and the projection element defines the projection. The name is the name of the place in the box.

In XML, it looks like:

```
<Box projection="v9" name="v10">
<northlimit units="v7a">v1</northlimit>
<eastlimit units="v7b">v2</eastlimit>
<southlimit units="v7c">v3</southlimit>
<westlimit units="v7d">v4</westlimit>
<uplimit zunits="v8a">v3</uplimit>
<downlimit zunits="v8b">v4</downlimit>
</Box>
```

For Lake Jindabyne in western Australia, it would look like the following:

```
<Box projection="UTM zone 55 south" name="Lake Jindabyne">
<northlimit units="m">5980000</northlimit>
```

```
<eastlimit units="m">647000</eastlimit>
<southlimit units="m">5966000</southlimit>
<westlimit units="m">644000</westlimit>
</Box>
```

And for the Duchess copper mine in western Australia, showing how the three-dimensional aspect comes into play (because the mine has a negative extension, below sea level):

```
<Box name="Duchess copper mine">
<northlimit>-21.3</northlimit>
<eastlimit>139.9</eastlimit>
<southlimit>-21.4</southlimit>
<westlimit>139.8</westlimit>
<uplimit>400</uplimit>
<downlimit>-100</downlimit>
</Box>
```

In addition, there is a time-related coverage element. Times can have names (for instance, the Jurassic period), or they can have delimitations in time (and in space; the Tokugawa shogunate lasted from the 17th century to the middle of the 19th century, and it affected Japan). The Period element in the Dublin Core has a start, an end, is expressed by using a scheme, and can have a name.

This example shows how to encode the Australian Football League grand final:

```
<Period name="1999 AFL Grand Final">
        <start scheme="W3C-DTF">1999-09-25T14:20+10:00</start>
        <end scheme="W3C-DTF">1999-09-25T16:40+10:00</end>
</Period>
```

These elements work fine for simple periods but will have problems representing more complex periods and occasions, where SKiCAL is probably a better choice. But by embedding these elements in your document, you will make it much easier for a search engine to find it and index it. That will mean that it can be used in services that you—or someone else—are creating.

Having created the service by encoding the information and using the DBMS to manage it, we will now move to looking at how to present the information—creating the user interface.

CHAPTER 8

The User Interface to Location Information Services

User interface design is important in location-dependent services—maybe even more so than on the Web. The services have to be intuitive to use because the business model is not based on numbers of views by casual users who might never come back. Rather, it is based on an established relationship with users who want to return and use the service time and again.

Who Needs a User Interface, Anyway?

First, user interface design should not be read with the emphasis on design, but rather on user interfaces. The kind of user interface design that flourishes on the Web—which mostly is based on paper design—does not work for mobile services. The form factor, for one, is all wrong. The intention is that the services should work on devices that are small enough to keep in your pocket. For the designer, this situation means rethinking what you have learned on the Web and learning from the established discipline of user interface design.

One reason for this is that the wireless Web is the Web, but it's different. For example, you can call a car a horseless carriage; it is that, but it is also a lot more, and different. The characteristics of mobile users are entirely different from users accessing the Web using fixed communications, according to several studies. They have a shorter decision horizon, do not want to wait for downloads, and have neither time nor space for information they do not want.

If you are in charge of the project to develop the location-based service, you will need to make sure that the design your team comes up with is not just a nice-looking shell on top of the database; you need to think about how it will make users feel that the service is an inseparable part of their lives. One way is to test it carefully; another way is to hire a human-computer interface professional to help you. But the first thing you should do is to follow the advice here.

There is no need for you to become an HCI professional. While I am sure you would be welcomed by the other practitioners of this illustrious profession, there are lots of useful things you can apply without having the basic schooling in computer science, psychology, and anthropology that HCI professionals enjoy. This chapter will give you a first insight into the basics; in the list of references there are a few other books you could read that will help you get deeper into the subject.

There are a few things you have to think about and a few other things that can be boiled down to rules of thumb that will help you create a better user interface. But first, let me note that the discussion about location-based data user interfaces is far from new. As a matter of fact, it has been talked about since GIS systems first were implemented. A lot of good solutions have been developed since then.

While you should not try to apply design lessons from the Web, you will find that because the services you are designing will almost certainly be Web services, you need to take the conventions that have developed for the mobile Web into account. This consideration is independent of which environment you intend to present your service on—the work in WAP, HTML; XHTML, GSM Services, Brew, and any other environment you can think about.

Mobile phone users do not surf the web, even if their phones have Web capability. That is the first truth you have to learn as a mobile information system designer. The users have a rather short menu of services they occasionally access, and they want those to be easy to use so they will keep coming back.

Mobile users have short fuses. If the information is not there the moment they ask for it, they become irritated and go somewhere else. This situation is true for regular Web pages as well as for location-based services. The Reuters rule of "three clicks and three seconds" holds even more true in the mobile systems, where the perception of "always on" helps create the perception that this statement also means "information always there."

There is a simple piece of advice that should go with any service that wants to be used more than once: Do not make the user think. The more intuitive it feels to use the service, the more usage the service will see. Nobody reads instructions, either. In the consumer electronics industry, users consider the gewgaw they just bought to be broken if it does not work when they take it out of the box and plug it in. It is the same way with services: If they do not work immediately when the user accesses it, then it is broken.

This knowledge is the simple secret behind the design of the iMode services. They are convenient to use. Always on, which you get with GPRS and 3G, is a big first step. But that the service is always on means that you must make it convenient to access—flip open the phone and it works (that said, there are applications that do not have a user interface, such as telematics).

Designing for the Small Screen

Phones, PDAs, and other handheld devices vary enormously in the capacity they have to present information. Different devices can also vary within a range. For this reason, systems exist such as CC/PP, the Composite Capabilities/Preferences Profile, which is intended to help designers understand the capabilities of a device (I discuss CC/PP a little more in Appendix D). This differentiation becomes both critical and useful when the display medium is a handheld device because they can vary so much and because the map (or user interface in general) can be generated dynamically, but there are significant differences between different printers. A plotter will give you a color picture, as will a color laser printer, but they will have very different capabilities in terms of how an image can be printed. Whatever the output capabilities, you must design for them.

Too many designs of Web pages are made for the screen of the designer and do not work where the user will see them (the best examples being Web pages with pixel widths adapted for the Macintosh pixel representation, which do not work when viewed on a PC, but it is becoming more frequent that people design for Internet Explorer and are too lazy to check how it looks in Netscape). Understanding the end-user device is necessary in understanding how your work will be received. Appendix G will give you some insights into likely developments of mobile devices during the next few years.

The real estate on the screen is minimal, and that circumscribes design rules from the regular Web. One useful way of designing pages is to put in all you could think about and then take away everything that is not strictly necessary for the user to do what he wants to do. Note that it is not what you want the user to do that is important; it is what the user wants to do. The biggest surprise that hits designers when they show their pages to users is that they do not use them the way the designer had expected at all. Just by being part of the design process, you have become tainted, if you will, by the "group think" about what is necessary and important. You will need to rethink this situation to create a service that will be used, and that will mean not only to "kill your darlings" as the advertising industry saying goes, but also be prepared to recognize that the customer is always right (this situation is probably true on the Web too, but there, the customers are the people who buy advertising space, not the user. So it might actually be said that pages on the Web are mostly designed for people who buy advertising).

Spending a little effort on making your pages self-evident to your intended users might seem like a hit on the project budget you do not want to take, but it will increase usage and so pay for itself. Unfortunately, the only way to prove that is to have two competing designs on the market, and no company will do that—so double-blind tests are simply too complicated for real life. That in itself is a reason to do a limited user test.

Also, the limited space on the screen does not allow for a lot of questions to the user, which the PC screen does—there is plenty of space around the error message boxes there (but not on a mobile phone). You need to make sure that the user interface interferes with the user interaction with the information itself as little as possible, and if you need to ask the user questions, you will need to do so in the sequence of interactions in a way that feels natural.

Also, it is important to get it right the first time. Once a user has found a service that works for him or her, he or she will tend to stick with it. And the switching costs in terms of mental effort are much higher once you have gotten used to something. If you are trying to break into an area where there is an established competitor, you will have to offer a much higher added value to attract users (or, if the Web can be used as a model, wait until your competitors overload their services so much that users feel the value that they get from them has decreased to a level where they are ready to change).

One simple rule that contradicts everything you knew from the Web is that typefaces must be larger, not smaller. The reason is because the user must be able to see immediately what he or she is supposed to do. If he or she has to squint, it becomes inconvenient and usage becomes lower (the easiest thing is not to bother with this size, because browsers set their fonts anyway and the user can always override your settings. Design your pages so that they work with a variety of typefaces and font sizes, not just the one the art director fell in love with that week).

Generally, you should also avoid large graphics—and not just large graphics, but multiple graphics. The reason is simple: even the fastest of the mobile systems have a significantly higher latency and lower access speed than what you can get out of fixed cable. That has a number of explanations, but the consequence is simple: anything that is large, and anything that requires more than one transaction (back and forth to the Web server), will mean delays. For instance, if you have five images on your page, the browser will require five separate requests (although in HTTP 1.1, there is no need to make five separate TCP connections). Each request has to be done even if the page is cached on the handheld, because that is the way the protocol works. You have to check that it has not been updated. If each request takes 0.5 seconds to get to the server and back, you now have 2.5 seconds of idle time on your hands. That is in addition to loading any text or other information.

Counterintiuitively, though, smaller is not automatically better on a small screen. Designing for mobile phones means providing something that is readable at arm's length. Shrinking traditional designs will just lead to clutter on the screen, as you can see in Figure 8.1. Instead, you need to think about what will be clear and easy to read—not necessarily in the alphabetical sense, though. One of the interesting things about maps is that we read them although they do not contain any letters. We read the symbols.

People do not read pages, nor menus or other instructions; rather, they scan them. This situation is especially true in pages that are collections of information. The user will quickly look for the choice that best answers his or her preconceptions of what he or she wants and not bother to really read anything.

As an aside, do not try to explain too much about how your application works. As a matter of fact, do not try to explain at all. Just make it easy to use. Most people do not care how their cars work (this statement is true for anyone—a programmer would be hard put to explain the automatic gear shift, as well). We want to get from point A to point B without having to think about how many sparks are set off per second. It is the same with Web services for location information. A user interface that is simple and intuitive and does not break conventions will make it more convenient.

Trying not to squeeze too much into a page is a consequence. It is better to shift to a new page when there is new information. Users also do not scroll. If the information is not on the telephone screen, it does not exist—even if it is only a click on a scrollbar away. It is more natural to flip than to scroll, so that is what users will do. The menus of the Lassoo system are much more accessible than the iconic system of the Lesswire CeBIT service, for instance, as you can see in Figure 8.2.

Users do not read through the available options and then select the one that is most appropriate for their needs. If you believe that, you are not only an engineer, but you

Figure 8.1 How a large-screen device becomes cluttered and unreadable on a small screen.

Figure 8.2 A text menu illustrated.

are married to one. In reality, users scan through the choices and pick the first reasonable one. This strategy is called satisficing, a term coined by economist Herbert Simon. We take something that seems to work and use it. Users have also learned that this process is not damaging the Web; you can always go back and try again.

This situation is especially true when we are in a hurry, and if you are standing at a street corner with the light about to turn red, you do not want to wait for the graphics to load. You want to know which way that restaurant is and get there fast. Seeing what is important in one single glance becomes paramount. That is why Figure 8.3 is a bad design example (and I am hacking on Lesswire again). How do you find the important things on this map of the CeBIT fair?

A much better way is to zoom in and show what is important and have a clear interaction model for how you should get back to the level where you were. Intelliwhere has done that nicely, as shown on its map in Figure 8.4.

Moving back and forth between levels is familiar to anyone who has used maps of different scales. In other words, you as a designer have to create a clear visual hierarchy on the page and between the pages (expressed, of course, in the navigation mechanisms you are using). The added control that you get with CSS and XHTML should be a great tool for performing this task. A visual hier-

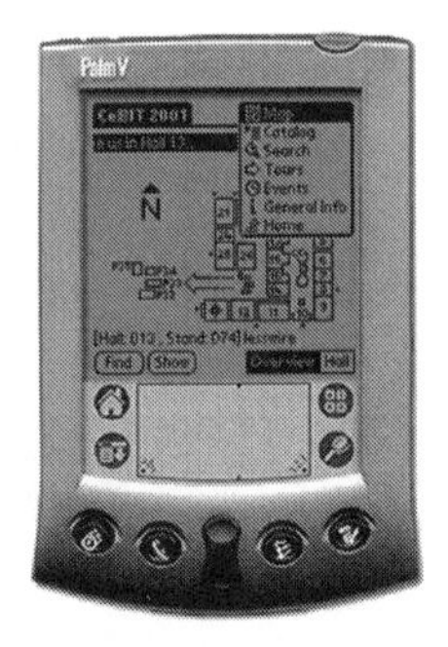

Figure 8.3 A trade fair map.

Figure 8.4 An Intelliwhere demonstration on an Ericsson R380.

archy is very simple: what is important is most prominent. You have to remember that prominence can be determined in three ways, though: through size, through color and luminescence, and through position on the page.

A very simple way of expressing hierarchy is to use colored plates or frames (not HTML frames, of course). You have to create them by using CSS, as well, because they would mean extra transactions otherwise. That said, doing so would help the user get a grip on what belongs together and what is important. Do not forget that space around an item makes it stand out. Some people seem to think that because they have a much smaller space on the mobile phone than on a PC screen they need to squeeze in more information (this same error occurs on the PC, by the way). Creating an open, empty space around something does not cost more if that is what you want people to do on that page. If you are uncertain about how to design it, take your regular Web pages and break them down into functions, giving each of the functions a new page.

What you have to remember when designing pages on mobile phones is that the position for the most important item is not necessarily where it would be on a larger screen, nor where it would be on a paper page. In a newspaper, we are (in countries that use the Latin alphabet) conditioned to start looking at the upper-left corner, then scan in a lazy zigzag pattern that looks like a shallow U, and then go on to the next page. In books and magazines, the layout convention is somewhat different, but these patterns are all learned. On a mobile phone screen, you have a very small space to play with, and the most important spot on that screen is the center. The next most important is the upper-right corner—in countries where we write from right to left. In Japan, the upper-left corner would probably be more important, and you would want to present the text (of important messages) vertically along the left border of the screen. This position has to do with the cultural conventions of writing, and while it is an utterly fascinating topic, there is no need to go into it now, other than to make you remember: the center is most important; the top-right the second most important.

Another thing you have to think about is how to connect things that are related. Make sure that things that are related logically are visually related, either by appearing at the same level in the hierarchy, in adjacent positions on the screen, or inside the same hierarchical elements. Users read this type of information every day without thinking about it. The user also does not think about whether the graphic conventions are clear enough and well established. Hierarchical elements have been used this way since Roman times, so this situation is nothing new. What is new is that the presentation is smaller than the palm of your hand.

Using User Interface Design Conventions

When you think about the conventions of the design, you have to think not only of the conventions in map interfaces, but also about user interface design conventions. If all your buttons are round when buttons everywhere else are square, it might look cute, but it will be hard to understand. If your links are underlined and purple and the followed links are blue and underlined, you will not only be flying in the face of convention, you will also seriously confuse the user. You have to follow the conventions in the environment you are executing, and because you are part of the Web framework (executing within the browser), that is what you have to take into account.

One problem you will have if you are not consistent with the established conventions is that users will find your application hard to learn. Intuitiveness is very much about being easy to learn, and the fewer new conventions a user has to learn, the easier it is to use. The reverse applies as well: the more new stuff the user has to learn, the more difficult it is to use. The tradeoff might be positive, such as when you learn a new word processing program because it enables you to become more productive by sharing files with your colleagues. But it might also be negative, such as when you do not want to learn a new word processing program because it does not make you more productive (any Word Perfect 4.2 user will remember fondly what I am talking about).

It is one thing to learn an entirely new application when the gains can be measured in money, such as in the case of a word processor. It is an entirely different thing when the gain is measured in pleasure and novelty. How do you measure fun? And that is the context where most location-dependent applications will be used, the market for location-based business services notwithstanding. It might start there, if the traditional models of new technology are anything to go by anymore, but it might also happen that the old models do not work any longer and that users are happy to jump into new technologies if the experience from iMode in Japan is anything to go by.

There are a number of conventions that have established themselves, on the Web and through other mass media during the 2,500 or so years that mass media can be said to have existed (and Gutenberg was not the creator of mass media, even if he was the creator of the printing press—there were woodcut prints in circulation in Europe at least 200 years before, and both church services and theatre performances can be seen as mass media, not to mention how long printing has been going on in China). These conventions have emerged not only from the Web, but also from other mobile services and from print design. If you do something that defies convention, it will be seen as counterintuitive and thus inconvenient—and harder to use. You should remember, though, that although some conventions can transcend different media, not all do. Headlines, for instance, are a convention that works well both in newspapers and on the Web. They do not work as well on TV (although news readers in German television actually do read the headlines first).

Conventions typically do not become conventions if they do not work. Very few media products are finished in the first try; they are tweaked and adapted by designers until they do work. Strangely enough, designers often seem to feel that they are smarter than millions of people who have been working for 2,500 years and either try to invent their own conventions or do not want to use the existing ones. In other words, there is an increased learning curve for the users—which means that the service will be harder to accept and harder to put in use. Involving user testing in this process is a simple way of ensuring that the products work well, by the way, and that the new conventions actually are better than the old ones (or not).One convention that is often violated on the Web is to make it perfectly clear what an action element is and what a decoration element is. By that, I mean, "What is clickable and what is not (and what result does it create)?" Underlined words in blue happened to be the way that Mosaic represented links, and that has become the convention for clickable links. Typically, they turn purple when clicked. Buttons should look like buttons, not like links—or—worse, like text. But put a label on them that says what they do (using an action verb).

Also, remember that pop-up menus do not work well on the screen of a mobile phone. If you move the cursor over a menu choice and 12 different choices pop up, odds are that at least three of them will fall off the screen (not to mention the menu obscuring everything else). It is better to rethink the information architecture and have sublevels for important sections. When you have more than three levels, make the user go to another page. Not only is the space on the mobile phone more limited than the space on the PC screen, but the attention span of the user is also most likely shorter.

One thing that helps a lot is if choices in fields are pre-filled. But they have to be pre-filled with text that is related to the user's situation and vocabulary. In a Japanese tourist service, it makes it more difficult for foreigners if the train table makes you choose between "upward trains" and "downward trains." You have to know that the direction toward Tokyo is normally referred to as "upward" and that away from Tokyo is "downward." Try to understand what labels the user puts on things before you invent your own.

Human cultures consist of shared context to a much larger extent than we appreciate. To understand one fact in a dialogue, you have to use seven additional, different, common-sense facts. In a dialogue with a person, you can get them through giving feedback. This process—raising your eyebrows, looking at your interlocutor with blank eyes, making a querying sound—is so deeply automatic that we cannot even think about it without making an effort. Needless to say, the computer does not get it when you look blank. In fact, it cannot even see you. Problems that require explanation occur when the assumptions about the context that the user and the software have are not the same. This situation can be true to different extents.

It is probable that the more you put in, the harder it is to learn. For this reason, the user only uses a limited set of functions (and only those that were most familiar, at that).

One good rule of thumb, however you design and whatever you design for, is never to trust the default values. Default settings are intended for the average user, but most users do not bother about the settings in a consumer market. How many people do you know who have set the clock on their VCRs? (If the answer is larger than zero, it is a sure sign you are an engineer.) They are useful to configure a device, but they might not be very helpful when you try to go from one set of representations to another. When you design for multiple devices, you will also have to care about a wide range of different device defaults, and there is no way for you to know that the ones that were set on your PC even were meaningful in the environment where the user will see your work without checking. Remember, it is not the display that makes the map; it is the cartographer.

It is very inconvenient to have to figure out what things are most important. Users want help, but they want help that is helpful to them. If the hierarchical elements do not correspond to their conception of what is important, they will feel that the service does not "talk their language," and the mental threshold of accepting and using it will be significant (maybe too large for them to bother).

Part of the hierarchy is expressing what is specific to the particular page you are showing now and what is common to the site. On the Web, navigation aids are slowly emerging, like logotypes and menu bars that help you get to the important parts of a site. On the screen of a mobile phone, you have much less space to design for this feature, so you might need to limit yourself to a single row of buttons or other navigation elements. The place for them is at the bottom of the screen, because that is the least-important place and because many mobile phones place their navigational elements there. The bottom is also the anchor in a hierarchy, and it enables the user to feel where the buttons belong.

There are some languages where the word for "more than two" is "many." Stone age languages though they may be, that is how we think. We get lost beyond three. Our short-term memory can only handle seven plus/minus two items, according to research findings. In other words, if the user gets so many things to think about, he or she gets lost (both the things and the user). "Three clicks and three seconds away" is the design slogan for the Reuters Web site, and especially in a location-dependent service where you are exposing a large database through a uniform interface, it makes sense not to have too many levels of hierarchy. The richness of the content should be evident from the searches that the user does, not from the number of documents you are showing.

Navigating Information Services

In the early days of computer games, the graphics were not very interesting. Developers had to concentrate on the puzzles, and that meant creating interesting instructions for the user. Navigating the virtual world of Adventure or Wolfenstein was very similar to how you would navigate the real world by using directions in a position-dependent system.

A problem in text-based adventures, as in location-based services, is that you have to describe the context very carefully. It is actually easier in the location-based services, because the user's position is known, and that way, a large part of the context that needs to be shared between the application provider and the user has already been established. Assuming that the position establishes the context tends to make us forget that you have to share the context with the user at all.

To understand—in other words, to interpret—the symbols that the user and the computer exchange, they need to share a context (for example, if the application is using Shift-JIS and the user is expecting ASCII).

It is easier to share context by using a graphical presentation, however. Reading text, or listening to spoken instructions, is linear. It does not allow for the user to concurrently take in a lot of data, which a graphical presentation does. This situation might not matter much in simple applications, but try describing the way from the Computer Museum in Boston to the American Repertory Theatre in Cambridge, and you have a problem that will take several pages of text at hand—especially if you want to give driving directions (well, until the Big Dig is finished, at least). Figure 8.5 is an example of how hard it is to share context in text, whereas Figure 8.6 shows that with a graphic presentation, it will be much easier to understand where you are (both examples from Webraska).

Another advantage of a graphical presentation is that it facilitates the visualization of the information (partly by setting and applying a shared context). In

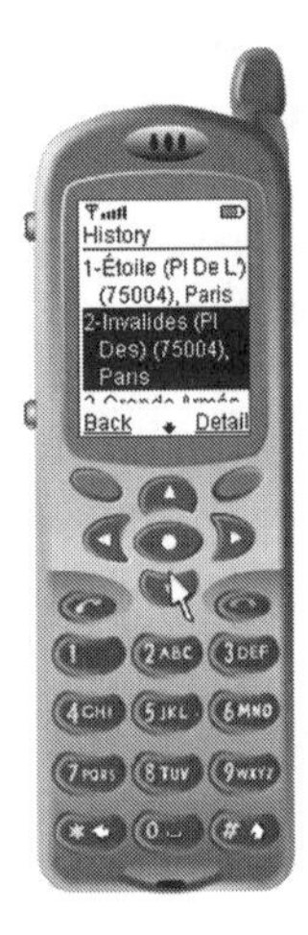

Figure 8.5 A text example on a mobile phone simulator.

other words, it is easier to see where you are on a map than in a long text. One reason is because the map relies heavily on an established set of conventions that are shared between users in most cultures today (maps look almost the same all over the world, but like any other visual representation of information, they are arbitrary—there is nothing that says that Muddy River should be blue on the map, for instance).

As a user interface designer, you should use those conventions and metaphors as much as you can. It will help you express more in your service, and it will

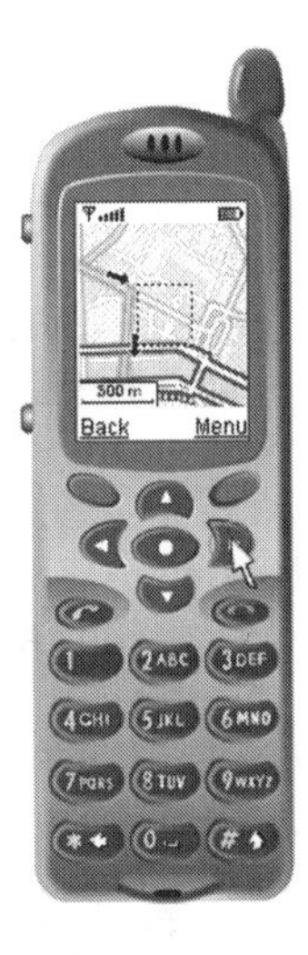

Figure 8.6 Same example as a map.

make it more familiar to the user—something very important when you are establishing a business—as you are when you are creating a location-based service today because there are none (well, very few).

The user interface is much more than the icons and menus on the screen. It is everything from the shape of the device where the user receives the service to the manuals that describe it. In the market today, few service providers have the luxury of controlling the end-user devices, however. The only mobile operator who actually does control all the aspects of the user experience is DoCoMo, which not only decides how mobile devices should work, but also how they should look, what the pages on them should look like, and how the manuals should be written. No other operator in the world has such an extensive control over the user interface and its design or any other aspect of the service.

While it is very unlikely that the service provider will have any control over the design of the end-user device (manufacturers of that have to make a business too, after all), they do have control over several other aspects of their services, which are often forgotten. One of them is help pages. Non-textual presentations are another.

There are two basic ways of navigating the Web today: searching and browsing. Some users will immediately enter things in a search box and never bother to click a link (even to the extent that they enter inappropriate things, such as the URI for the services they are going to in the search boxes of other services). Others will follow links forever, never making a search. For them, it is easier to have a hierarchy of links to follow so that they can create a mental model of how they are navigating the site. Which behavior you mainly use depends on temperament and a lot of other things. There is no need to go into it now, except for saying that there need to be two ways of addressing your service: through a search interface and through a list of links.

For the link-based navigation strategy, getting around a Web site is much more similar to navigating a store than it is to reading a newspaper, for instance. When we want to find things, we use our physical sense of location.

It also makes it very important to have a home page that exposes your entire service and where you can set a bookmark as a user and easily come back. Needless to say, you need to create a navigation bar or similar tool that helps users get back to the home page. It gives them a fixed starting point in their surfing. Knowing that there is somewhere you can turn makes you feel safer, and that makes it more likely that you will continue to use the service. But you should regard the home page as the navigation bar for the entire site.

There are four questions you need to be able to answer when you look at a home page of a site. If you cannot give immediate answers, you must start rethinking the design. Try it on your grandmother, and it is even better if she is not computer savvy. The five questions are as follows:

- What is this?
- What can I do here?
- What do they have here?
- Why should I use this service instead of some other?
- Where and how do I start using it?

If you can get the answers the users will give to those four questions to be the same as your intentions for the design of the site, then you have succeeded. But if users give different answers from what you wanted to hear, remember that the customer is always right and you are not.

The conventions for the Web are continuing to evolve, but on many (if not most) large sites, you will find a navigation bar at either the bottom or the top (some have the graphical version at the top and the text version at the bottom). In the extremely limited real estate on the telephone screen, it is important that you do not waste what little space you have on navigation elements. Nor should you have too many graphics, because the user is rarely willing to wait for the downloads. If you were using the Web around 1994, when 2,400 bits per second modems were state of the art and 9,600 bps modems were just coming out in stores, you will realize what graphics downloads feel like on a mobile phone.

The navigation bar should be present on the pages that the user can access within your site, and it should help them find their way and get away from where they are if they want to go somewhere else. The only exceptions are pages that automatically take the user away when used, such as forms and the home page. Using tabs is not a bad model because they also let you show where you are in relation to other pages in the rest of the site very simply.

When you do the user testing, take some nonsense text from a foreign language Web page and fill in where your text and labels would have gone. That way, you can really judge whether the interface says what it is supposed to say without the text (if you try and cannot design the page so that it is possible to navigate without understanding the text, consider doing away with the design and keeping the text. Life is full of choices like that).

Search interfaces need to be designed so it is clear how to use them. For instance, if you have a form that the user needs to fill in to get the coordinates of an address, there needs to be a button that activates it. How you place that button, and what its label says, is important. If users are thinking about searching, they will look for something that says "Search." It is useless to try to be fancy and call it something else. Search is what you do—you do not "enter keyword" or "find it" or something similar. If it is a search, call it a search.

Useful Functions on a Navigation Bar

- **Home page link**
- **Search form link**
- **Links to the main services**

There is no need to have more. But there are a number of questions you should be able to answer by looking at the navigation bar alone:

- **What site is this?**
- **What page (in which section) am I?**
- **Which are the major sections of this site?**
- **What can I do here?**
- **Where can I search for information?**

If it is not clear from the context (for example, the page is called "Shops on Motomachi by name"), then it should be made clear that the search is only within the site, not on the Internet. Also, remember to provide a link interface for people who would rather browse than search (for instance, a list of the shops; these are generated from the same database, as we will talk about in Chapter 9, "Maps as User Interfaces"). Remember that you will need to show the user where he or she is, for instance, by graying out the search page link in the navigation bar. That is a way of telling the user where he or she is and that he or she cannot go to a different service by clicking that link. There is no need to be too subtle, especially on pages where it is not quite as evident where you are as on the search page.

The names of the pages in the menu should be the same as the page titles. That makes it easy to find them and easy to identify when you have found them. That also goes for other links in the site that connect back to pages in the site: They need to be easily identifiable. Think of page titles like street signs. In Web pages, you might have to repeat them a couple of times, but in a mobile application, there is no room for that.

If it does not work, do not be afraid to change it. You will be able to follow up on how users access the service through your log files, but that information is useless if you do not do anything with it. You can do it in either of two ways: continuous redesign or rerelease.

Continuous redesign means that you continuously change small pieces of the layout and structure to other things that you have tested and shown to work.

Re-release means that you gather all the changes you need to make into one big batch, and then you do them all at once and throw a release party for the media and advertising agencies. Quite frequently (at least, during the dot.com boom), the main reason for the re-release seemed to be the party.

Which philosophy to choose depends on how you choose to present yourself and how you think your changes will be most easily explained to users. The re-release model is not bad if you can combine it with an advertising campaign and some kind of direct communication with registered users. But it is wasted if the only people you are communicating with are media people and people from advertising agencies.

The center of the screen is most important; the top-right is the second most important.

Personalization

This book describes a way of working that will change the way traditional system design works. But there are two important aspects that you will need to consider in your design, which will affect the way it works. The first is that maps can be created dynamically by using personalization information obtained from the user at the same time as the position (or before). The second is that in a computer display, there are new options compared to print—most important, probably, animations. They have to be used carefully if at all, but they have the potential to make your maps tremendously more useful than they would have been if you just presented a static raster image of the printed map.

Personalization is about creating services that customize the end-user's experience for the individual subscriber. Personalized content is considered a key success factor in mobile device usability due to the limitations of the user's display. Relevant information must always be only a click away, because any exist-

Useful Features in the Mobile Web

- **Menu bar at the bottom of the screen**
- **Menus are maximum three levels deep**
- **Change context after three levels**
- **Minimize the number of objects to minimize the waiting time**

ing wireless device is not as flexible as a PC screen. Personalization is the difference between a usable and an unusable application.

In the m-Commerce and mobile location services arena, personalization is critical because research studies (such as Durlacher's Mobile Commerce Report, 1999) have shown that every additional click required from the user reduces the transaction probability by 50 percent.

Remember that you are creating an experience for the user, and part of that experience is solving a problem. It might well happen that the user does not know that he or she has a problem until he or she has used your service and found that it added value to his or her life, like most of us did not know we had a need to follow hypertext links until the Web came along. It helps your design if you identify that problem early on in the process, for instance, by specifying use cases that your service is addressing.

However feature-rich your application is, it is likely that most users are similar to users of most other applications and are happy with just a few of the features you provide. One thing that has happened as applications are being developed for the Web is that because they are executing within a framework, they have to have fewer features. In a way, each service can be specialized to do the things it is supposed to do, and the features are accessed through the framework application, which is the Web browser.

In other words, you can optimize the particular functions that are going to be used in your service. They are the ones that you hope users will use frequently, and the easier they are to use and access, the less the barrier from the user for their use (and the more likely that your company will be profitable).

When push comes to shove, the important thing is that you are designing pages to create a user experience—and an experience you expect the user to pay for, at that.

If we do not understand what is meant, we try something and see whether it works. We muddle through. Users will do the same if they do not understand what you mean in the service. But muddling through in a service where you have external pressures like screaming kids, an angry wife, and a bunch of foreigners talking very loudly and gesticulating in an incomprehensible language as the light is both green and some kind of purple at the same time, all while people drive on the wrong side of the road and you can feel that Montezuma wants to take revenge on your stomach, can be quite stressful. You do not have time to try a lot of stuff. You just want to go away. And that is when your service should work.

As an aside, there is a scope for great synergies between the personalization and user interface here. Because all pages are generated from a database anyway, you could make the service express itself in the terminology that is most familiar to the user.

Users on the Web usually try to find something. Even if they are just browsing around, it is rarely totally aimless. This situation becomes even more acute if you are using a location-dependent application (a process known as wayfinding, but my book *Designing Wireless Information Services* by John Wiley & Sons, 2000, is about that).

Identifying the Stakeholders

The users of your service matter a lot when you are considering your design and how it is supposed to look. But the end users are not the only ones who are interested in the service (and might actually not be the only user group you need to take into account). Other groups also have a stake in the system that you create. Most notable is perhaps the systems administrators, who will have to maintain the system that you have designed. They might not be the only data maintainers, however. In many cases, data will come from other organizations than the one that is providing the service, and somewhere someone is feeding the data into the system. Either you have to think about interfaces to existing systems or you have to think about a user interface that enables them to do their job in a simple way (or possibly both).

Then, there are the owners of the information. They might have sold it to you (or given it away), but they can still have a requirement on how it is used. For one, they have a moral right to how it should be used.

This situation makes them, along with end users and system administrators, one of the groups that has a stake in the system—a stakeholder. There can be a quite large number of these, but you need to identify them to be able to design the system appropriately.

Other possible stakeholder groups are the owners of the service, the advertisers (if any), the provider of the location information, and possibly the provider of the communications services (who might be the same in the case of network-based positioning services). Any other sponsors, hopefully represented in the project steering group, are also among the stakeholders.

All stakeholders have something they want to get out of the service. Ask them to tell you about it, and summarize it in a way that can be analyzed in terms of what the system should do; then divide them up so that each contains a single task. These are requirements. Fulfilling requirements play a very big role in formal development processes and are often used to measure the performance of a project. How many requirements did you fulfill? In complex systems, the requirements specification can become a specification in itself, and starting with the requirements is actually not a bad way of developing systems.

Documenting the requirements helps you understand what the system should do. Documenting use cases helps you understand how it should do it. A use case is a description the system should work. Normally, you document the typical ways the system should work from the point of view of the user; rare cases (edge cases, at the edge of the Bell curve of a statistical normal distribution) can be documented if they limit the functionality of the system (for instance, if there is something that might happen in certain circumstances that should be avoided). Very often, the edge cases hang up an entire development process. Documenting them is worthwhile not just to know whether they should be avoided, but also how they can be avoided in the development process.

Use cases do not have to be formal descriptions. If you use them to evaluate the performance of the system, it actually helps if they are not. They can be scenarios, little stories that describe what would happen in different situations. It is from this use case document that you create a formal specification. We will discuss this subject more in Chapter 10, "Pulling It All Together: LBS-Enabling Your Web Site and Developing New Applications," but the use case documents are something you should discuss with the steering group of your project and with other stakeholders. The formal specification is not something you should show them; it is something to be used inside your project and that you should use as part of the project documentation. You can also submit it for review to a systems expert together with the requirements and use case document, which is good practice because any problems are usually spotted at the specification stage (and not when the development has finished).

Designing a user interface for a tourist application for the small town of Grönköping is very different from designing a user interface for a large application to visualize the sewer system of a major city. It is not only a difference in how you express the functions, because you will be dealing with very different audiences with very different contexts and expectations, but it is also difficult because they represent two ends of a scale. On the tourist application end, you have casual users and you have to decide what to show (essentially design a visual presentation). On the other end, you have users who are exploring a complex database and its mapping onto reality. They want to know what they can know, and they probably represent a group that uses highly developed visual thinking in conceptualizing their questions (as most sanitation technicians do).

In the presentation end of the spectrum, the user executes well-defined tasks that are routine. The interface needs to be highly structured to be accessible, and the graphics can be presentation-oriented and rather complex as long as they enable simple operations (it's not that the graphics should be complicated; rather, that they can contain a lot). In the exploratory end of the spectrum, the graphics need to support exploration and assist the user in tasks that are not routine and that can be ad hoc. The interface needs to be flexible, and the graphics must support analysis. The engineer studying a presentation does so to find patterns, identify problems, and establish relationships. The presentation, on the other hand, needs to be simple and needs to support a limited set of tasks that the user can execute.

Usability Testing

It is not hard to test whether your site is usable. Just get a few people together, show them, and ask. Well, there are a few more refinements to the method than that. But you do not need to hire a psychologist, although it might be useful to have an HCI professional as a facilitator and report maker.

Human factors specialists typically try to test applications in real situations. In other words, gathering performance data, recording the users' thinking about the tasks (by asking them to think out loud), and by studying the errors that they make.

In general, tasks can be differentiated between strongly structured and unstructured. The level of abstraction can determine the structure.

Different users use different concepts. Geographic features, especially when represented in a computer, are objects with location, magnitude, identity, and time (plus a host of other parameters). If you have two features, you have a distance between them, and with that, you have proximity. Features can be similar or different, and if you have more, you have dimensions like density, pattern, and dispersion (clustered in an area, uniform over an area, and so on). With a large set of occurrences, you will have concepts like regionalization, dominant and subordinate relationships, and paths and routes.

How the tasks are performed depends on how we acquire the spatial information. Wayfinding seems to work better if we are given sequential instructions, such as directions. Getting an overview of the occurrence of something in an area is easiest to do with an overlay on a map. When people are navigating, they like to know where the decision points are and what they can do (and have to do) at these points.

But the individual preferences are strongly influencing this situation and might change it—as might education. Map reading is an abstract skill, and doing it well requires practice and training. At some level, all people probably share some level of conceptualization and visualization of space. But different groups might have different concepts of space. Some modalities are better for some users (regardless of, for instance, disabilities). How can users be distinguished by the mental models they use, rather than by cruder measures such as their abilities or the type of device they are using?

This reason is precisely why location information and personalization are so tightly tied. Computer learning can work to distinguish between different categories and groups of users and help them get the precise presentation that is best for them. The tasks that they perform can be modeled by using cognitive task analysis in a hierarchy of increasingly complex tasks. Grouping people who perform similar tasks in related groups is a first step toward creating a machine-learning system based on personal preferences.

Basically, the test has two functions: it exposes users to your site without prejudice and records their reactions, and it exposes your staff to the reactions of users.

The last is not the least important. You need to expose your staff to the reactions of users to make sure that they understand how to redesign functions that will be hard to use. As a group that has been working hard on a Web site for a long time, you will know every nook and cranny of the site, and you will not have any idea about how it will feel to people who see it for the first time. You need to get that opinion, because you will pretty soon expose it to people who have never seen it before, and if it does not work for them, it will not work for the people whom you want to be your customers, either.

The equipment you need for the tests is very simple: You need the platform you will run the service on, or a reasonable simulation of it, and you need a video camera with a recorder. Some people recommend that you tap a live feed into a room where the developers can watch the test in real time, and while that might be fun (and they will find it frustrating), the social function of it is limited. It is better to edit a report together from the video recordings and give a half-day seminar about the user interface as part of your talk. That will usually serve as a better eye-opener. But there is nothing that stops you from doing both; after all, knowing that there is something going on will make the developers feel more involved.

That said, you do not need to test hundreds of users to get an idea of where the problems are. On the contrary, a very small number of users can expose the problems. It has been statistically proven that with only five users, you can expose 85 percent of the problems. But there is a twist: If you first test with a small group of users, correct the problems, and then test again, the users will find another batch of problems. This way of approaching it makes it possible to highlight the areas where repeat users will get problems, and because you want users to come back, that is important. If you can afford it, repeat the tests with a fresh batch of users on the fixed site, as well.

The test users do not have to be scientifically selected. While focus groups enable you to get a reliable opinion about what your prospective customers think, behavior in general rarely varies that much between groups of individuals. Of course, if you test your site in China for release in the United States, you will get some problems (as you will if you do it the other way around). The users need to have a basic idea about how to use the technology. But for a site that is not geared toward a specific group of users who use a very specific vocabulary, it is sufficient to test John Doe from the department of sanitation technology. As a matter of fact, the broader the focus of your site, the less the participants matter (unless you introduce some other bias in your selection, such as testing it on white male computer engineers with considerable programming experience aged between 30 and 45). Instead, take anyone who is reasonably inexperienced in the site. It works.

It is possible to recruit the users on an ad-hoc basis by sending e-mails on lists, putting up notes in the building or campus where your offices are, or putting out a flyer on a nearby campus (students are an excellent test audience; they typically want to make a buck, have flexible hours, and are knowledgeable in technology). You can offer them incentives, such as $50 per test (or the chance to win a trip to Hawaii, if you can afford it). If your users are specialists whose time is valuable, you will have to be prepared to pay for it.

You do not have to be an expert in usability to facilitate a test. All you need is a good protocol to follow (those are listed in the books about usability that are in the reference section), reasonable social skills, and to not be too involved in the product. Seeing users who are not involved in the product try to use it can be very revealing. But it is important that the facilitator does not try to help them do things the way it "is supposed to be done," but rather let them walk through the site themselves. As a facilitator, your sympathy should be with the participants. If you are a facilitator, run through the test first.

Creating a report that helps develop the product is the most important task of the facilitator. Using the video recordings is one way; creating a written report that highlights the way the user perceives the product is another.

But it is not certain that the user test comes at a point where the product can really be fixed. If the data modeling and database programming has to be redone, and the templates have to be changed, it can be quite costly to redevelop everything. It is better to do a conceptual check at an early stage. That can be done by using a simulation—PowerPoint or even paper cards.

There are a few things that you will typically find out when you do user tests. One is that the users do not understand what you meant with it. They look at it, and they do something wrong. Or, they do not understand what they are supposed to do at all.

Another possibility is that they are looking for something that is not there. What you do when you are creating a site is that you set users' expectations by the way you present the information. That means making sure that the expectations of users are fulfilled by the information you present. You might have other expectations than the users, but they are the ones that count, not you.

It is hard to measure usability, like any qualitative dimension. The easiest thing to measure is, of course, failures. If it does not work, the result is 100 percent negative, and that is it. You can measure usability along a few other dimensions, however.

Effectiveness:

- How many tasks, goals, or objectives were achieved during a predetermined time period?
- How completely were these tasks achieved?

If the test is for a tourist application, the time period should be fairly short because the tasks are likely to be simple. Setting a goal—for instance, "Find the Westerlundska Konditoriet and get the route there from here," is appropriate because you can measure how far along that task the user has come. This process might require a couple of trials because it is hard to say beforehand how long it takes for an average user to manage a task. It is also a better idea to let the user concentrate on the task during the test and not have him or her speak his or her thoughts, because this situation is likely to be distracting. Note that the tasks must be specifiable in terms of goals.

Efficiency:

- How quickly was the task achieved?
- How many errors did the user make?
- How much of the time was spent on relevant tasks, and how much time did the user spend sidetracked?

Acceptability:

- This measurement is essentially subjective, where the user can judge the progress on a scale. The question is how the user perceives the system.

Learnability:

- How fast did the user learn to do a task?
- How much training was required?

You can measure the speed of learning by grouping a series of goals that has common tasks and seeing whether the time to reach the goal is getting shorter—as it should if the user is learning to perform the tasks. The optimum amount of training is none, but the effectiveness of this goal can be measured by providing a group of users with different levels of instruction and then letting them perform the learnability test.

Creating Help Pages

Help pages for services tend to be scant at best. It seems that service providers assume that the only help the user needs is in handling the application (and providing help for that is the application provider's problem). If they cannot navigate the service without assistance, tough.

If you start providing a service with richer interaction, however—for instance, animations—you had better provide help pages as well. The very least you will need to do is to tell the user how to switch off the rich presentations. This situation will probably, if the experiences from the Web are anything to draw on, not lead to a rapid increase in people who switch off the features of the service. But it will lead to a higher level of satisfaction for those users who intended to do it but could not find out how.

Help pages are contingency measures. The important thing to remember is that the user does not look for them until he or she has a problem. Those problems might not be something that the designer ever believed would cause a problem (see the section about testing). But the important thing is that the user perceives that he or she has a problem, and if it is not solved, he or she will not have a positive association with the site or the service.

For that very reason, the person creating the help pages should not be the same person who is responsible for the design of the service. Help is about explaining how something works, and putting a trained technical writer on the job will not only give you excellent feedback on what you have done, but it will also help structure the help pages in a way that is, well, helpful.

Good help pages give overviews of the most frequent functions, not only how to use them step by step, but also what can go wrong and how to solve the problem. Creating all the error conditions that can be involved in a large software program is next to impossible, of course. That is why there are bugs in Word. In a small project, it is easier, but it still requires careful thinking about ways that the user can imagine for using the service that the designer has not thought about.

When you design a service, you make assumptions about what will be the most frequently used services. Those should receive extra attention in the help pages if for no other reason that the more people there are using something, the larger the purely statistical probability that some of them will have a problem.

If you can control the error messages in your service, you might want to connect them to the help pages by using a hyperlink—assuming that the problem is not with the browser. It might actually make more sense to have the help pages in a separate program so they can be started when there is a problem with the application that your service is executing in (for instance, the Web browser). Make sure it is clear how to access it in that case. Put it on the system menu or whatever similar function there is.

It is also impossible to make sure that the user understands what happens. Make sure that the error messages are informative. "General Protection Error 101" is not very informative to a general user who wants to know what he or she did wrong (neither is "404 Page Not Found"). Use the frame of the browser to provide information about what is going on. A rotating slash and the word "searching . . ." will decrease the impatience of the user to no end. Make sure the messages are not debugging messages for programmers but are adapted to the expected user group's terminology, though.

If it is possible to have screen dumps in your help pages, it facilitates the use. But you should always provide a text description, as well. In devices where the memory is limited, this option might be the only one that exists. The mere act of telling the users how something works is likely to help you find problems, by the way.

When you explain functions in the help pages, make sure to use the same words that are used on the labels in the application. It does not help if you say "press the OFF button" when the button itself says "shut down." Also, make sure that you do not talk to your developers but to the prospective users of the service. It is very easy to fall into a jargon that is not reflected in the community of users, especially if they are very different—like GIS developers and users of a tourist application.

Handling Advertisements

It is very tempting not only to charge the user for the access to your service, but also for someone else to access the user. This procedure is called advertising, and it is such an ingrained part of our society that we hardly realize that for most of their history, newspapers did not have advertisements (they started to appear in the form we know them around the 1840s, but the first newspapers as we know them were published in the 17th century). On the Internet, the business model of advertising-financed services has become so pervasive that people hardly realize that someone is paying for the service, and it is not the user.

The way the advertising industry is set up provides a number of unfortunate consequences for user interface design that are surprisingly little discussed but that have a vast impact on position-dependent services. The first is that because advertisers do not care about the quality of the impressions (the number of hits), the advertising carrier—the service—tries to maximize the number of exposures and users with any means possible (well, not quite, but almost any means possible). In other words, designing a maximum number of advertisements into the pages, designing the pages in such a way that advertisements are impossible to avoid, and designing the site in such a way that the user is exposed to a maximum number of advertisements as he or she is using the service is essential. Because there is solid evidence that users unconsciously avoid looking at advertisements and do not like them, this situation will continue to be a losing battle for the services and advertisers.

There have been a number of trials with position-dependent advertising, and it continues to crop up in business model discussions about location-dependent services. After all, if you knew the user was heading your way, you might tempt him or her into your store instead of the store of your competitor. If he or she knew what you were offering is the thinking. You can see an example of an advertisement on an iMode phone (the first, actually, since iMode had been free from advertising up until September 2001) in Figure 8.7.

If the assumption is that the user has a mobile phone, then there is no space to put advertisements along the edges or around the fringes. Advertisements will hog the center, and the user will be more irritated than not and stop using the service—in which case both the service and the advertiser has lost. Another problem is that the small screen requires the designer to think about the message and how it should be exposed to the user. The smaller screen and the comparatively less graphic presentation is also something they are not accustomed to (and something they regularly complain about).

The only viable solution is to alternate the advertisements with the information, like a slide show. If this feature were optimized, it would probably be easier to do through a specialized PDA that received a broadcast of maps and advertisements and threw away what was not required at the position of the user. As it is, because the position information is part of the personalization of

Figure 8.7 A test advertisement in iMode.

the service, the adaptation goes beyond the position (adaptation to different devices is already a reality in the WAP world) and probably includes other personal information items, as well.

If the device is running Java, it becomes possible to create a presentation that is a little more interesting. With Java, you can animate the presentation. While Java has turned out to be more efficient for server programming than screen animation, there are several comparative (and competing) solutions, like Flash from Macromedia and SMIL Animations for the SVG data format from the W3C. SVG is interesting because it is possible to create transformation sheets that accept data from the database and turn it into an XML presentation, which can be more highly personalized. This task is, of course, possible by using Java as well (but much harder).

Animations can take several forms. You can roll the advertisement over the map after a period of inactivity of the user (which, if there is a position information source such as GPS in the terminal, can be determined in terms of the user not moving—possible to do by using network-based positioning as well, but more expensive). You can create an icon that animates and appears on the screen of the user when an object is approaching (the logotypes of many fast-food and coffee chains are likely to appear in this way). You can create pop-up bubbles and balloons that contain information (being careful not to position them over the icon that marks the user's current position).

A simple solution that many service providers seem to shy away from is to have two subscription levels: one for users who want to pay for the service and escape advertisements and one for users who want a cheaper subscription and are willing to accept advertisements. The argument that the users who deselect the advertising are the ones with money who the advertisers really wanted to reach might have some truth in it, but on the other hand, the way human nature works, greed is more likely to make even those users want to save and therefore select to get advertising.

If you decide to accept advertising into your site, not only do you have to think about how you would expose it to the user (as we have been discussing here), but you also have to think about how the data is modeled in the database. An advertisement is likely to cover an area, and advertisements for chains are likely to have two sections: one that is general and covers everything and another that is specialized and covers only the specific area where the store is located. Modeling the coverage of an advertisement requires some careful thinking that has to go hand in hand with its design. The question you must answer is what do you want to achieve? Do you want to draw users that are not in the general vicinity to your store, or do you want to advertise in the general vicinity of your competitors' stores? For instance, if you are an e-tailer of something and your competitor has lots of bricks-and-mortar stores, do you want to place advertisements that pop up when the user comes close to the store and tells him about your superior offer?

This feature might not be motivated by pure competitive thrust, actually, but by the general concept being evoked when the user passes by a landmark. When a user is passing by a physical object that is prominent, they are likely to be reminded of the general concept. This statement is true not only in the dimension of position, but in a time dimension, as well. Landing at Chicago's O'Hare airport in winter, you are more likely to be attracted to a travel advisory from Florida than if you are sitting on a beach in California.

How you create the metadata that lets the service select the correct advertisement is part of the data modeling. Each ad has to be designed in a way that is most efficient for itself. But there are a few guidelines that will assist designers in creating ads that are more efficient.

One of the most important rules of thumb you have to remember when designing for personal devices like mobile telephones and PDAs is that the screen belongs to the user. Any intrusion by anyone, be they a service provider or the device itself, will be seen as an irritation (if anyone likes the "A general error has occurred. OK?" sign that Windows occasionally presents, they are welcome to raise their hands). And anything that is presented without permission will be seen as an irritant (and any associations connected to it will be negative). This news is, of course, bad for advertisers. If you pop things up on the screen of the user, they might actually become *less* likely to do what you wanted them to do in the ad.

Instead, you have to create buy-in. You have to get the user's permission, either explicitly or implicitly. If the user knows that his or her mobile phone will display ads but that he or she pays a dollar less per month for the service, he or she is likely to think about the ads in a different way than if they all of a sudden started to pop up on the screen.

I am not the only person who believes this point of view, either. Very few tests have actually been done on the effectiveness of advertising in the new medium, but from September 2000 to January 2001, the American company SkyGo conducted a consumer market study in Boulder, Colorado. The company studied permission-based alert-type advertising using WAP phones (Ericsson R280 LX cellular phone) in the AT&T Pocketnet network with more than 1,000 users and 50 advertisers during four months. Of course, the company did not test push advertising but permission-based, opt-in advertising.

The company concluded that consumers perceive relevant, compelling, and convenient information as content—*not* advertisements (a finding that is substantiated by other research). The permission-based models will be accepted by consumers and will be effective for advertisers, because as much as 64 percent of all ad alerts were opened (the consumer clicked the advertisement title to view the advertisement), and 3 percent of participants made a purchase as a result of viewing an advertisement. 19 times more "delivered ads" were viewed than "searched advertisements"; 31 percent of consumers used "search all advertisements," and 7 percent of consumers said they would use wireless coupons at the point of purchase.

One way of creating buy-in is to give the user something back. Instead of just preempting the user's personal space, you provide a few pages of useful information. This situation is especially true for mobile services where there is likely to be information that might not be accessible from the current service but that the user could serendipitously use. For instance, while "Public Toilets by Bank of Kanagawa" might not create the associations you would want to create, the idea is right—and all it takes is a link at the bottom of the screen. The trick is to find something useful and interesting at the same time.

Infomercials on TV sell things that are not particularly useful, using methods that used to be reserved for Tupperware parties. But what you want to create here are infomercials, which associate your services with something on which the user has a positive view. A local bank could, for instance, create a series of historical notes about the area where the user is, associating itself with the history and creating a positive association with the location (something that would not work in most of America, of course). Another way is to find something that the advertiser can become unique in providing so that the user associates it with the advertiser whenever he or she sees it or thinks about it. Advertisements are as much, or maybe more, intended to create a positive association with the brand as they are to sell products, and this feature is something you can leverage in creating position-dependent services.

In a multimedia terminal, it is also very easy to play sound as part of the advertisement. It is actually relatively easy to play sound on a regular Web page as well, but it is a much-underused feature for no other reason than it is irritating. It is not only the screen of the device that belongs to the user; it is the mental space surrounding it—which includes the sounds. Anyone who lives in a country rich in mobile communications will know that telephones will ring at the most inappropriate moments. Imagine the effect of a packed bus where everyone's mobile phone almost simultaneously started playing a jingle. It would probably lead to an increase in mental health care bills as well as a large number of devices thrown out on the street, not to speak of lawsuits for noise pollution. The more crowded our environment is, the less we want to intrude in the private spaces of others (which, by the way, shrink when we are crowded together). Sound can be fun, but only if it is activated at the user's choice.

Another question is the multimodality of metaphors. While there has been some work done on "earcons," or sound bites describing a concept (like an

Rules of Thumb for Efficient Mobile Advertising

- **The screen belongs to the user.**
- **Before you do anything, make sure you have the users permission (either implicitly or explicitly).**
- **There are many ways of advertising on a small screen, and they do not have to be traditional advertisements.**
- **Using animations and overlay planes can make advertising information more appealing.**
- **Consider infomercials as a goodwill measure.**
- **Use sound, but wisely.**

icon does visually), there is lots of room for trials by companies that want to distinguish their user interfaces. Adding sound is very easy in a Web-based interface (you use the object element), but there is nothing that says anything about how sounds should be selected. There might be two paths, actually: finding generic representations of a concept, such as the din from a restaurant; or a specific representation of a brand or business, such as the signature jingle from a restaurant chain. How to perform this task in a way that is unobtrusive is another matter. If the user is irritated by the user interface, then he or she will not continue to use the product.

When he or she does, it means that the user is looking at a map. And that is what we will look at in the next chapter.

CHAPTER 9

Maps as User Interfaces

Traditionally, a map has to be read just like a text in a newspaper or a report to management. Maps are supposed to be objective, emotionally void representations of space. But maps can be filled with emotion and can as easily be used to persuade as they are used to objectify. If the red line on a map that points you to a goal gets thicker as you move toward it, or if the goal itself is represented in high color on a background that has been grayed out, it will be perceived very differently than if it is just the point where the path ends.

Presenting Maps

The first thing to think about is what happens when the map is delivered to the user. It has to be presented in appropriate scale. What the user believes to be appropriate is, of course, impossible to predict. However, the scale can be adjusted according to a few parameters, which can be included in the service request from the user.

If the user is not satisfied with the initial presentation, he can change it (and the zoom in or out can be stored in a preference profile). The user preferences are one factor in determining the scale. In the mobile environment, you also have to weigh in the capabilities of the terminal and the network, such as actual bandwidth (in Japan 3G, theoretically capable of 2Mbps, is constrained to 64Kbps in the downlink). Then, there are the display capabilities, in what speed the user is travelling, in what kind of environment the user is located, and other parameters that have to be weighed in.

Typically, map databases contain a large number of geographic objects. Not all of these are interesting to the users. Even normal paper maps deselect some of the information available, and this has to be done to an even higher degree in maps that are to be used for mobile applications.

A location-dependent map service should include the following geographic information:

- A background map containing just enough information to enable the user to orient himself
- The position of the user (the mobile station)
- Application-specific information, which may be restricted according to the user's choice, such as restaurants, train stations, or other mobile subscribers (friends and so on).

It may be reasonable to expect that the *service provider* will show the location of the nearest branch of the company or its sponsors in the map. For example, a map service provided by a petroleum company will most likely show the location of its gasoline stations, regardless of what information the user requests.

It is also likely that specific graphical input devices are used in some applications to enable the service provider to have better control of the interaction with the user. Such graphical input devices can be checkboxes for determining what information is to be shown, what buttons need to execute certain scripts, and so on.

A user may want to do a number of possible operations on a map. Zooming, already mentioned, is changing the scale while maintaining the center of the map. This means that the level of detail in the map (or the level of detail displayed in the map) must be changed if the display is not to be cluttered. Panning is a different type of operation, in which the user can scroll sideways and up and down into different areas outside the displayed map. It is also possible to do this automatically, provided that the map stored in the mobile station is large enough and the position information is continously available.

Panning can be done continously, in which the next area of the map moves into view without interruptions, or discretely, in which the map is updated frame by frame. This is the only practical solution for raster maps, but for vector maps, the continous pan is very easy, and this is something that users are likely to demand. Possibly, they can be combined, continously sending frames that are cached in the device that lie outside the borders of the current frame.

The map frame can itself be larger than the current display, especially if it is a vector graphic map. The user can also rotate the map, perhaps in accordance with the orientation of his mobile station. If he has a vector map, the user can search for text elements in the document that has been transmitted. This would enable the user to pinpoint objects with much higher accuracy than if he has to search visually on a raster map. The alternative is for the user to send the data to the service provider to do the search and to get a new map in return, something that is likely to take some time due to network latency, and, therefore, be a source of irritation for the user—which will detract from his propensity to use the service.

Vector maps have traditionally been produced in different planes, and the planes contain information of different kinds (for example, natural geographic features in one plane, man-made geographic features in another, cultural features in the third). In XML terms, you can think of them as top-level elements encapsulating different aspects of the contents and presentation. Although this is useful in selecting features that are displayed in paper maps, the amount of information is too large to transmit over the mobile network and still maintain the interest of the user. Instead, a plane with links to salient features—using Xlink and Xpointer, or directly by giving the URI—will allow the user to click on a feature and receive the information about that point. This has to be predicated on the preferences of the user and the current query; a user who asks about hotels is not very likely to be interested in information about the city hall. If there is a lot of information it can be linked through a menu system, although given the limited display of the mobile phone, this is probably already too much information.

Maps also contain a great deal of metadata that is usually conveyed at the edges of a map, such as who created it; how it was compiled, from which sources, and when; and which projection and datum has been used. As already mentioned, planes of information containing information that is salient for the user may not be displayable. In a small display it is most likely not possible to display this information. Even a copyright marker might steal space from something that would be useful to display, so consider doing without it. It might not matter, because it is rather unlikely that the user will take the map and download it to somewhere other than her primary display (which is often a piece of printed paper, as things are now). It is also not formally necessary, because copyright does not have to be registered to be valid, even in the United States. Put metadata in meta elements instead, where they also become usable for a computer.

Many maps are used to visualize statistical relationships. These relations are invisible, but a visualization can make them usable in real life—as the famous example of the first use of statistics mapped to a map to find a well that was spreading cholera illustrates. This process is not easy because you have a tradeoff between the statistical representation of data and the geographical dimension. In generalizing and simplifying data distribution to get it to work on a map, you might as well distort the data in a way that removes or obscures the very relationships you wanted to display.

For example, if you have a choice about the number of the categories you want to use to display the data, too few categories might obscure the contours of the data distribution. Too many categories will make the data drown in categorization and are unlikely to reveal any existing spatial patterns in the dataset. Remember that the human short-term memory works on the "seven plus minus two" basis, and you should never use more than seven categories, because they will be difficult to distinguish for the reader. Fewer than three categories will probably mean that there are alternative ways of displaying the information that are more efficient than trying to project it onto a map (such as a table).

Which ranging method you use will also affect the way your data is perceived. If you use discrete steps you will get a very different effect than if you use a continuous scale, for instance. The larger the steps, the bigger the difference. Which method you use depends on what in the data you want to highlight.

Which diagram type you use will also result in a very different understanding of the user of the data. A histogram will give a totally different picture of the distribution than a pie chart. Representing data as absolute or in relation to something (the total data set, a selection of it, or a different data set) will give very different results. For instance, air population in a city will seem negligible if it is displayed in proportion to the country as a whole but might appear very large if it is related to the air quality in the nearest national park. In other words, it also might make more sense to have two different diagrams displaying the data.When you see a map, you automatically think about it in terms of distances. This is related to the problem of the adjacent map frames just discussed and is used in buffer analysis to study which geographic objects are located completely or partly within a certain distance from another geographic object, for example, a point, a polyline, or a polygon. An example is a situation in which a user wants to find a certain service within a defined distance from his location or from a certain object, for example, restaurants within a 2 km distance from the conference center or from a certain highway. Area computations can also be a desirable function in some situations. Many professional users may find a lot of different applications for area computations—for example, real estate agents, surveyors, farmers, and different official authorities.

Buffer analysis currently has to take place in the server because the computation is most likely too heavy for the mobile station. It is also possible to use the functions of the database to do buffer analysis and other distance-related computations, as we saw in Chapter 6, "Providing Databases and Doing Searches."

Animations are a special case, because although the map uses two dimensions to represent three, animations have the potential to add a fourth—time—which often is represented in compressed form (waiting for the growth of the Himalayas to display in real time would take millions of generations, and your PDA would be obsolete before anything had happened on the screen). This situation is true even in the case of wayfinding, in which you can play a scenario that displays how you should go from Place de l'Étoile to Montmartre on the screen, for instance. Of special consideration here is that such displays might not necessarily be limited to maps. When riding the subway, you are essentially in a different space from the world above, so it might be more interesting to display a map from where you are to the subway entrance, a video sequence on how to enter the subway (including, of course, instructions for how to buy a ticket from a French cashier), which direction your train should go, which line it is, the name of the station where you should get off and how it looks (including a video sequence shot from the direction you are coming), and a video sequence on how to get to the right exit from the station, plus a map of the walk from Place Pigalle to Montmartre (or taking the funiculaire).

Developing this system is part of developing a position-dependent application, as I have talked about in previous chapters. The user interface designer's job is to knit all of this information together in a meaningful presentation. It involves, as I am sure you gathered from the previous example, a lot of thinking differently. But it will be tremendous fun as well, and it will be a great business for any Web design company that understands how to do it right.

Maps and Objectivity

While a map should represent the geographic reality, all maps are selections from that reality. Maps can show only roads and not buildings. They can abstract away unidirectional roads; they can show only roads that are accessible to 18-wheel trucks. They do not have to be objective, which is often perceived as equivalent with boring. A map can be fun and useful at the same time.

Tourist maps prove this point. They often show the buildings along the main streets in a layout that enables the user to see the facades of the buildings flattened along the road. Needless to say, this representation of the construction of those buildings is not very accurate, but it enables the user to see how the street looks from his or her perspective. In a way, the map has become subjective.

Maps distort space in order to display it. They constitute a model of the world that sacrifices real dimensions to fit them all into one display format. This model is so ingrained in our way of thinking that we no longer reflect on the remarkable idea of representing altitude as curves. Depending on your background, it is likely that you will perceive space in different ways. To the naive user, space is equivalent to the map; to the GIS professional, the space represented by the map is really a swath of fields that interact to form loci, where the models meet reality. Specialized interfaces will not be meaningful to the general public, who will require a representation that follows custom as closely as possible (unless it is intended to be alienating).

The same philosophy can be applied to maps in location-based services. Instead of showing a supposedly objective view, why not show the world from the subjective view of the user? Few cities have Helsinki's total three-dimensional object model of all buildings, so it is hard to show the view as you would see it outside your car window as you go through for example Mobile, Alabama, or Hong Kong. But you can add dimensions to the map without violating its integrity as a map.

Another unclear aspect is representing maps outside the visual dimension. How does a map sound? How does it feel to the touch? The second question is probably easier for most people to imagine (especially if you have vision-impaired acquaintances), but why should a map not have sound? Operators' echolocation equipment in submarines put an aural picture of the world in the operators' heads. How do we translate the visual map to that aural representation?

Both maps and GIS systems represent space in a format that is appropriate for a specific task. There is a hierarchy of complexity inherent in the different levels of spatial cognition that people typically use:

- The lowest level is "just the world," a pre-cognitive level. A small child will perceive the world in this way.
- The next level is a perception of spatial relations, something you get into when you start using language. Language describes the relationships between the person and the objects in the world (including other people).
- After that comes mapping, starting with the mapping of concepts to persistent representations. This step developed into writing systems as well as maps.
- Beyond the geographic map, you have representations of objects in the world as models and graphic expressions of these models. This kind of map emerged during the Middle Ages in Europe but really did not come into wide use until the 19th century when statistical representations were overlaid on a geographic representation.
- Finally, the representation of abstractions using spatial visualizations (for instance, organization charts, which usually do not have the CEO sitting physically on top of the associates).

These representations, while easily mappable onto a historical scale, also represent different levels of complexity in the use of the visual language of maps. Any representation of space is subjective, and maps—especially when used as third-party communications devices—can be used for persuasion as well as representation. In addition, in location-based services we often have a need to represent objects on the map more clearly than the generic map symbols will let us. The user's current position can be a cross, a polygon, or a circle; it can flash, be in a different color, or be a combination of things. There is not yet an established convention for this (although "X marks the spot" might work for

Americans, it would not work as well in, say, Japan). And when there are conventions, you have to be careful that they fit into the local culture. Directions are usually represented by arrows, but arrows are drawn differently in Japan and China from the West.

There are a number of questions surrounding metaphors for user interfaces in geographic systems. Evaluation criteria are one. Before deciding on good metaphors, you have to decide which criteria determines what is good. This situation comes back to the user group. The communication of the metaphors to users by the designers and the methods that are integrated into the design process will also be important.

When taking the view that the way objects are represented are mappable to tasks, you should remember that there is no one a priori preferable representation that is closer to the truth than others. Any computer interface expresses metaphors, and while the metaphors of spatial representation are different from the metaphors of word processing, they do not in themselves have an inherent scale that lets you say, "This map is always true." We already touched upon one simple reason—that the perception of space is always subjective. You cannot tell precisely what another person is thinking, and while maps are one of our most firmly established cultural conventions, different cultures and different individuals will perceive the user interfaces differently for different tasks. Real-world objects also tend to be fuzzy—not easy to represent on the crisp map.

As a cartographer, your job is to represent reality and make it understandable by simplifying and symbolizing it. Use the traditional means of cartography is well understood after thousands of years of mapmaking. But using animations in maps, something that becomes simple with the event of electronic representation, can help users make sense of your information—even on the small screens. Zooming in on a map from a larger scale to a smaller scale, centering it on the position of the user, panning out at the same time as you zoom out, adding context sensitive annotations that appear as you move the pointer over them—all of these actions become simple, but that does not make it easier to use. It is often easier, as a matter of fact, to confuse people by explaining things than it is to help them understand.

This knowledge becomes especially important when the process is not about simple wayfinding, but when your map is a part of the visualization of a dataset—a thematic map. If you want to track the growth of Tokyo over the Kanto plain, the flow of sewage in the favelas of Sao Paolo, or the spread of myxomatos among Australian rabbits, you have a time series that can be visualized as part of the position of the user and that can be related to a place by using a map. But it does not affect the map itself. If you want to visualize the growth and decline of mountains, the spread of vegetation types at different climactic extremes, or the expansion of Tokyo into the mudflats of Tokyo Bay, you can. Finding the right representations of these data series in relation to the map will be a major problem for designers in the next few years, and unfortunately there are few rules of thumb that you can follow. They will develop as time passes, and you might be part of developing them.

Designing Maps

Maps are probably secondary to writing in symbolic representation of information in society. They record and store information—not geographical features and their distances, but also other types of information, like statistical information, natural resource information, and property information. Representation of the information in relation to spatial patterns also helps make it available for analysis by helping the users visualize it in a larger context. The visualization also means that maps are a way of communicating information, especially information that is hard to communicate verbally. Maps use a special visual language for describing spatial relationships. The communication is based on subjectivity. Maps do not have to be objective; they can instead be used to communicate opinions just as any other medium and be used for persuasion or propaganda as desired.

Cartography is not like other types of graphic design, however. It is a highly specialized discipline that requires training to master. But it is not technical and not harder than, say, designing detergent boxes—only specialized in a different way. Cartographers use visual resources such as color, shape, and pattern to communicate information about spatial relationships. In many ways, it is similar to language, where we use sounds according to a set of conventions to communicate. It is definitely similar to visual design of other types that also use shapes and colors to communicate. If you do not use these conventions and basic principles, your maps are likely to be misunderstood or even cause confusion.

Depending on the type of map you are making, you might have to take into account the coordinate system, map projection, and scale in which you are representing objects. Cartography and graphic design are actually similar in many ways, but cartography is about representing abstractions and generalizing real-world phenomena and their representation. The world as represented on a map is not only simplified, but also a symbolic representation. When creating a map, you must consider how much to simplify the situation being depicted and how to symbolize the relationships being represented. If you do it within the framework of conventions, it will be accessible for anyone who shares the conventions—which is pretty much everyone in today's society. Conversely, if you go outside the conventions, the map will be incomprehensible as a map (however good it might look).

Cartography is a form of communication, so the measure of a good map is how well it communicates the information it is intended to represent to its readers. Too often, only the aesthetic appeal of a map is considered when weighing its communicative value. However, how you look is only part of what you communicate. Instead of looking at how things should look, you should start a project where you intend to use maps to determine what the map will communicate and to what audience. The goal of the communication also has impact on the design of the map and on the communicative and cartographic strategies used.

Who will read your map will also affect how it should be designed. While maps follow convention, there are aspects that you can change depending on the audience (especially the background of the audience). Schoolchildren will have a different level of experience than retired generals, for instance. If your map is intended to shore up the construction of a new sewer system, it will look very different from a map that is used by aviators to find their way by landmarks.

Who is as important as where. Location-dependent services will be used in very distracting conditions, and the maps themselves will be used on small displays and in principle only once. The context of use will influence the design. You want details to be clear, but you also want them to be readable on a small screen at arm's length. The light can be bad (too much or too little), so the contrast needs to be good and in colors that are not easily hidden in glare. Any letters you use should be bold and have high contrast against the map and not be placed where they might interfere with the display of features (neither should any other symbols you introduce on the map; for instance, the position marker of the user). The position marker needs to be different from other symbols used on the map, as well. It is often represented as a circle or a dot, but if it is represented as a torus (a dot with a hole in the middle), it is easier to see where the user is—especially if you put a fine cross over the hole (as in Figure 8.19). Other symbols are sometimes also used, like a star, but stars are often used to display the location of other points of interest (like restaurants).

Be careful that you do not overuse them or your map will look like an advertising for Lucky Charms cereal (for non-American readers, a kind of cereal with marshmallow shapes included). Also, try not to use symbols that are regularly used in maps. If you do, you will confuse users. The same goes for other symbols that are used in other contexts, such as traffic signs.

Another question is, "What data and other resources are available?" Because good maps are available from a number of sources, it might not be very likely that they draw their own. But if you do, you have to consider how and who draws it and what experience they have. Even if you are a good illustrator or designer, you might not be a good map designer.

Making maps is the art of generalizing and simplifying features in the real world so that they are represented within the framework of the map in such a way that they are comprehensible to the audience. Note that not everything that is in the real world needs to be present in the map; only that which is salient is necessary. Maps for tourists generally do not attempt to show the local orienteering landmarks, and geological maps do not display the names of the shops in cities if they even display the houses (for readers who have never orienteered, it is a sport in which you are competing to find your way through an unknown terrain by using a map, a compass, and your feet—great fun).

Deselecting the detail is part of the art of making maps. The amount of detail depends on the scale of the map, of course, because a pocket map of a country will not be able to display the amount of detail the local survey map does. You eliminate some details to draw attention to others. Adding just a little information might lead to confusion.

There are a few elements that are basic on all maps. These include the title of the map, the legend, the map body, and a north arrow (if it is not clear that the map is following the convention that up or top is north or if the map can be tilted, as is often the case on phones and other small screen devices). You can include some other things as metadata; for instance, who made the map, who is responsible for its contents, the original date of production and the latest update, the scale, which projection is used, and information about the data sources. Some of the elements will be present in the same whatever the theme; others will not be included on all types of maps. And, as mentioned, some might be contained in metadata describing the map. Conversely, if you are buying maps from a professional map producer, do check for that information. In many areas, a map that is 10 years old will be horrendously wrong.

A map is often a work of art. Its design is a matter of balancing the different elements of the map and prioritizing them on the screen. That which is most important should be shown most prominently. That can mean making them larger, putting them in the foreground of the design, or placing them in positions of importance—in cultures that use an alphabet that is written from left to right, the top-left corner. Less-important elements should go to the bottom right. In cultures that write from right to left, the positions are mirror reversed.

Designers of small maps often get trapped by the desire to make the image exactly proportional (or might be forced to it by the raster image they are using). But there is nothing that says that all dimensions have to be exactly to scale. If the most important dimension is a road that runs vertically on the screen, you could give the vertical dimension higher priority and squeeze the horizontal. Maps already have this feature, because any projection of the surface of the Earth onto a piece of paper will mean that there will be dimensions that cannot be represented properly and that are built in by convention.

A map is, however, not simply a utilitarian representation of an area. It is also an illustration of that area, and that means it does have an aesthetic dimension. Many service providers who want to use maps forget this fact, and many mapmakers do not think about it because the conventions of creating maps are so ingrained. On the other hand, overdoing the aesthetic dimension will mean that the facts the map is intended to represent will be obscured.

Within the frame of the map, you have to create a balance between the elements. In most cases, this balance is guided by the proportionality that the projection of the Earth's surface onto the paper enforces. In other cases, it is guided by the desire not to crowd them together unnecessarily or not to create large blank areas (except where that is used to express priority). You can also align the elements along the frame of the map (or along the perhaps imaginary guiding lines that exist in a map). Remember that the human eye is very good at spotting patterns, and if you can align objects that are important you will have created an easy way for the eye of the user to find them. If you worry that this image will be a distortion of nature, you would probably be surprised at how minor the adjustments need to be for you to perceive objects as ordered into patterns.

Creating a map is no different from creating any other illustration. The process is no different from the process of making, say, a cereal box. You start with some simple sketches, which create the building blocks for how you will use the space available. Making a few of these will allow you to consider different layouts before the composition of the detailed map begins. Computer systems sometimes have functions for this feature already built in, especially those that are specialized for drawing maps.

Remember, though, that if you break the conventions for how things are placed on the map, you had better have very good reasons for it. You need to have good reasons for how you place each element, and you need to be able to defend it in a discussion with a trained cartographer if necessary. Think carefully about the function of the element as you place it on the map. Does it have to be in the foreground, or does it work better in the background? Is it important for the reader to see it in order to enhance comprehension? For instance, on a field map, if your customers are ramblers (hikers in America), the paths will need to be in the foreground and the vegetation type might be a background fact. If the customers are plant biologists, on the other hand, they will require a very different level of representation of the biota in each location.

As the example shows, the world is full of interesting things. One of the necessary things in making a map is excluding features that are not necessary for the audience to comprehend it. In some cases, it is easy—"Is that rock gneiss or igneous granite?"—but in other cases, the choices might be harder. Do not be afraid to test a paper prototype (which you can print out from your computer) on prospective users. It will make it much easier for you to see whether you have understood the priorities of the users.

If you are making a map of a shopping street, make an agreement with a store to honor gift certificates (for a few Euros) you print on your map. Make them save the maps, record the time the customer came in, and collect them later. Then print out 50 or 60 copies each of three or four different layouts. Hand them out to people who are a reasonable distance away (at the other end of the street). The shop will be glad for the extra advertising and possibly new customers, and you will be surprised to see how easy it becomes to disqualify one or two layouts (probably those that held the ideas you liked best, but that is life).

There are four levels of representation of real-world phenomena on a map, going from the simple to the complex and lower to higher cognitive processing, as outlined in Table 9.1. You will often hear them referred to in discussions about maps (and other data representations as well).

Several of the relationships can be used together to enhance each other. There might also be several different visual representations of them that coincide, and you might have to select the ones that both communicate the relationship the best way to your audience and make the map work in the environment it is going to be used.

Maps are visualized non-verbal communication, which means that adding text to them is somewhat problematic. Anchoring the features on the level of the map in other levels of information enhances the relevance of the map tremendously (if the station to which you took the train is marked, you can use your mental model of the world at one level to identify it in the mental model repre-

Table 9.1 The Four Different Levels of Representation

TYPE OF DATA	FUNCTION	EXAMPLES
Nominal	Grouped into qualitative categories	A road as distinguished from a river
Ordinal	Grouped by some quantitative measure	Small, medium, large cities; width of rivers according to water flow
Interval	Arranged according to some scale, along which addition and subtraction has meaning.	Temperature in centigrade, distance from starting point, "scenic route"
Ratio	Like interval data, but uses a scale which begins at a zero point (which is non-arbitrary), at which no features are present	Elevation, distance, population size

sented on the map). If you have doubt, ask anyone who has tried to use a Japanese map without understanding Japanese to find an unknown destination. Reading maps is as hard as reading text, and while it is possible to identify the location by using the map, it is much easier if you can highlight some relevant geographic features using text.

The factors that limit the readability of the text are the same factors that limit the readability of any text. You have to think about fonts, font sizes, and the placement of the text in relation to the features. Conventions for these already exist in the world of cartography, of course, so if you want to change conventions, you had better have very good reasons (such as, the traditional placement of a label would obscure the object you wanted to highlight).

If you are adding anything beyond simple names, consider how you do it. To have captions and annotations that are formulated with care can make a large difference in usability (and hence, in the amount your customers spend on your service). In a service where you annotate a map with driving directions, for instance, the directions will need to be written in a way that is relevant to the user. But they also have to be short. And you need to have specific directions for all the different starting points a user can have. It helps to have a writer do the captions for you instead of trying to be a writer yourself.

The placement of the text also relates to its orientation. If some of your labels are horizontal, some are vertical, some read from top to bottom, and some read from bottom to top (which is possible in Japanese and Chinese, for instance), you will confuse the user no end. Make sure that the lettering points out the feature it relates to, but also make sure that it does not cross any of the boundaries of the map—in which case what it related to would become confusing. Lettering also shows a shape, but make sure it relates to other shapes on the map in a way that enhances readability. If you are labeling an area, take care where you place the label so it is clear that it relates to the area and not to any of the features surrounding it and that it is clear how it relates to the area. Color can be helpful in placing labels. If you have a field that has a name but also a vegetation type that you need to highlight, use one font family, size, and color for the name (typically Times in black) and another for the vegetation type (for example, Arial in green). Take care with the contrasts. If you label a green field in green, the color should not be so distinct that it distracts from the reading of other features of the map but not so light that it disappears in the color of the field itself. Note that the alignment of the text can also be used to convey information. Text that is pertinent to the theme of the map can be placed where it aligns with the frame of the map, for instance, text that does not can be aligned with the features.

Handling letters and other characters is an art unto itself, and we will not go into it here because there are good books both about the theory of typography, the computer science theory involved in character representation, and about how to write text. Do remember, though, that less is more in typography. You are actually involving the non-verbal communications centers in your readers when you ask them to read a map, and the verbal communication of the captions and labels works with a different part of the brain. Confusing the issue with a number of different fonts will not help readability. Traditionally, typographers try not to use more than four different fonts and font sizes on a given page of print. Use different fonts and font sizes only when you have a good reason to, such as when convention dictates it (for instance, while cities of different sizes are often represented with symbols of different sizes, their functions [county seat, capital] are often represented by different typography [capitals have their names in, well, capitals]). This feature is just one aspect of how text is used on the map, actually. In effect, the typography is used to group information in the text into categories that reflect the theme of the map.

Having said that, we must also state that no guidelines are absolutes. If you encounter a situation where the placement of labels implies a conflict, use your judgment—or ask someone who has more experience than you in designing maps.

Bringing attention to objects means lifting them into the foreground or putting them in the background. Differences between these can be critical to design and comprehension. The limits between some types of geographic features (and socially constructed elements) must also be made clear and unambiguous. For instance, if it is not clear what is land and what is water, you will have a large number of wet users. In some countries, displaying the right of way of a path becomes critical to the hiker not being shot (which could happen in the United States).

Foreground and background are separated by using color, value, and patterning. The foreground and the background are largely a matter of convention, but there are relations to the physiology of humans as well—for instance, in the fact that using the same shade of red and green on a map without patterning might make them invisible to some users who are colorblind. We also perceive shadows in a very nonambigous way, and adding shadowing to a map can create an illusion of relief and highlight directions, for instance.

Databases, Maps, and Visualizations

The database is one representation of a geographic model of reality; the map is another. Using XML technology, you can go from the abstract representation in the database to the visualization of the map information in two primary ways: Virtual Reality Markup Language (VRML), specifically the GeoVRML version, and Scalable Vector Graphics (SVG).

GeoVRML

The *Virtual Reality Markup Language* (VRML) was designed as a three-dimensional language based on HTML. The group that is standardizing it, the Web3D Consortium, is working on an XML version. But meanwhile, there has been work towards representing geodata by using VRML, essentially enabling the creation of three-dimensional maps.

GeoVRML provides the capability to embed latitude/longitude or UTM coordinates directly into a VRML file and to have the browser transparently fuse these into a global context for visualization. There are additional functions for numerical precision, scalability, and animation.

An additional problem is how to browse large areas in a virtual geography. In VRML, you "fly" over the terrain, which brings new problems compared to browsing a link space (which has extensions in other dimensions). The designers have decided that the velocity at which users can navigate around a world should depend on their height above the terrain. For example, when flying over the coast at a height of 100 m above the terrain, a velocity of 100 m/s could be considered relatively fast. When approaching the Earth from outer space, however, a velocity of 100 m/s would be intolerably slow. What is fast obviously depends on the scale in which you are looking at things. To enable smooth navigation you also need to maintain a constant pixel flow across the screen. A simple linear relationship between velocity and the user's elevation above an ellipsoid such as WGS84 seems to provide an acceptable solution, which also is easy to calculate.

The second issue is following the terrain. The Earth is not flat, so if we fly over the surface of the Earth, we would expect to follow a curved flight path through 3-D space. But in VRML, the default navigation methods (such as WALK and FLY) work along a linear path that is parallel to the plane of the representation. It would be more correct to maintain a particular height about the surface of the Earth.

You represent the geographic characteristics in VRML by using a number of new nodes that have been defined to fit into the VRML standard. They are outlined in Table 9.2.

With GeoVRML, you specify each point in terms of its coordinates, the elevation, and other geographic parameters together with the normal VRML parameters. Using GeoLocation, you can reference a traditional VRML model to a point on the Earth.

VRML has been accused of being verbose, and it is far more verbose than, for instance, XHTML. It attempts to represent a far more complex space, however, and the rendering is as complex as that of SVG. It should not be expected that we will see VRML viewers on mobile devices anytime soon.

Table 9.2 The New Nodes in GeoVRML

NODE NAME	DESCRIPTION
GeoCoordinate	Build geometry using geographic coordinates
GeoElevationGrid	Define a height field using geographic coordinates
GeoInline	Inline a file with control over when to load and unload the data
GeoLocation	Georeference a vanilla VRML model onto the surface of the earth
GeoLOD	Level of detail management for multi-resolution terrains
GeoMetadata	Include a generic subset of metadata about the geographic data
GeoOrigin	Specify a local coordinate system for increased precision
GeoPositionInterpolator	Animate objects within a geographic coordinate system
GeoTouchSensor	Return the geographic coordinate of the object being pointed to
GeoViewpoint	Specify viewpoints using geographic coordinates

Transforming other data formats to VRML is relatively straightforward. (The basic way to do it is the same as transforming GML to SVG.) Rendering tends to be less detailed, though, because VRML views are really meant to move through.

Following is an example that displays a standard VRML cone at the latitude/longitude location of Lossiemouth, Scotland. The Cone is rotated through 180 degrees about the X axis to point straight down rather than up. The GeoElevationGrid node gives an underlying reference model of the Earth, albeit a coarse one. The cone is 500 km high and 200 km above the surface of the Earth, so it can be seen at a global scale.

```
#VRML V2.0 utf8

EXTERNPROTO GeoLocation [
  field SFNode   geoOrigin
  field MFString geoSystem
  field SFString geoCoords
  field MFNode   children
] [ "urn:web3d:geovrml:1.0/protos/GeoLocation.wrl"
    "file:///C|/Program%20Files/GeoVRML/1.0/protos/GeoLocation.wrl"
    "http://www.geovrml.org/1.0/protos/GeoLocation.wrl" ]

EXTERNPROTO GeoElevationGrid [
  field        SFNode   geoOrigin
  field        MFString geoSystem
  field        SFString geoGridOrigin
  field        SFInt32  xDimension
```

```
  field        SFString xSpacing
  field        SFInt32  zDimension
  field        SFString zSpacing
  field        MFFloat  height
  exposedField SFNode   color
  exposedField SFNode   texCoord
  exposedField SFNode   normal
  field        SFBool   normalPerVertex
  field        SFBool   ccw
  field        SFBool   colorPerVertex
  field        SFFloat  creaseAngle
  field        SFBool   solid
] [ "urn:web3d:geovrml:1.0/protos/GeoElevationGrid.wrl"
    "file:///C|/Program%20Files/GeoVRML/1.0/protos/GeoElevationGrid.wrl"
    "http://www.geovrml.org/1.0/protos/GeoElevationGrid.wrl" ]

EXTERNPROTO GeoViewpoint [
  field        SFNode     geoOrigin
  field        MFString   geoSystem
  field        SFString   position
  field        SFRotation orientation
  field        SFString   description
  field        SFFloat    speed
  exposedField SFFloat    fieldOfView
  exposedField SFBool     jump
  exposedField MFString   navType
  exposedField SFBool     headlight
  eventIn      SFBool     set_bind
  eventIn      SFString   set_position
  eventIn      SFString   set_orientation
  eventOut     SFTime     bindTime
  eventOut     SFBool     isBound
] [ "urn:web3d:geovrml:1.0/protos/GeoViewpoint.wrl"
    "file:///C|/Program%20Files/GeoVRML/1.0/protos/GeoViewpoint.wrl"
    "http://www.geovrml.org/1.0/protos/GeoViewpoint.wrl" ]

Group {
  children [

    Background { skyColor 1 1 1 }

    GeoViewpoint {
       geoSystem [ "GDC" ]
       position "51.5122 -40.0 10000000"
       orientation 1 0 0 -1.57
       description "Initial GeoViewpoint"
    }

    Shape {
       appearance Appearance {
       material Material { diffuseColor 0.8 1.0 0.3 }
       texture ImageTexture {
```

```
            url "http://www.geovrml.org/1.0/doc/images/earth.jpg"
         }
         }
         geometry GeoElevationGrid {
         geoSystem [ "GDC" ]
         geoGridOrigin "-90 -180 0"
         xDimension 11
         zDimension 11
         xSpacing "36"
         zSpacing "18"
            creaseAngle 1.05
         height [
              0 0 0 0 0 0 0 0 0 0 0
              0 0 0 0 0 0 0 0 0 0 0
              0 0 0 0 0 0 0 0 0 0 0
              0 0 0 0 0 0 0 0 0 0 0
              0 0 0 0 0 0 0 0 0 0 0
              0 0 0 0 0 0 0 0 0 0 0
              0 0 0 0 0 0 0 0 0 0 0
              0 0 0 0 0 0 0 0 0 0 0
              0 0 0 0 0 0 0 0 0 0 0
              0 0 0 0 0 0 0 0 0 0 0
              0 0 0 0 0 0 0 0 0 0 0 ]
         }
      }

      GeoLocation {
        geoSystem [ "GDC" ]
        geoCoords "57.7174 -3.286119 200000" # Lossiemouth, Scotland,
elev 200km
        children [
           Transform {
          rotation 1 0 0 3.1415926
             children [
               Shape {
                 appearance Appearance { material Material { diffuseColor
1 1 0 }}
              geometry Cone { bottomRadius 100000 height 500000 }
            }
          ]
           }
         ]
      }

   ]
}
```

GeoVRML is a way of creating a richer data representation than is possible with other markup languages. Its emphasis is on presentation, not features in the data (which is why it is in this chapter). For most use cases, however, the presentation works better if it is flat, because that is how maps are usually presented.

SVG: Vector Graphics in XML

Most images on the Web, and on computers in general, are raster graphics (or even bitmaps). Raster graphics are displayed on a raster, a grid of coordinates in a given space. The image is formed by pixels fixed to a grid (and instructions describing whether they should be lit up or dark and in color or not). An even more restricted form of the raster image is the bitmap. BMP, TIFF, GIF, and JPEG are examples of raster image formats.

The difference with vector graphics is that although rasters describe the image in terms of dots on the screen, a vector graphics format describes the image in terms of lines (and shapes formed by the lines).

A vector graphics file can be scaled and modified (as anyone who has ever hacked a Postscript file knows). That cannot be done as simply with a raster image file; it is hard to modify without loss of information (although there are tools that convert from raster format to vector format and back again without a loss of information).

Vector graphics are actually quite old. In the 1960s and 1970s, the display standard was vector graphics. The Visual Display Units (VDUs) of the time drew lines on the screen that were controlled by the central computer. The screens were green and were nothing more than glorified oscilloscopes.

The difference between vector graphics and raster graphics has nothing to do with the way the image is formed; rather, it depends on how the image is described. The vector graphics file is also smaller, which makes it more suitable for mobile use and scalable. The relationship between the lines that define the image can be changed to make it larger or smaller without changing the image, something that is not possible with a raster image. The disadvantage, however, is that a vector graphics format requires more processing capacity to be displayed when the description of the vectors are transformed to instructions for the display. (The raster format can be displayed without transformation.)

Most maps are designed in a vector graphics format, and when they are used on the Web they are transformed to raster images. The alternative would be to transform them into a standardized vector graphics format. There are actually a few such as these, but what is most interesting in this context is the SVG format from the W3C, because it is a vector graphics format in XML.

As you already have discerned, XML is important as a universal data format. SVG is rendered by an XML application, which uses the data from the XML parser to render the image. It works a little like a BASIC interpreter, but it is not an interpreted language because the entire file has to be read by the XML parser first and then forwarded to the SVG processor. (We will discuss more about XML in Appendix B, "XML: An Introduction.") The SVG image is drawn by the SVG renderer in the order that the objects occur in the document. The document itself consists of elements in an XML hierarchy. Objects occurring early in the document will be drawn on top of objects occurring later, but it is also possible to make objects wholly or partially transparent. In the client, the image is rasterized by the application and output to the screen. This makes it possible to adjust the anti-aliasing to the presentation scale.

Scalable has two meanings in SVG. The first is that it is possible to change the scale of a vector image without distortion or degradation—that is, not being limited to a fixed pixel size. This is common with other vector graphics formats. The anti-aliasing, which smooths the appearance of the graphics, is adjusted to the scale that is used. The second meaning is that the technique itself is scalable, which means that it is possible to use it in a wide variety of applications.

One way of leveraging the scalability is to make calculations on map data, presented as SVG. If the relationship between the image coordinate system and the geodetic coordinate system is known and the image coordinates are accurate enough, it is possible to calculate real-world length and areas of objects in the map.

This is most easily done using scripts. The algorithms are made simpler if the image coordinates are the same as the geodetic coordinates. Lengths of paths for instance, can be calculated with repeated use of the Pythagorean theorem, provided that no arcs or curves have been used. There are also theorems for calculating area using the coordinates of the corners.

An alternative to using scripts is to include this functionality in the browser. This probably will use the resources of the device more efficiently, because a browser showing SVG images must have the ability to calculate lengths along paths anyway. One reason for this is the browser uses length calculations to make it possible to render a text at a specified distance from the starting point of a path.

The text in an SVG file is actually stored as text and not as a set of pixels, so it becomes possible to search an image for certain text strings. This can be useful when viewing complex images such as maps. It is also possible to select, cut, and paste text. Searches can be done on the actual SVG document, which of course is XML, so that you can search for features in images by their names. SVG uses a set of fonts called Webfonts, which are defined in CSS level 2. These also contain the most frequently used cartographic symbols, and it is possible to create new symbols for use in SVG documents. Symbols can be cached in the client, which speeds up rendering. Because the format is scalable, the symbols can be resized; their orientation can be changed; and the style can be adjusted to match the rest of the vector image. You can assign style to individual characters, and texts can be aligned along paths. This can be useful in map applications if the text is the name of a curved or irregular feature such as a river and if it is desirable that the text follows the shape of the feature.

The styling of the text is included in the SVG document and not handled by CSS as in the case of XHTML. Examples of styling properties are "font-family," "font-size," "font-weight," "fill," "stroke," and "text-decoration." The text can be rendered straight or along path elements. If the text is to be be rendered along a path, the path should be referenced by an xlink:href attribute within a textPath element contained in a text element. The direction of the path is important because it influences the direction of the text.

Two styling properties that are especially useful when working with text along paths are the properties "startOffset" and "dy." The property startOffset controls how far from the starting point of the path the rendering starts. The property dy makes it possible to place the letters higher or lower with respect to the orientation of the letter. For example, if one letter is oriented with the bottom to the left and the top to the right—that is, it lies down—a positive value for the dy property moves the sign to the left.

If only a part of a longer text should be influenced by some specific styling properties, for example, if one word should be underlined, an element called "tspan" can be used. This element makes it possible to cut up a text in pieces in order to alter the appearance of one piece from the rest of the text.

SVG does not allow the use of several characters—for example, the Swedish characters å, ä, and ö—in the document text. If you want to use these characters or some other forbidden character, you have to escape them. The characters in the preceding example—that is å, ä, and ö—are written as "å", "ä", and "ö".

An SVG document can be used by itself, or embedded in other XML content, using XML namespaces. It can also be embedded in HTML, although with XHTML, there is no reason to produce HTML any more. SVG image objects can be used as hyperlinks or they can be connected to scripts, something that can be used for interactivity (for example, conditional hyperlinks). Hyperlinks can refer to other elements in the same SVG document or external documents using URIs. It is also possible to make animations with SVG, either with scripts or with a special animation element in the SVG document. Animations use the SMIL Animation function, which works in a similar way to Macromedia's Flash. It is also possible to use different script languages with SVG, provided the renderer can handle it. The most frequently supported are ECMAScript and JavaScript, which is a previous version of ECMAScript.

The execution of scripts is triggered by events. This is specified as an attribute to the element, which receives the event. That element can be a graphics element receiving a click event from the pointing device (in which case an attribute has to be set to make sure that the action occurred at the event), or the whole SVG document receiving an event when the loading of the page is complete. Using event attributes links only one function to one type of event at a time. If you want a certain function to get triggered by all events and received by a certain element, the function can be registered as an event listener to that element.

All elements inside the SVG document are reachable from the scripts via the W3C Document Object Model (DOM). The objects in the SVG document are manipulated primarily by changing the attributes of elements. For instance, if you want an object to disappear, you change the style attribute for the object from "visibility" to "hidden."

When animations are created with scripts, a timer function in the script language is used, making it possible to call functions after certain time intervals. These functions define the behavior of the animation by, for example, moving or resizing objects. A special element, SVGZoom, also occurs when the scale of an SVG document is changed. The intention is that it should make sure that no important information is lost, for instance by triggering a script that handles the generalization of the map. When the event occurs, it is possible to determine what the initial zoom level was and what the new zoom level is.

In SVG it is possible for scripts to adjust the style and visibility of single objects in the map. To keep track of which objects are visible on different zoom levels, similar objects intended to be visible at the same levels could be grouped using the g element. The scripts can change the appropriate style property for the whole group with a few commands. Typically, the style property "visibility" is changed dependent on the zoom level.

The SVG element "animation," which is another technique for creating animations, has inherited its animation functionality from Synchronised Multimedia Integration Language (SMIL). SMIL includes a specification for general-purpose XML animations. The SVG animation element also has some SVG-specific functionality.

All the content of an SVG document is contained in the root element, which is called svg. The content is mainly structured as graphics elements and container elements. The graphics elements of SVG are text elements (for example, letters and symbols), raster images, paths, shape elements, and the use element.

Symbols are used for representing objects of special interest to the user with simple well-known shapes. An example of a well-known symbol used in maps is the cross, which often is used to indicate the location of churches. There is a special element in SVG, called "symbol", that is used for defining symbols. This element creates graphical templates, which are instantiated with the element "use", which references the symbol with a URL. Other elements can be instantiated with the use element, but when the symbol element is referenced it creates a viewport and adds semantics to the document. When using symbol elements it is possible to add styling properties, which have not been specified already in the symbol template.

The basic shape elements are:

rect (draws a rectangle)

circle (draws a circle)

ellipse (draws an ellipse)

line (srtarts at one point and ends at another)

polyline (draws a line between many points, typically used for open shapes)

polygon(draws a line between many points, from the last point a line is drawn to the first point so the shape is closed)

The path element is not a shape element, but it is important for constructing shapes. This element draws the outline of the shape using a current point as the starting point. The current point can be thought of as the position of a pen on paper. Different functions are used for drawing lines with this pen and by moving the current point. There is also a function for moving the pen without drawing a line, which can be used in order to create doughnut holes in closed shapes. A closed path resembles a polygon, and an open path resembles a polyline. The path element, however, is even more flexible. Except for the possibility to make doughnut holes, the lines between different current points can be drawn as elliptical or circular arcs, or cubic Bézier curves.

Lines can be used to represent many different types of objects. Examples are streets, rivers, property boundaries, and power lines, which have very different properties. The parameters in SVG that can be altered in order to create different line patterns are color, thickness, dashes, symbols, and the course and multiplicity of the line.

The color and the thicknesses of lines are controlled with styling properties in SVG. The color is specified in the property "stroke", and the thickness is specified in the property "stroke-width". Dashes are controlled by a styling property called "stroke-dasharray", which holds an array with numbers specifying the lengths of the dashes and the space between them. The array "10 5 20 5" specifies a line with a 10 units long dash , a 5 units long space, a 20 units long dash, a 5 units long space, and from the beginning again with a 10 units long dash, and so on.

If you want a symbol to occur along a line, you can partly accomplish this with a styling property called "marker". The marker property references an object that shall be used as a symbol by specifying a link with a URL. The link can either reference an object within the same document or an object in another document. The chosen symbol is inserted on the locations of the points building up the line. This operation can be quite complicated, as well as computationally demanding. Making the pattern of the line change direction requires altering the coordinates specified for the line by inserting points that describe the pattern.

Line patterns with many lines can be generated by specifying additional lines parallel to the original line. However, this requires computation of the positions of the points used for creating the parallel lines, and with an even number of lines, the result is that no line has the coordinates of the initial line. The initial coordinates can be needed when trying to determine, for example, whether a terminal is located on a road by relating positioning information to the line representing the road.

A different technique that may be possible to use is the styling property "mask". When a mask is specified for an object, the opacity of the mask influences the object. If no opacity is defined for the mask—that is, it is completely transparent—the influenced object is also completely transparent with respect to the appearance of the mask. If you draw a line with the style property "stroke-width" set to a width that corresponds to the outer boundaries of the two lines, you can define a mask based on the same coordinate data as the initial line. The mask has the width of the preferred space between the lines, and the opacity value is set to 0. The last step is to specify the mask in the styling property "mask" of the initial line, creating the illusion of one line being two. This is, however, not an optimal solution.

In some cases a set of lines should look like an area. A common example of this is streets on large-scaled city maps. The effect is achieved by using homogenous equally colored lines that are wide enough. This technique avoids the seams that would appear if the lines were styled with different colors or as multiple lines.

However, it is often preferable that the area built up by the lines is outlined in order to create a contrast to the surrounding areas. This effect can be achieved in SVG by drawing the same lines twice. The first time the line is drawn a bit wider than the second time. Because of this, when the first line is drawn over by the second a bit of the first line is still showing, creating a boundary for the area. It can also be drawn with a different color or contrast to enhance the outline effect.

The endings and corners of lines can be affected by styling properties that define whether they shall be rounded or not. When the property affecting the line endings is used, it also affects the endings of the dashes within a dashed line. The property affecting the corners affects the line where it changes direction. The marker element can also be specified to influence the start- or endpoint of the line.

Different areas are usually separated by using different colors or patterns. The color of an area is specified in the styling property called "fill". Area patterns are important in cartography partly because many conventional area fills in maps use well-known patterns for different geographic objects (for example, blue for water), partly because patterns can be used to indicate another dimension of area information. An example of using two dimensions of area information is using color for indicating land and a stripe pattern which is possible to see through, to indicate the extent of nature reserves in the same map.

In SVG an area pattern is defined in a pattern element, which creates a square with the pattern inside. The pattern is defined by graphical elements contained in the pattern element, which are drawn in the square defined by the pattern element. When the pattern is attached to an object, the square is repeated over the area of the object in order to create the pattern. Patterns for areas are specified by referencing the pattern element. This is done in the fill property, also used for specifying the area color if no pattern is used.

Although SVG is a true vector format, it also has support for some raster effects, such as drop shadows. It is possible to insert raster images as objects in the vector image, for example, if you want to use an aerial photograph as a background to a map. The inserted raster image is resampled so that it is possible to scale it to fit the image context. In SVG, raster images can be inserted into the vector image with an element called "image". The difference between referencing an SVG file with an image element and referencing SVG content with a use element is that the image element can reference only a whole file and not single elements contained in the SVG document. The image element also always defines a new viewport and a new document object model for the inserted image.

The specification for SVG requires that tools for SVG support at least the image formats PNG, JPEG, and SVG. Both PNG and JPEG are raster image formats. In the image element, the file containing the image to be inserted is referenced with an xlink:url attribute. The xlink:url attribute and the width and height attributes, which specify the extent of the image after insertion, are required attributes for an image element.

Alternatively, you can use the filter element in SVG. Examples of filter effects possible to define in a filter element are "feBlend" and "feOffset". The filter effect "feBlend" is used to blend two objects in different manners, and the filter effect "feOffset" is used for moving the input image and is used for creating drop shadows. Different kinds of lightning effects, blurring, and so on, are also possible.

Transparency usually is used in maps when two graphical objects covering each other should both be visible. This can happen when a graphical object is expanded outside its physical boundaries, in order to emphasize its importance. An example of this is a road that is expanded and accidentally covers a symbol indicating a house. If the road is completely opaque, the house has to be moved away from the road in order to be seen. This is a complicated task if it is to be carried out automatically. Therefore, it is easier to solve the problem by making the road partly transparent.

The transparency is controlled with four opacity properties, which are styling properties. These properties influence different elements in different manners and make it possible to create gradients of opacity. The opacity properties specify alpha values, which are used when compositing color values of transparent objects that cover each other. The alpha value ranges between 0 and 1. An alpha value of 0 means complete transparency; an alpha value of 1 indicates that the object is completely opaque.

The container elements in SVG are elements used for structuring the document and containing graphics elements or other containing elements. The g element is a container element intended for structuring graphics. It can contain other g elements and shape elements. This creates a hierarchical structure of the shape objects, making it easy to define a style attribute (or other properties) for many objects at the same time. Title and description properties can be attached to groups of elements called "title" and "desc", which both are possibly included in the g element.

In a map application, the g element could be used to group all shapes representing public buildings. The title attribute of the g element could be set to "Public Building," and the desc attribute, representing a description, could be set to a text holding a definition of what a public building is (such as "City Hall" or "Hôtel de Ville"). An SVG viewer may show the title attribute as a tooltip when the pointer device moves over the element. The style of the g element could be set to one property, for example "red fill". It then will be easy to change the attributes for all public buildings at the same time.

Other container elements are the element for creating symbols and the "a" element used for hyperlinking.

The initial coordinate system of an SVG image is *cartesian* and has its default origin in the upper-left corner. The area where the image is placed is called a viewport. The viewport is the extension of the SVG image on a computer screen or in a containing document. The size of the viewport is expressed in the units of the containing document or system units, such as screen centimeters or pixels. The preferable height and width of the viewport can be specified in the SVG document. However, it is possible for a containing document or application to override these declarations.

If the viewport is smaller than the size of the image, only a part of the image is visible when using the initial coordinate system. This is because the initial coordinates of the image have the same scale as the coordinates of the viewport. Using the attribute "viewbox" makes sure that the whole SVG document is visible regardless of the size of the viewport. A viewbox defines a square, expressed in the image coordinates. When the image is placed in the viewport, the viewbox makes sure that it fits exactly in the viewport without changing the scalar relation between the x- and y-axes.

To make this possible, the image is stretched or shrunk equally along both the x- and y-axes until the top and bottom, or the right and left sides, touch the sides of the viewport. For example, if an image with a viewbox 1000 pixels high should fit into a viewport 500 pixels high, the side of an image pixel would be equal to half the side of a viewport pixel.

You can also change the coordinate system itself (and not just the presentation format, as with the viewbox). This is done using transformations, which are attributes that can be attached to both container elements and graphic elements. The possible transformations in SVG are translation, rotating, scaling, and skewing. A transformation attached to an element affects all elements contained in the element. It is also possible for transformations to be nested, for instance first skewing all elements in a container element and then rotating the text elements. There are also techniques for applying many transformations to one element.

Thematic maps are used to show the distribution of geographically dependent data such as statistics and inventories, demographic data, and pollution rates. You can create thematic maps in several ways. One is to let the different saturation of the color used to paint different areas correspond to different values of a certain variable. This kind of effect can be created in SVG by letting the value of the variable influence the color property of corresponding areas. The relation between color and value could be specified in a function that is called either when the SVG document is created, for example in an XSLT routine, or by scripts executed in the client.

A way of visualizing statistical data is to use small charts inserted in the map on appropriate locations. This kind of object is also best created with automatic routines, such as functions called by an XSLT transformation. This is because if many charts are drawn in the same map, they probably look quite similar, and once a function has been written, it can be reused when creating charts in other maps. The size of the representation becomes a problem here, however, as there is a risk that these charts become too small to be visible if displayed on a mobile phone.

Color can also be used to demonstrate cartographic priority, that is to illustrate the importance of a certain object or object collection. Putting the object or objects on top of all other objects by changing the drawing order makes them more conspicuous. Both these steps are easiest to do when transforming data from a GML representation to an SVG presentation.

By grouping elements it is possible to let the user handle the map as a set of *layers*. A layer is put on top of the others by moving the node corresponding to the g element containing a layer to the last position in the document tree.

SVG should feel familiar to graphics designers. Color, gradients, and patterns are painted onto the screen. You can fill and stroke shapes and text or apply other graphic techniques like masking and opacity. Nobody is likely to try to write SVG because it is very complicated and would not be useful as there are already good drawing tools available, and it is being included in several of the most frequent drawing tools (although so far not in mapmaking tools).

The interesting part is that while the storage format is text-based, neither the user nor the designer should notice, and because it is an XML format it can be transformed to and from other XML formats by using XSLT. For instance, geographic feature descriptions in GML can be transformed to SVG for display.

That said, so far there are no SVG renderers for mobile devices or PDAs (like Palm and the Ericsson r380). The working group in the W3C is working on a mobile profile, however, and SVG viewers are expected to appear in mobile phones in a year or two.

Converting Databases to Maps Using XML

In Chapter 7, "Data Formats for Geography-Related Information," I discussed markup formats for geographic information based on XML. Because XML formats can be transformed to other data formats by using XSLT, you can take a database encoded in GML (or another format) and transform it to a map. If the target device has a vector graphics viewer, you can transform it into vector graphics. If it has a raster viewer it becomes slightly more complicated, and you have to transform the data to raster data. That process essentially involves rendering the document onto a virtual screen and reading the raster image into a file. Specialized software performs this conversion from vector formats to raster formats, so if your output has to be a raster format that is what you should use.

Converting GML to SVG is one of a few possible techniques for presenting GML data. It is simple to display GML data with SVG for several reasons. SVG is a vector format, just like GML, which makes it possible to preserve the structure of the geographic objects. Second, SVG is an XML language, just like GML, which makes it possible to use XSLT for transforming between GML and SVG. Any additional XML data included in the GML documents from other XML standards can be included in the SVG document.

To transform a GML representation into a map (regardless of the output format), you will need to add graphical styling to it. Typically, this process involves the interpretation of the GML content by using graphical symbols, line styles, area or volume fills, and often some sort of transformation of the geometry of the GML data into the geometry of the visual presentation. This process is similar to the way you add styling to an XML document by using XSL and CSS.

When the format transformation between GML and SVG is executed, you can do other transformations and manipulations of the data. This can be, for instance, coordinate transformations that make sure that all geographic data references the same SRS. Cartographic generalization and spatial queries can be performed as parts of this transformation.

After the abstract graphical representation is finished, the map needs to be rendered. Rendering happens when the map format is drawn as a series of pixels on the screen. There are viewers for the most popular formats available for those handheld platforms that have sufficient memory, processing capacity, and an operating system that can handle it. The more modest mobile phones will not be capable of handling this type of presentation.

The general process for this transformation is illustrated in Figure 9.1.

Note that there is no need to do this process on the client; it can be done in the server, and you can send the raster image to the client. This situation depends on the capabilities of the client. The same transformation sheets (XSLT programs) can be used if they are written with this feature in mind. Ron Lake of Galdos software is really the pioneer in this area and has demonstrated the process several times. My previous book, *XSLT* (John Wiley & Sons, 2001) does not demonstrate this particular case but tells you how to use XSLT.

When you write XSLT transformation sheets, you create a program that reads the input document and replaces some XML elements and adds other things to the output document. An XSLT processor such as Xalan or Saxon (or Java classes produced by an XSLT compiler) takes the input document and turns it into an output document. (Which, by the way, is *not* the same as the input—you not only transform the document, but you also create a new document that is transformed.)

The principle behind XSLT transformation is to locate the GML property—for instance, the extentOf attribute for each FLOODPLN_POLY element (which is

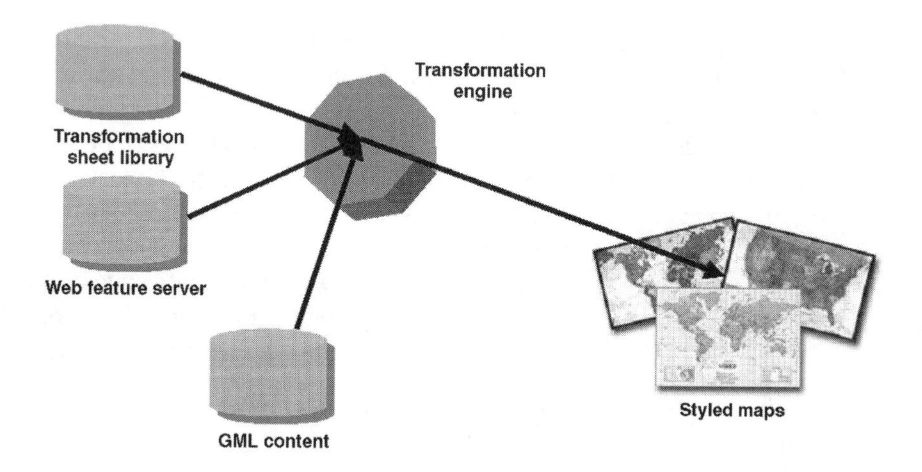

Figure 9.1 The general process of transforming an image.

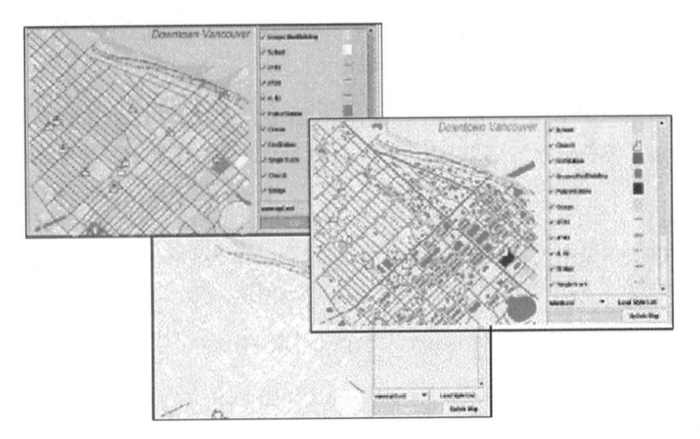

Figure 9.2 Some simple maps from the same data source.

a member of a feature collection) in the GML dataset—and to create a corresponding SVG graphical element (here, a path element) for the inner and outer boundary elements of each located feature. The fill color, stroke color, and stroke width of the polygon boundaries are established in the stylesheet. You can also build in tests of presentation parameters—for example, determining whether output should be in world user coordinates or in pixel coordinates.

For some transformations it is necessary to extend XSLT. This situation is true, for example, for coordinate conversions. XSLT can be extended (how depends on the transformation engine you are using). There is no need for a large number of extensions just to perform arithmetic and string operations that are difficult in XSLT. Galdos Systems, Inc., and Ionic Software SA have proposed a standard set of XSLT Extension Functions for Map Styling using SVG as the target graphical presentation language.

Some style editors are specialized for this purpose, and I provide one on the CD-ROM that comes with this book. To create a transformation, you need to locate the GML data source, fetch its schema, select a symbol library, and match the symbols to features (for example, I want roads to be represented as red lines), then the software automatically creates the stylesheet, which performs the transformation. All you need to do is load it into your transformation engine and send the data you want to transform into a map into the transformation engine. Using the map style engine, you can create several different representations from one source document. The simple maps in Figure 9.2 were all generated from the same data source by changing the map stylesheet.

The difficult part is fitting your data to the map, which is what we will look at in Chapter 10, "Pulling It All Together: LBS-Enabling Your Web Site and Developing New Applications."

CHAPTER 10

Pulling It All Together: LBS-Enabling Your Web Site and Developing New Applications

If you already have a Web site, you have shaped the way your users will expect you to present yourself. This may have happened by chance, or it may have been a throroughly designed move. Either way, change is difficult, as you will recognize if you have ever tried to change the layout of a magazine or the graphical profile of a company. The presentation sets expectations.

Location-Enabling Your Web Site

One of the tricks in Web site design is recognizing that the user's expectations are not the same for the mobile as for the fixed presentation. But you must still present the user with a consistent experience if the user is to understand that it is you with whom he is dealing. Changing the user experience means that you have to convince the user once again that he wants to deal with you; if you retain the user experience, the new presentation can carry over into the old.

Let us assume you are the owner of four candy stores in Middletown, and you have decided to make your stores available to mobile users who are looking for candy. You can use HTTP content negotiation with CC/PP to make sure that the right users get your information. This, of course, also means that you can use the same URI for all your information, which is handy if you want users to remember it when they are at home or want to use your delivery service. One of the things that is sure to turn users off is if they have to write a long string because your system is unable to make their life convenient. Also, it reveals that you do not know how the technology works and that is pretty bad, especially if you are a consultant and want to impress your customers by proposing a clever solution. Besides, it sends the wrong signals. Having one URI for all information tells the user that Middletown Candy has the same basic offering. Having different URIs for the different versions is like having different versions for the different stores, which may be right in some cases, but which is wrong more often than not.

In the case of Middletown Candy, it means that www.middletowncandy.com senses if you are a mobile or fixed user, and if you are mobile, the site adapts the presentation to your position. Middletown is very small, and it has a very simple structure, as you can see in the map in Figure 10.1.

This makes it very easy to provide location-based services (assuming, of course, that the infrastructure is there). There is nothing magical in providing services this way, only a lot of elbow grease and sweat, plus some quick tech fixes. You could do it today, if you could get the position information from the mobile network operator (or the user) in a way that you could use.

You do not need to change the external interfaces to the Web site at all. What you do need to change is some of the internals. If a request comes to your

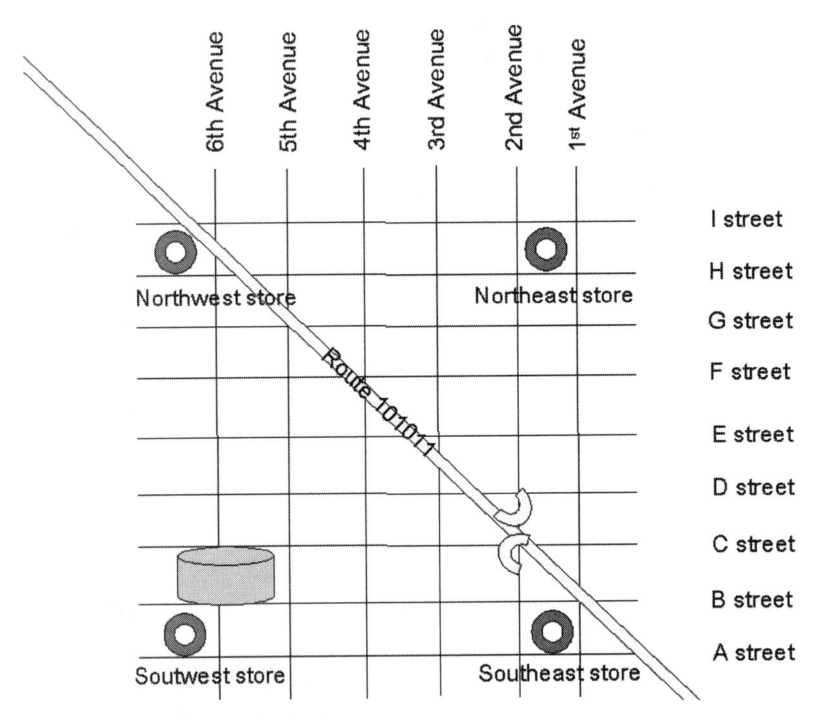

Figure 10.1 A map of Middletown.

server at www.middletowncandy.com, you first need to find out if it comes from a mobile or fixed user. This has consequences not just for obtaining the position, but also for what information you should present, as discussed in Chapter 8, "The User Interface to Location Information Services."

Today, there is one reliable and one slightly less unreliable, but more useful, way of doing this. The reliable way is to look at the user agent header of the HTTP request. This field contains the brand name of the software with which the user is accessing your site. It has been misused in the past (when Netscape was the dominating browser, Microsoft labeled theirs "Mozilla"—the name then used for the Netscape browser in the User Agent field). In sites that present different versions of pages to different browsers, or even block different browsers, this field is used to determine which browser is making the request. If vendors cheat, or simply do not fill in it, you are in trouble, however.

The other way of finding out if the user is mobile is to look at the address. This will become progressively more complex as time goes by, but currently, the user in the mobile network has to pass through a proxy (irrespective of which system he uses, WAP or iMode or Jsky, which are the three existing mobile Internet systems used commercially in the world today). The proxy is likely to reset the address of the user to its own, and it is also likely to strip out any cookies you try to send back—not that the mobile phone has cookie capacity, at least not currently.

If the user is a fixed user, direct her to the normal Web page. Here, you can have animations of dancing taffies, lollipop cancan, bonbon polka, or whatever you fancy that might catch the attention of the customer and make her feel good about the site—and buy candy.

A user on a fixed line does not mind waiting for something good. The psychological setting of the surfing makes that both possible and inviting, and while he is downloading the dancing lollipops, he can read the latest Dilbert comic.

The mobile user has no such luxury. He has one small window on the world, and that is it. If the information is not there before three seconds have passed, then you have lost him, and that means lost a customer as well.

Before you present the mobile user with his information, however, you need to make sure that it is appropriate—both position-wise and when it comes to formatting. If the user is a mobile user, then you can use the CC/PP profile describing the device capabilities to optimize the presentation (of course, one simple way of selecting which branch of your information tree to send your user to is to check for any profile present at all first, and then, if there is a CC/PP profile, proceed). This is one of the advantages of the WAP 2 framework: It does include CC/PP, as well as the Location Attachment function.

After you have found a profile describing the device (and cached it), you need to check whether the position was enclosed (as a Location Attachment), or whether you have to retrieve it by sending a request to the positioning service. Having done that (which may require that you have a previous relationship with the provider of the service), you can compare the position and the data you are providing (there may be an intermediate step here where you have to translate the coordinate systems, but that is trivial).

This requires that you determine the polygons in which the position is relevant. The simplest way to do this is to take a point (for example, the address) and draw a circle around it. This can be done by geocoding the addresses and assigning a circle radius to each address—essentially, the reverse of the uncertainty area. If you have categorized the addresses, you do not have to assign coverages to each address, and if you know the size that is relevant for each category, you could assign different sizes of coverage areas for different categories—for example, the walking distance for a sandwich shop in the city center at lunchtime or half the state for a rubber jewlery specialty shop in a major mall.

That, however, may mean that you will get holes and overlaps in the coverage of your system, and you will need to define algorithms to resolve this. It is simpler to define polygons that cover the area completely, but this may require that you take demographic variables, such as the population density, and aspects such as which roads lead to the stores and how accessible they are into account. Designing these *regions of interest* for each store (or other location of business) is likely to be a very interesting craft for future information architects, and while there has been a lot of work done in this area, it has not been focussed on location-dependent systems. You can find an entire discpline in the marketing industry, if you want to apply it.

You may also have a behavior for the case where the position is not within the coverage of a store. Then, you can either direct the user to the default home page or send him to a menu, for example. What you absolutely do not want to do is give the user an error message, the HTTP 404. The likely effect of that is a loss of confidence in the service, and as an extension, in the brand of your Web site and your company itself. Users have a tendency to believe that you are doing what they think you do, not what you think you do. Providing a credible alternative is a very easy way of getting out of this situation, for instance putting up a page saying something like "You are not within the coverage we have defined of any of our stores at present. Please select the store you would like to visit from the menu."

The important thing is the tone, as in all marketing communications. You do not want to suggest to the user that he is the one doing something wrong, and you do not want to suggest that your company is not available—even if it is not present in that location. I talked more about this in Chapter 8, "The User Interface to Location Services."

To return to our Middletown example, that is not a very large city, and the coverage areas of the four stores is rather easy to put onto a map. Taking demographics into account, such as the location of the football stadium and the turnoff from the highway, you get something like Figure 10.2.

In applications such as the Oracle location services database, you can draw the polygon on a map and give it a name, and it will be stored with the appropriate coordinates in your database. There are various other tools that let you do this as well. In essence, however, what you do is declare the coverage areas. This can be done in another way, as we saw in Chapter 7, "Data Formats for Geography-Related Information"—by encoding them in the markup language that is used to represent the database.

If you have a geographical editor, this will be done automatically, as part of the output from the system (just like you can choose to have your data model output as Oracle, DB2, standard SQL, UML, or XML in major data modeling tools).

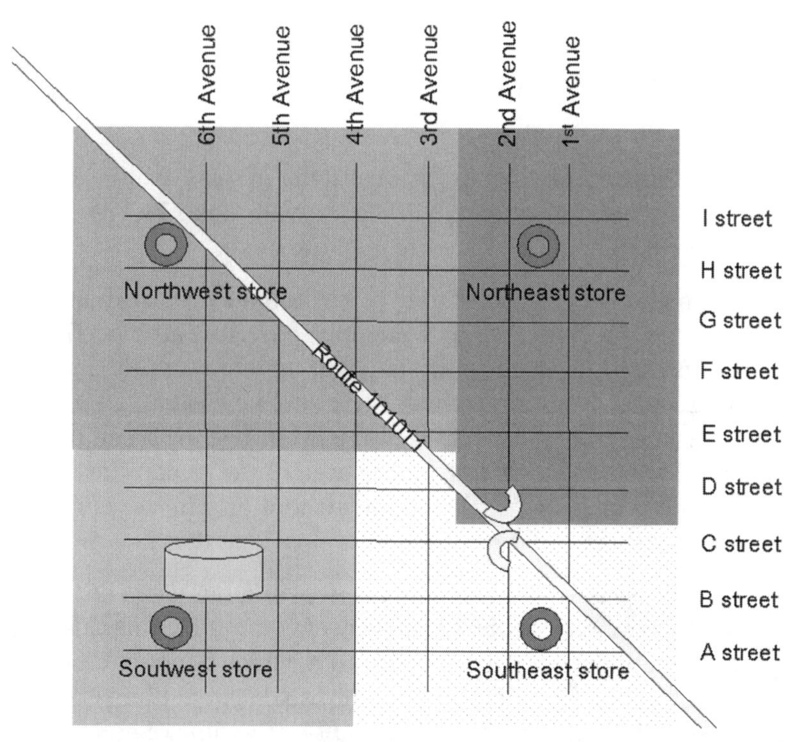

Figure 10.2 The coverage areas of the Middletown Candy stores.

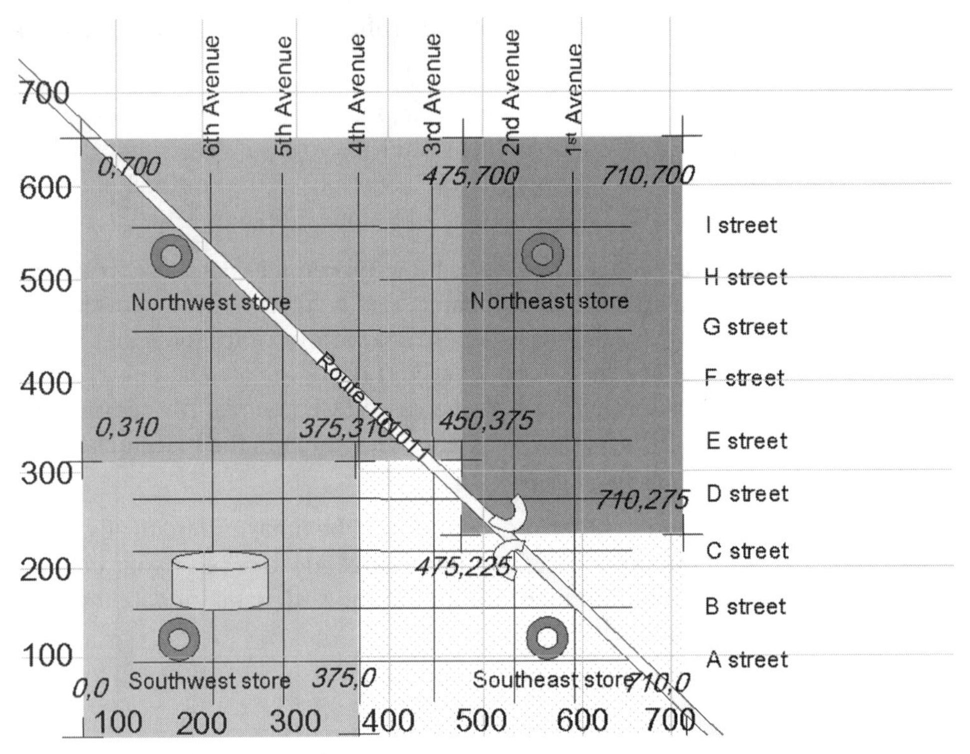

Figure 10.3 Adding a Coordinate System.

If we put Middletown into a coordinate system, in this case the World Famous Middletown Coordinate System (WFMCS), it will look like Figure 10.3. I have also assigned coordinates to the corners of the boxes.

If the user position is an uncertainty polygon, you need to have software in your server that matches that to your polygons. You also need a resolution mechanism in case the uncertainty polygon overlaps two or more polygons. This can be as simple as using a majority rule, like assuming that the polygon with the highest degree of overlap is the relevant one; or it can be some other, for example if a major artery falls in the area of the uncertainty polygon, it is likelier that the user is on that than somewhere else in the polygon, and she should be directed toward the store that is closest to that artery.

The uncertainty areas are assigned by the network, but users have corresponding areas of interest. As yet, there is no way of communicating this to the server (except as a component of the CC/PP profile, which is not standardized). This area is four dimensional; there are the four usual spatial dimensions, plus time (you are less interested in a lunch restaurant by the time of your afternoon coffee break) and are predicated on the entire personal preferences profile of the user (such as whether she likes Chinese food or prefers Italian). This discussion is much larger than we want to go into here, and I touched on it in Chapter 6, "Providing Databases and Doing Searches," and Chapter 8, "The User Interface to Location Information Services," so we will move on.

When you have finished encoding the coverage areas of the stores, the system generates the GML (or other encoding you have selected). If you are a real hacker, you can do this manually in any convenient text editor, but that is not something that any sane person will do more than once. The problem of maintaining the data and avoiding bugs in a data set that is as big as any GML description is likely to become is also insurmountable, so here is where you really need to use computers for what they are good at: keeping track of data.

Your GML will look something like the following (for the northwestern store):

```
<candyStore ID ="MC-NW">
               <description>Middletown Candy Northwestern Store -
Yummy!</description>
               <interestRegion>
                    <Polygon srsName="wfmcs:0000">
                         <outerBoundaryIs>
                              <LinearRing>
                                   <coordinates>
                                          0,310
                                          450,375
                                          475,700
                                          0,700
                                          0,310
                                   </coordinates>
                              </LinearRing >
                         </outerBoundaryIs>
                    </Polygon>
               </extentOf>
</interestRegion>
</candyStore>
```

Remember that you have to repeat the starting coordinate of the linear ring to close the polygon. That is why it is both at the start and at the end of the coordinates.

When you have compared the position of the user to the region of interest and determined which store is the most appropriate, you can compose a response page. There are, however, several steps you can take while composing the content. One is personalization: If you have data about the customer, you can include them in the page. For instance, "Hi, Bill! So, you want some of those yummy jawbreakers again, huh? Well, we have a special offer in the Northwest Store for you! Here is how to get there from where you are," is much more inviting than "Map of Route to Northwest Store from your location." Yes, the real estate of the mobile phone is limited, but if you open a dialogue with the user, you may make him to click on a link for the map when he gets the first page.

Another set of relevant parameters you can use is the CC/PP profile (the WAP UAPROF and other device capability data), such as screen size and so on. By using this, you can create an optimized presentation. There are systems that let you select the appropriate stylesheet for transformations automatically based on the CC/PP profile, so you can generate the map (or route description) that will work best on the user's device.

It is possible to realize the software that performs the functions described here in many different ways. You can use an application server. Or, you can do it using a Web server and scripts. For instance, you can build it as XSLT stylesheets using Xpath to address the documents containing coordinates and other parameters; you can write it in PERL, or Java, or the language of your choice. You can realize it within your DBMS, or as file structures. You can use a commercial product, or open source software. The steps that follow are logical steps, and because the implementation will be so different depending on which architecture you select, I will not go into implementation details.

If you want to get fancy, you can calculate the route and the time it will take the user to get there from her present position. If you know the traffic situation (assuming that it is available in a structured format, which it may be because radio stations in Europe use an XML-based format to send out this information), you can also take this into account. That may mean that if the road is blocked in the route the user would need to take from the origin to the destination, then it may actually be faster to go to one of the other stores. To determine this, you first have to calculate the route and then check the traffic information. Of course, if this is something you intend to do regularly, you will set up a system that reads it often and downloads the information into a database, but we will not go into that now.

Note that depending on how much you take variables, such as demographics and traffic patterns, into account your maps may be very different. To cover Middletown it is not enough to have four maps. If the user is eastbound on Assumption Artery, then he should be directed to store number 2, but if he is westbound, he should be directed to store number 4. This means that you need at least two more maps to cover these cases. If you start to analyze a real situation, you will find that there are close to an infinite number of possible situations, and that you will need to take all of them into account, which means that it is probably easier to generate a map; after all, you can always generate a vectorized map and rasterize it.

This also places requirements on your information architecture. Because we know (as I discussed in Chapter 1, and again in Chapter 8) that the user satisfaction gets much higher when the user gets to the information he wants without any delays, always directing users to the home page is not meaningful. This means that you now have a Web site with at least two different home pages. I say at least, because you might want to create a separate home page for each store (which could then be presented randomly to fixed users and depending on position for mobile users). In a small town like Middletown, it is likely that the users know the location of an address (and given the naming scheme of the streets, they can conclude it if they do not).

Depending on what is most important, the site could have different features depending on the time and date. If one store is close to a movie theatre, and the time of the request is evening, it is likely that the user wants movie candy, so you could feature that. If there is a college football team that plays in a location near a store, you could sell candy in their colors on game days. You do not have to push the data to the users to sell more. After all, why would they look for the site if they did not want candy?

This means that you now have a conditional architecture; which pages you serve depends on when it is and where the user is. Designing this can be complicated, because you want to reuse Web pages, but the simplest way is to design each set of homepages and the following two or three levels as a mini-Web site. The user will not, after all, go deeper if he is satisfied.

Instead of one linear Web site, you now have six (the other two stores will have the same offering independent on the time). And, they interlink at the third level, when they drill into the catalog (it is very unlikely that the mobile users will go this deep, but you still have the fixed users to consider).

This sounds very complicated, if you think of Web sites as electronic brochures. If you are a designer who is used to creating object reuse diagrams in UML, you can do this with your eyes closed and one hand tied behind your back, while you drink a can of Jolt Cola. When you add conditions to the pages and start regarding them as information objects, you can apply object-oriented design to them. They can inherit elements; you can order elements into classes; and you can give the objects behaviors (for example "start appearing at 5 PM in store 3, and stop appearing at 10 PM, except Sundays when the appearance starts at 3 PM. Replace the "licorice twist" element with "caramel popcorn"). It does not have to be more complicated than that.

Of course, it does as soon as you add more objects and more behaviors to the objects. An object-oriented design will let you keep track of what is happening, and it is really the simplest way of designing a system of semi-static objects, which is what we have here. If your site is entirely dynamic, a process approach may be better, as I discussed in Chapter 6, "Providing Databases and Doing Searches."

Buying Databases and Maps

If you do not want (or cannot) create the data for your location-based application yourself, you will have to buy it. Or, you have to buy part of it, for example demographic information. A number of sources will provide you with both map data and database data, but the format and the terms will be very different. In the United States, for instance, all data produced by federal agencies is in principle free for use by citizens. In other countries, data may not be free, and depending on whether anything has been added to the data, it may be the product of commercial providers. Basically, you have to weigh the tradeoff: Is it worth getting the data from someone else, or can you create similar data yourself cheaper?

You need to consider a number of things when you acquire data: What is it intended for? Who guarantees the quality? Which geography is it intended for? Which coordinate system does it use? What is the file format? Who maintains the data? Also, remember to think not only about what your data needs are now, but also think about what they will be in the future.

If you intend to use your data for engineering purposes, the data has to be accurate. If you are going to use it for a tourist application, the accuracy may not be so important. The data quality also becomes a factor if you are going to use your data for serious applications. You need to be sure that all the objects in the database are correct.

The geography is important because the data must match the location for which you intend the service, not only when it comes to coverage, but also when it comes to the types of data needed. If you are planning an application for rural Kansas, having data about skyscrapers may not be required.

Map databases available from a location-dependent map service may contain a huge amount of geographic information. They may contain streets, public buildings, residential buildings, objects of cultural interest, restaurants, hotels, bus stops, parks, or tourist information just to mention a few common examples. Other examples of more specialized nature may be locations for spotted

Things to Consider When Acquiring Datbases

What is the database intended for?

Who guarantees the quality?

Which geography is it for?

Which coordinate system does it use?

What file format is it in?

What is the data model?

Who maintains the data?

unusual birds, a meeting spot for a skating club, or the location of a broken pipe in a city. These may be grouped as different layers, for example objects of a certain type (organized in elements). A map containing all this information would be considered very messy and unclear. Both the capability of defining whole layers to be presented on the map as well as the choice of specific objects should be provided.

You should also consider the data format: first the file format and then the data model. Both are addressed if you require the data to be in GML, but so far, almost no commercial map providers deliver GML data. The most widely used format is the ARC/Info format. MapInfo and ESRI are two other frequently used formats, but this is highly dependent on your application server. File format may also affect another important factor: the size of the file. GIS files can be very large, if the coverage area is large. What is the delivery medium? How can it be subdivided to be manageable? Does the data layering help?

The coordinate format is also a factor that you need to take into account. If the data does not use the same spatial reference system as your other data and your application, then you will need to make sure that there is either a conversion function, or or that the data is converted into the correct format before it is delivered to you (and you may want to check the conversion algorithm, as well). Which coordinate system and datum are used must also be checked in advance. If you are at the equator on the zero meridian, this may not be a problem, but if you are closer to the poles, this will matter. And if you are going to use the data to display which cities you have offices, it may not be as important with a correct datum as if you are going to use the data for surveying.

How you intend to display your data must also be taken into account. If you want to show it on a mobile phone screen, it should not have too much data; or at least, a function to select which data is to be displayed, like you get if you use SVG. You need to be able to turn the data layers off and on as required. If you intend only to use the data as an image, you may be content with a raster image. The difference between vector data and raster data we discussed in Chapter 9. For raster maps, you need to know the scale, as well. This may not be required in vector data. Remember that you may need to buy several maps on different scales, if you want to allow the user to zoom in or out of your initial map.

Remember, the work does not stop when you have created the database. There is a maintenance cost, as well. Someone has to keep the database updated, guarantee data quality, and deliver the updates. This means that whom you choose as data provider becomes important. If you would rather go for a cheap solution, which may not be there tomorrow, you can just buy any old map from anywhere. If you want the map to be maintained for longer, you need to find a reputable data provider. Another thing that is important is, of course, is that the data provider listens to you.

Building a New Application

Building a new application is somewhat different from enabling an existing Web site to handle location information. You have considerably more freedom when you are creating an entirely new user experience and do not have to take the old look and feel of the Web into account. As an example, let us create a simple tourist service for Paris. I am sorry for the quality of the pictures; the publisher did not want to pay me to go to Paris to take them. This example is not entirely realistic (that would not be quite as fun), because I am going to assume that there is positioning coverage everywhere, including in the subway. If you have ever traveled by subway, you know that is simply not the case. But in the future, there might be Bluetooth micropositioning beacons available or something similar (even where there is mobile phone coverage in the subway, it might not include positioning, as we discussed in Chapter 2). I am also going to ignore the challenges of buying tickets from French vending machines (let's just assume that the user has previously bought a carnet—a bundle of 10 tickets —for now).

I am trying to show this information on the screen of a mobile phone, but it is an imaginary device because these services are not available in Paris as of yet (although there are map and direction services, better in France than most other countries, thanks to companies like Webraska). Let's divide it into steps based on what happens in the user interface.

Going from Place de l'Étoile to Montmartre via subway:

1. User makes a request for the route (see Figure 10.4).
2. Locate user (or we can simply assume that the position of the user is sent as part of the request; for instance, in a WAP Location Framework Appendix).
3. Initiate periodic triggering of the user's position.
4. Present map of surroundings with subway entrance marked (see Figure 10.5).

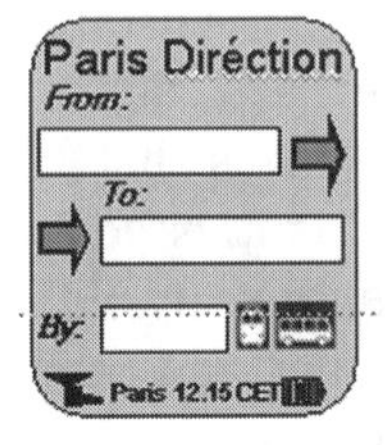

Figure 10.4 A request screen for a location-based service.

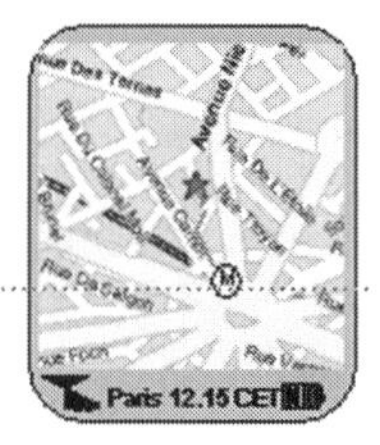

Figure 10.5 Map of the neighborhood with a route to the subway station.

5. Show subway timetable and indicate the train that it is likely that the user will be able to take given his current position, mode of locomotion, and the distance to the subway station. Also, indicate the direction and end station of the train (see Figure 10.6).
6. As the user comes closer to the subway entrance, move the map so that the position marker stays centered in the screen but the movement of the user becomes apparent. Add a red line to the user's previous path and a green line to the path he or she is taking.
7. As he or she comes within 200 meters of a subway entrance, flash the position of the entrance on the screen. Show a photo icon at the bottom of the screen for the user to select whether he or she wants to see a picture of the entrance (see Figure 10.7).
8. As the user enters the subway, show the screen in step 4 again, updated for the current time.

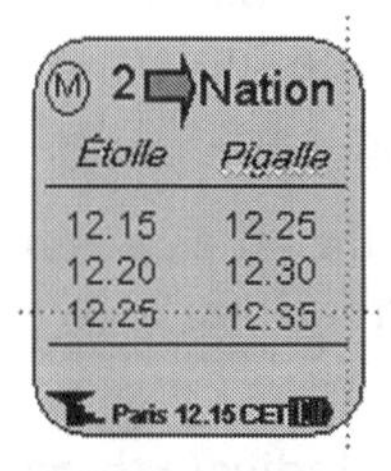

Figure 10.6 Subway timetable with line and direction.

Figure 10.7 Picture of the subway entrance so the user can identify it.

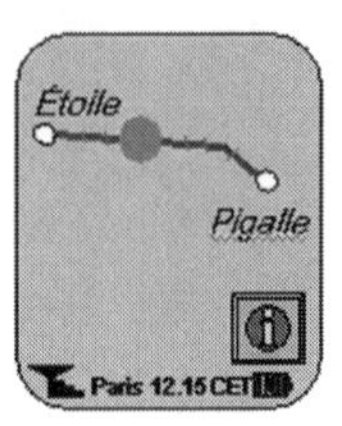

Figure 10.8 Showing the progress of the user on the subway train.

9. Show a map of the subway station with the position of the platform the user is supposed to get on marked with a green X. Show the current position of the user with a red dot. Insert the possibility for the user to click a photo button to get a photo of the turnstiles that he or she is going to go through.
10. Show the timetable screen again.
11. As the user enters the train, show a video of the train approaching the destination station.
12. Show a map of the subway line. Indicate the current position of the user with a red dot. Indicate the target station with a flashing green X (see Figure 10.8).
13. As the train approaches the destination, show the video of the train approaching. Make a sound signal.
14. As the user exits the train, show a map of the destination station. Show the current position with a red dot and the destination exit with a green X. Show the path from the current position to the destination exit. Display the exit number and the name of its closest features on the screen.
15. Move the dot along the path as the user moves through the station.
16. As the user exits the station, display a street map. Mark the current position with a red dot and the destination with a green star (see Figure 10.9).
17. At the bottom of the screen, create two buttons that indicate there are two different ways to get to the top. Include hyperlinks to the funicular.

Figure 10.9 Display of a street map of the user's path.

Figure 10.10 Display of the destination of the user with a link to information.

18. (We will assume that the user chooses to walk.) Show the path to the top of the bout de Montmartre. Include clickable links to the Sacre Coeur cathedral and general tourist information.
19. As the user progresses toward the destination, show the position marker moving along the path. Show the path taken in red and the path left in green.
20. At the destination, show a screen with an image of the cathedral and a clickable link to additional information (see Figure 10.10).
21. Terminate periodic positioning of the user when the destination is reached.

As you can see, this situation brings about a user interface problem not only in that you have to show where the user is and where he or she is going, but you also have to illustrate that he or she is setting off an action (the periodic positioning) that will be present on the screen during the entire time he or she is using the application. Because the information stays in the application server and does not go to the mobile phone, you cannot represent it with a local information icon (like the field strength indicator on the mobile phone screen), but you have to represent it with an animation that indicates that a process is going on. What that icon should look like is another question.

Developing the Database Structures for a Position-Dependent Application

From Chapter 5, "The Application Server," you remember that how the data flows in the application depends on whether it is a triggered application, in which case the MPC will return periodic positioning messages, and the application server has to forward them to the appropriate application (which will take the appropriate action). Or, if it is a request-response application, the user would generate a request that starts the data flow.

Now, let's look at an example. Actually, it is really a generic database example, as I am sure you will note—the position dependency comes in as a geocoding of the data elements. Deciding which elements will be used is certainly part of the development process for a position-dependent application, however. It is relatively easy to see what data we need to create the application, once we start breaking it down, and then we can put it into an ERB diagram and build the databases from the descriptions of the data structures.

Let's go back to the example we had before and the steps based on what happens in the user interface.

Going from Place de l'Étoile to Montmartre via subway:

1. The user makes a request for the route.
2. Locate the user (or we can assume that the position of the user is sent as part of the request, for example in a WAP Location Framework Appendix).
3. Initiate periodic triggering of the user's position.
4. Present a map of the surroundings with the subway entrance marked.
5. Show the subway timetable and indicate the train that the user will likely be able to take given his current position, mode of locomotion, and the distance to the subway station. Also, indicate the direction and end station of the train and show the ticket price on the screen.
6. As the user comes closer to the subway entrance, move the map so that the position marker stays centered in the screen but the movement of the user becomes apparent. Add a red line to the user's previous path and a green line to the path he is taking.
7. As he comes within 200 meters of a subway entrance, flash the position of the entrance on the screen. Show a photo icon at the bottom of the screen for the user to select whether he wants to see a picture of the entrance.
8. As the user enters the subway, show the screen in Step 4 again, updated for the current time.
9. Show a map of the subway station with the position of the platform the user is supposed to get on marked with a green X. Show the current position of the user with a red dot. Insert the possibility for the user to click a photo button to get a photo of the turnstiles that he will go through.
10. Show the timetable screen again.
11. As the user enters the train, show a video of the train approaching the destination station.
12. Show a map of the subway network. Indicate the current position of the user with a red dot. Indicate the target station with a flashing green X.
13. As the train approaches the destination, show the video of the train approaching. Make a sound signal.
14. As the user exits the train, show a map of the destination station. Show the current position with a red dot and the destination exit with a green X.

Show the path from the current position to the destination exit. Display the exit number and the name of its closest features on the screen.

15. Move the dot along the path as the user moves through the station.
16. As the user exits the station, display a street map. Mark the current position with a red dot and the destination with a flashing green X.
17. At the bottom of the screen, create two buttons that indicate there are two different ways to get up to the top. Include hyperlinks to the funiculaire, the cog railway.
18. (We will assume that the user chooses to walk.) Show the path to the top of Montmartre. Include clickable links to the Sacre Coeur cathedral and general tourist information.
19. As the user progresses toward the destination, show the position marker moving along the path. Show the path taken in red and the path left in green.
20. At the destination, show a screen with an image of the cathedral and a clickable link to additional information.
21. Terminate periodic positioning of the user when the destination is reached.

Before we start drawing the diagram, remember that the tool you use only does the formal analysis. It does not identify the objects for you; that is still a semi-magical process you have to do yourself. Let's walk through the example again to see which objects we have:

1. The user makes a request for the route.

   User request

2. Locate the user (or assume that the position of the user is sent as part of the request, for instance in a WAP Location Framework Appendix).

   User position

   User mode of locomotion (vector, if available)

3. Initiate periodic triggering of the user's position.

   Triggered positioning request

4. Present a map of the surroundings with the subway entrance marked.

   User position

   Area map

   Subway entrances

5. Show the subway timetable and indicate the train that the user likely will be able to take given his current position, mode of locomotion, and the distance to the subway station. Also, indicate the direction and end station of the train and show the ticket price on the screen.

User position

Desired position

Subway timetable

Correct line related to the current position and desired position

End station

Calculation of time to reach the station

Current time

Subway timetable

Time for next train

Ticket price

6. As the user comes closer to the subway entrance, move the map so that the position marker stays centered in the screen, but the movement of the user becomes apparent. Add a red line to the user's previous path and a green line to the path he is taking.

   User's previous route

   User's future route

7. As he comes within 200 meters of a subway entrance, flash the position of the entrance on the screen. Show a photo icon at the bottom of the screen for the user to select whether he wants to see a picture of the entrance or not.

   User's position

   User's position marker

   Subway entrance position

   Subway entrance marker

   Photo icon

   Photo of entrance

8. As the user enters the subway, show the screen in Step 4 again, updated for the current time.

   User's position

   User's desired position

   Subway timetable

   Correct line related to the current position and desired position

   End station

   Closest station to desired position

Calculation of time to reach the station

Current time

Subway timetable

Time for next train

Ticket price

9. Show a map of the subway station with the position of the platform the user is supposed to get on marked with a green X. Show the current position of the user with a red dot. Insert the possibility for the user to click a photo button to get a photo of the turnstiles that he will go through.

   Subway station

   Correct platform

   Closest station to desired position

   Route to correct platform

   Link to photo

   Photo of entrance

10. Show the timetable screen again.

    User position

    User desired position

    Subway timetable

    Correct line related to current position and desired position

    End station

    Closest station to desired position

    Calculation of time to reach the station

    Current time

    Subway timetable

    Time for next train

    Ticket price

11. As the user enters the train, show a video of the train approaching the destination station.

    User's position

    Closest station to desired position

12. Show a map of the subway network. Indicate the current position of the user with a red dot. Indicate the target station with a flashing green X.

    Subway map

Current line

User's position

13. As the train approaches the destination, show the video of the train approaching. Make a sound signal.

    User's position

    Closest station to desired position

    Sound signal

    Video of train approaching the station

14. As the user exits the train, show a map of the destination station. Show the current position with a red dot and the destination exit with a green X. Show the path from the current position to the destination exit. Display the exit number and the name of its closest features on the screen.

    User's position

    User's desired position

    Station information

    Station map

    Correct exit

    Exit name and number

    Path on map

    Features closest to exit

15. Move the dot along the path as the user moves through the station (same as Step 14, actually).

16. As the user exits the station, display a street map. Mark the current position with a red dot and the destination with a flashing green X.

    User's position

    User's desired position

    Area map

    Subway entrance

17. At the bottom of the screen, create two buttons that indicate there are two different ways to get up to the top. Include hyperlinks to the funicular.

    User's position

    User's desired position

    Area map

    Mode of transportation

Walking

Funicular

Link to Funicular

18. (We will assume that the user chooses to walk.) Show the path to the top of the Butte de Montmartre. Include clickable links to the Sacre Coeur cathedral and general tourist information.

    User's position

    User's desired position

    Area map

    Path to desired position

    Link to tourist information

19. As the user progresses toward the destination, show the position marker moving along the path. Show the path taken in red and the path left in green.

    User's position

    User's desired position

    Area map

    Path to desired position

20. At the destination, show a screen with an image of the cathedral and a clickable link to additional information.

    User's position

    User's desired position

    Photo of cathedral

    Link to tourist information

21. Terminate periodic positioning of the user when the destination is reached.

    Terminate positioning

What do we need to do in the way of database development here? We could structure the data in such a way that we have four primary data sources (apart from the user's position itself): the map database of subway stations, the subway timetable, the map database of the city, and the database of video and photo images of the subway stations. We also have an external data source, the tourist information.

Of course, we can structure the data in a different way, as well (probably more than one different way, actually, but let's be content with one different way for now). Data can be ordered according to location type, for example. In that case, we would have one entity containing all the station information, including the maps of the stations, the photos of the entrances, and the videos of what it looks like coming into the station. We would have one timetable for the entire network, including destination information and final destinations. And, we would have a table with the maps of the city.

Which of these you should choose depends on a lot of variables, including how the data is created (for instance, the city maps might exist as raster images, or they could be generated as vector maps of the city from a database of geographic features encoded in GML); how the data is maintained (in other words, who maintains it; if you have a previous database that you are extending, so you must take existing data structures into account; organizational reasons; and so on.

It is worth noting that in principle, we have only one state in the database for this example, and it will be true for a large class of other position-dependent applications, as well. We are only retrieving items, not manipulating their values—that can be done at the user interface (alternatively, you could imagine an older Web browser receiving a series of images of the maps, but these would most likely be generated by a separate piece of software in the server and not retrieved from a database). It would be different if the changing position actually meant changing something in the database—for example, updating a driving log. This situation is one major difference between tracking and retrieving applications. The objects we fetch from the database have one thing in common: their position. In other words, it is a good idea to use the position as the foreign key value for the database.

Scenario-Based Development

In this chapter, I have told little stories that have been the basis for the creation of a slightly more formal use case description. If you are familiar with object-oriented development methods, you will have recognized the technique. What you may not have done is personalize it to such a degree.

A formal description, either in UML or some other language, is not necessary before you start the coding. Ambiguities in software specifications are almost always to blame for conflicts between the vendor and the customer, however, so it makes a lot of sense to be as clear as possible on what you are developing and how you proceed to make sure that the development is consistent and fast.

What is not usually taken into account is why. The exception is the Nordic user-centric development methods, which insist on user involvement in the development process. They work fine when you are developing for a concrete group of people, who can come into the room and test the applications. They do not work well when your system is intended for a user you may not even know by name, except from his credit card.

In those cases, scenario-based development can be very useful. Your advertising agency (or marketing department) probably has an idea of your model customer and maybe has already constructed a sociographic prototype (perhaps without knowing it). You can also find out a lot from your existing customer databases—if you have them. One of the major differences between online stores and brick-and-mortar stores is that the physical store does not keep a record of each user. If somebody walks into Wal-Mart and buys a gallon of milk for $2.50, that company has no way of knowing that this is the same person who was in their store in the next town three days ago and bought five sixpacks of beer.

Customers value this privacy and trust the store to ask before it tries to find out about them. Violating that trust is a surefire way of losing your customers.

Scenario-based development works best when you anchor it in a well-known reality, which is why I selected the name of one of the most well-known store chains in the United States in the privacy example I just cited. Create a person you can identify with and use a location you know to check whether your system will work. If you imagine yourself in the shoes of the person you are telling a story about, it is very easy to spot any logical faults that might come back and haunt you later.

The important thing for your future development is that it fits with the specifications. Web sites are, however, notoriously badly specified and almost never developed according to some process, rather as a haphazard advertising agency adventure. With database-backed development, some order has come to the industry, and with the creation of location-based services, the requirement for order will be larger still. Starting the creative end in a scenario and going to a formal data model is an efficient way of satisfying all the stakeholders in your development process.

It is also very easy to get buy-in from outside stakeholders (like investors) and comments from prospective users on scenarios, but very hard on technical specifications. They start to fall asleep before the third line of UML, and they turn the sequence diagrams upside down. But, even your grandchildren understand stories. Remember, even though the *customer* may be a large company with a professional development department, the *users* are probably ordinary people, and if you are developing a new service, that is definitely true. You need to approach them in a way they can understand and believe.

Transporting Data

A problem in applications intended for presentations on mobile phones is the size of the files. Network latency is already a problem on the Web, which of course is compounded by retrievals from several locations into one presentation.

Even at regular modem speeds, however, the file size of some Web pages may be a problem. At the low speeds available in the mobile networks, it becomes like the Web was in 1994, or maybe 1997. Despite the theoretical transfer rates that all manufacturers and mobile providers enthusiastically quote, the practical data rates for GPRS may be only 28.8 Kbps, and for IMT-2000 only 64 Kbps (the latter because the Japanese operator NTT DoCoMo has determined that it should be).

Using SVG, you can do a few things to minimize the transfer time. One simple way to create good transference is to minimize the size of the SVG files. The SVG language has many possibilities for doing this. Two elements that can help in reducing the size of the vector image are the path element and the use element. The support for Bézier curves in the path element makes it unnecessary to tesselate curves into polylines when converting from tools used for CAD, Computer Aided Design. Besides, the coordinates in a path element are denoted in a much abbreviated form.

The use element makes it possible to use a graphic that has been defined once in the file an unlimited number of times in the document. Typically a symbol element is referenced. The strict handling of text fonts makes it unnecessary to draw the outline of characters, which is normally done to preserve the exact shape of texts. When using the SVG fonts, the texts can be sent as character data instead, because there is no risk that the fonts will be exchanged due to a of lack of support in the presentation system. In fact, the whole CSS styling mechanism helps in keeping the file size down (especially because the stylesheets can be cached).

A compression tool for text files, such as gzip, could be used and reduces the file size considerably. However, the text compression tool is also needed for the decompression, and the tool may be unavailable in the system receiving the SVG document, such as a mobile phone (where the processing would require more capacity than the processor in the mobile phone has available).

A way of making it seem like the file is downloaded faster is to use progressive loading. Especially those elements that the user could be expected to interact with should be placed at the top of the file. After that, the important elements, giving an orientation of what the image looks like, are downloaded. In a map application these objects could be for example roads and borders between built-up areas and other areas. This may require that the server takes this into account, which it of course can do if it gets a CC/PP description of this module from the client.

The transformations normally will be done in the server, at least in the current state of the mobile terminal industry (and the network capacity). As I mentioned previously, the mobile phones are rarely capable enough to render anything but text and raster images. However, some mobile phones have more capability. What they are, and how they get them, I will talk about in Chapter 11, "Location-Based Services in Terminals."

CHAPTER 11

Location-Based Services in Terminals

The handset—the mobile telephone, PDA, or special-purpose receiver—is not a very good place to deploy services. The Internet end-to-end model did not take into account that you might want to present a highly interactive, vividly graphical service in a device that has a smaller processor than a Furby and less memory than the average technical calculator.

As of this writing, the mobile terminals on the market do not have any implementations of hardware or software to support any of the terminal-based positioning solutions, for example, GPS or E-OTD. However, in order to fulfill the accuracy requirements defined by FCC, if UL-TOA is not to be implemented, software and hardware to support one of the terminal-based solutions (or both) have to be added to the mobile terminals.

The better accuracy of the position information obtainable, the more applications for location-dependent services are conceivable, and the more useful they get. Therefore, some users may not find E-OTD sufficient with respect to obtainable accuracy for the services they are interested in. For these users, support for Assisted GPS in the terminal should satisfy their need for accuracy, which is the most accurate positioning solution for mobile terminal application. A terminal with a GPS receiver will, however, cost a bit more, and some users may not want to pay that extra amount for the accuracy enhancement compared to E-OTD.

Nor is the software required for the more interesting applications implemented. Web phones—at least, in the foreseeable future—will receive a ready-made presentation from the server. While they will probably have some limited capability to execute Java code, 128KB of memory is not something that enables you to build particularly impressive services, although SVG viewers are available that will work in a mobile phone. Still, there is a programming environment for them that executes on their even punier companion processor, the SIM card. There is also one pioneering device that has more programmability: The Benefon Esc.

Combined Mobile Phones and GPS Receivers

Many GPS receivers have map services built in, but very few can communicate the position that they receive to another service to receive a response that is adapted to the request. An example of a device that can perform this task is the Benefon ESC, which is a combined GSM mobile telephone and GPS receiver. This device has a large enough window to present a position-dependent application—for instance, with a map interface. That is an exception, however, and will most likely remain that way for as long as the price of mobile phones is a primary buying argument.

GPS receivers can be made small enough to be integrated into a handset, and several manufacturers have demonstrated prototypes. Just having the GPS receiver in the handset, however, means that you have to perform a cold start every time you want to retrieve a position. You need a way of getting additional data about the satellites to the GPS receiver. From a cold start with the almanac, it typically takes two minutes to determine its position. A warm start takes about one minute; a hot start (with the last fix less than one minute old) takes about 15 seconds. Without the almanac, however, the time it takes to acquire the satellites is approximately 12 minutes.

There is also another concern. When you are using a protocol like the WAP Application Framework of LIF API, you send the position request along with the identity for the terminal that is to be positioned to the application server or MPC. All of the traffic concerning the positioning, including the execution of triggered requests, can be done in the fixed network connecting the base stations to the MPC (because you are asking for information that the network has collected anyway). But when you have the positioning in the handset, you have to transmit the position over the radio channel.

That might not seem so problematic, but consider the numbers. If 720 users in a cell on a GSM network are sending position information messages of 256KB every three seconds, that means they are using 400 kbit/s. The capacity of the cell is 2160 kbit/s. So, they will be using some 18 percent of the capacity of the cell for signaling.

Traffic modeling is more complicated, of course. Because the traffic tends to be bursty, there are fixed allocations of capacity that are required for certain traffic types and so on. But that simple calculation shows that just signaling would use up a significant portion of the traffic, assuming that the return messages with the personalized information are the same, independent of the technology used. Another matter is pricing. Nobody knows how the tariffs will be set in the end, but given the charges of some European operators for the GPRS traffic (the packet data traffic on the GSM network), it would become extremely expensive to keep sending your position.

Benefon Esc

One of the first mobile phones that has a built-in GPS receiver is the Benefon Esc, built by the other Finnish mobile phone company, which in a sense combines the best of two worlds (Benefon is actually based in Finland, the same as market leader Nokia, but it has never reached their market volumes and remains specialized in niche applications). It is shown in Figure 11.1. While it uses GPS and has all the mobile communications possibilities of a mobile phone, it actually loses out because the GPS function cannot handle assisted GPS.

Strictly speaking, with the Benefon Esc you are not developing for the mobile phone; instead, you are developing for the system. The adaptation of the data is done in the service center (in other words, the application server), and the phone is providing coordinates that are used to adapt the data. The service center is a specialized application server (although it should not be too hard for application server developers to provide this kind of functionality if it becomes popular). The phone has a built-in GPS receiver, but it can also

Figure 11.1 The Benefon Esc.

receive position information from the GSM network (or, strictly speaking, the service center can).

To enable development by others and to allow service providers to supply information, Benefon has released the protocol, the Mobile Phone Telematics Protocol, which is used between a service center and a terminal. It uses the *Short Message System* (SMS) of GSM as a carrier format, but that is not an absolute binding—other carrier formats could be used if required. The format has two parts, text and binary, so the conversion should be simple. The communication between the service center and the mobile terminal is binary, but when communicating with a regular GSM phone, it uses text format. Functions enabled in the protocol are primarily intended to handle emergency calls and security and safety applications.

In addition to the message part of the protocol, it also has a numbering for a series of images that can be used as icons. They include circles, squares, arrows (relative to the compass directions), and a lethal danger symbol in addition to icons for buildings such as banks, hospitals, kiosks, camping sites, and photo shops. It also includes religious buildings such as synagogues, mosques, churches, and temples. There are icons to represent different modes of transportation (possibly only Finnish developers would have imagined a snowmobile as a standard mode of transportation) and travel symbols such as electric outlets, sight, shower, toilet (and again, betraying its Finnish roots with symbols for lean-to shelter and orienteering control point). Sports symbols include golf, riding, ice hockey, and swimming (but not bandy).

The commands are based on characters already implemented in GSM and use ASCII text as the base format. Messages in the protocol have a header that is always constant. All field lengths except the header and CMD include one field separator character; in other words, the preceding underscore. The field separator is the underscore (_) character. All messages from the service center or normal to a GSM phone start with a question mark (?), and all messages from a terminal using the protocol terminal start with an exclamation point (!). If information is not available, the corresponding field is filled with "-" characters.

The basic commands of the protocol are listed in Table 10.1. They are what are supported by the Esc telephone, but Benefon also has several other models that can provide functions that are more complicated.

The idea is that the service center sends one of these commands to the mobile terminal, which then responds with a message reporting its status and position. It looks like Table 10.2. The second-row figures are the length of the fields, the third row is an explanation, and the fourth is an example.

The Benefon protocol does not use HTTP but SMS for data transfer, which makes it somewhat different from Web-based applications. The terminal can

Table 10.1 The Basic Commands of the MPTP Protocol

COMMAND	EXPLANATION	DIRECTION
CRO	Create route: MPTP terminal can send create route request to SC.	MPTP terminal to SC
EMG	Emergency	MPTP terminal to SC
LOC	Location information	SC to MPTP terminal to SC
NMR	Network measurement report	SC to MPTP terminal to SC
RSE	Route sending	SC to MPTP terminal to SC
RWP	Request waypoints	MPTP terminal to SC
STO	Stop trigger: This will stop only those triggers, which are activated with the easy version of tracking TRC.	SC to MPTP terminal
SWP	Send waypoints	SC to MPTP terminal
TRC	Easy version of tracking	SC to MPTP terminal

send back a position in response to a request from the service center. The request from the service center is sent as a response to a request from the user.

The TRC command is intended to set up a tracking service based on the number of tracking messages and the interval between them (the tracking messages here go from the phone to the service center).

The system can also transmit a route to the terminal by using the RSE command. The route points can have names (instead of numbers) in addition to coordinates. Up to four route points can be sent in one message, but by using the "part" field, messages can be chained so that up to 99 messages can be sent (because the field length is six and has to contain the number of the current message, a slash, and the number of total messages). The last point of the route has a name and an icon (one of the encoded icons that are part of the protocol). The terminal can also request a route and a waypoint from the service center.

The terminal can also be used to send network measurements to the service center (and it is likely that this market will be big for it in the short run, because there are many network operators who want to measure where their competitors' cells are located). The network measurement includes the country and network codes, the cell identification number, the time advance, the transmission power, and other related data.

SIM Toolkit

Traffic is one factor to consider, because it might make GPS receivers in terminals unsustainable. Another factor is the processing capacity. The mobile station

Table 10.2 Position and Status Reporting Message

HEADER	COMMAND	PART	MODE	BATTERY STATUS	POSITION SOURCE
1	3	6	5	5	4
	According to Table 9.1	Indicates which part of the command this message is	Indicates terminal operating status. Three values are possible: norm (normal); emer (emergency); and test (test mode)	Battery status in percent of fully charged, three figures	Two possible values: ps (GPS receiver); gnet (GSM network)
	LOC	10/12	norm	025%	gps

POSITION FORMAT	POSITION	TIME STAMP	SPEED	DIRECTION	DATA
2	25	20	8	7	60
Three possible values: 0 (unknown); 1 (WGS84); 2 (UTM)	Position information. If the position is not available, the network position can be sent. If no position is available, filled with blanks.	GPS time	Calculated from GPS. Can be blank if not available	Calculated from GPS. Can be left blank if not available	Optional
2	N68.28.43,9 E027.27.02,4	31.12.1999; 12:13:15	40 km/h	90deg	

in a mobile telephony system might contain pretty powerful signal processors, but as a computer it has a long way to come close to the current PDAs. Processing power, and memory even more so, costs money. All manufacturers of mobile equipment do everything they can to shave off as many cents as possible off the price, which of course is good for the consumer because the mobile phone becomes cheaper. But on the other hand, the consumer does not get a device that is capable of doing very much. Even the smart phones, like the Nokia Communicator and the Ericsson r380, do not have much processing power available when compared to devices like the Compaq iPac or the HP Journada. The operating system might be better, but there is still a ways to go until you can do the same type of processing as on the PC.

Even in the most humble cell phone, however, there is a computer that can be used by applications developers. It is the SIM card, which is a smart card used to process calls and manage the traffic from and to the telephone (and to make sure it is logged into the billing system with the correct identity in GSM and IMT-2000, that is). Systems like TDMA, CDMA, and PDC do not have a smart card built in.

Compared to the processor in the telephone, the processor on a smart card is laughable. But performance is not the issue when developing handset applications; the main presentation format is text, after all. The main issue is access to information, and in GSM and IMT-2000, that is provided through a standardized set of APIs, the SIM Toolkit (there is also a Java virtual machine available for smart cards, but it does not have the interfaces to system functions on the smart card that the SIM Toolkit has). SIM is an abbreviation for Subscriber Identity Module, and the telephony part of the handset interacts with the SIM to receive the identification of the user. Other information, like the telephone number of the user and the calling party, is also stored on the SIM (and transferred from there to the telephone book when requested). It is the fact that the SIM is the point of interaction for data services with the telephone that makes it interesting as an application platform. The SIM card is the receiver and transmitter of all SMS messages, and it also has access to all other information that pertains to the users of the telephone system (including the cell information).

The original SIM cards had fewer than 10 kbytes of memory, but the cards will soon have more than 100 kbytes of memory. They have also been given more functionality than when first released in the 1980s. The standard is set through the *European Telecommunications Standardization Institute* (ETSI), which sets the standards for GSM. It is the ETSI SMG9 that defines the box of programming tools and protocols that comprise the SIM Application Toolkit (the number is GSM 11.14).

The SIM card controls the mobile telephone, and it receives all the information that comes to the mobile phone. This information includes the network and cell identifiers, which are the base for location-dependent services based on the SIM Toolkit. It also controls what is displayed to the user and can communicate with the network by using the SMS, which was introduced for this type of communication even if it is now mostly used for messaging between teenagers.

Today, there are a large number of applications based on the SIM Toolkit deployed. Among the applications developed using the SIM Toolkit, Internet access, mobile banking, and location-dependent services seem to be among the most popular. The SIM card can also download new applications (if they are encoded as SMS messages).

Internet applications using the SIM card are based on a browser that is located on the card. It is, in turn, based on the WAP standards. The most popular applications are those that are dependent on security, however, such as banking, online brokerages, and payments. The card is more secure than the rest of the phone because it not only manages things like telephone numbers and the like, but it also contains the encryption key for the communication (all GSM communication is encrypted, including the voice calls).

Location-dependent information based on the SIM Toolkit means that the card requests the area information based on the cell ID to the operator and then receives messages containing the information.

There is also a standard Java Card API for SIM cards. The Java-compatible SIMs can run a number of programs independently. The standard is also used in the credit card world. With Java, the card can also be reprogrammed in the field. A Java card runs a *Java Virtual Machine* (JVM). The card handles applications according to the standard behavior for virtual machines, but it is also possible to upgrade the software in a SIM via over-the-air data communications technology or other techniques, such as downloading software via *point-of-sale* (POS) terminals in the network operator's retail outlets. The operator can also remotely update the cards without the user requesting it (because the assumption is that the user does not own the phone; the operator does). Only changes to the applications need to be transmitted.

Smart cards are also integrated with WAP 1.2 (and 2.0, where it is replaced with TLS, the same standard as is used on the Internet) standards. There is a smart card-based application known as a WAP Identity Module, or WIM. The WIM handles the Wireless Transport Layer security between the WAP gateway and the mobile terminal. Several different cryptography algorithms are available. The card can also be used for application-layer security, using a digital signature and non-repudiation techniques.

The SIM card is a GSM standard, but there is a similar standard for CDMA called the *Removable User Identity Module* (RUIM). In IMT-2000, there is a similar function: the *Universal Subscriber Identity Module* (USIM).

Programming with the SIM Toolkit is not simple, and you need to access the operator's network, not just the SIM. The assumption in the system is that the operator owns the application, the network, and the SIM card. So, instead of going deeper into it here, I will include the specifications on the CD-ROM that comes with this book.

Java in the Mobile Phone

When Java became available in Web browsers, it was hailed as the biggest thing since shaved ice. It was supposed to change the user interface of the computer forever. Well, a lot of water has run under the bridges since then, and Java has established itself thoroughly as a server programming language. It turns out that what you can do as a developer is very limited—far more limited than was originally thought. The standard MIDP and CLDC implementations do not actually enable more than very simple animations and games. It is when you get additional functionality, such as sprites for games (which the Japanese company Jphone has in its Java phones), that you can start doing really interesting things.

The first company to manufacture a mobile phone using Java was Nortel, but that phone hardly sold at all—being released before the standard was ready. The first network operator to release a phone with Java was Nextel in the United States, which released a phone based on the MIDP profile. Next came NTT DoCoMo, which released a phone based on K-Java—but it has a very special operating environment that does not allow the Java applet to interact with the network other than through the browser. Jphone in Japan came next, and it released a mobile telephone with the full MIDP profile (and some additional goodies, such as three-dimensional sprites for games programming) in 2001.

The Java libraries in the mobile telephones consist of the *Connected Limited Device Configuration* (CLDC), which is based on the *K Virtual Machine* (KVM) and is implemented on top of device operating systems and provides the interface between specific phone operating systems and higher-level Java technology-based applications. The J2ME *Mobile Information Device Profile* (MIDP) is layered on top of the CLDC and includes APIs for the application life cycle, HTTP networking, persistent data storage, and *graphical user interfaces* (GUIs), as shown in Figure 11.2.

CLDC is a basic set of libraries and Java virtual machine features. It must be present in each implementation of a J2ME environment on highly constrained devices, such as cell phones and pagers. CLDC takes care of the telecommunications-related aspects of the system, such as pausing applications when the connection drops, switching over to the telephony application when a call comes in,

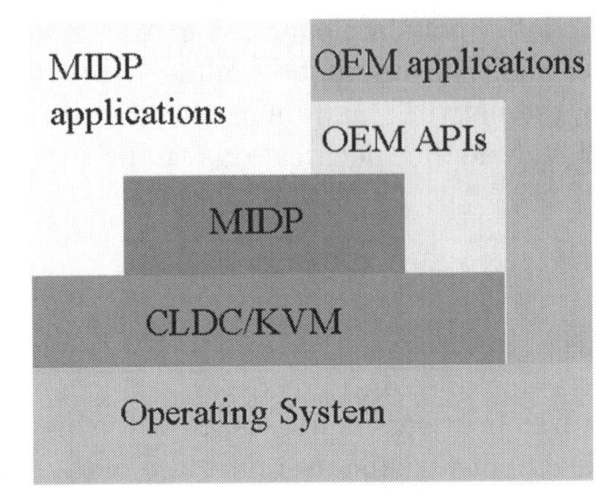

Figure 11.2 How the Java libraries in the mobile phone belong together.

and so on. MIDP adds supplementary libraries that provide APIs not handled by the low-level CLDC, such as the user interface, database, and device-specific networking. It also contains HTTP so you do not have to develop your own browser or your own protocol stack. It also contains TLS security over TCP, which enables end-to-end security.

When you use Java technology phone applications, you'll be able to either leverage an existing WAP infrastructure or, by using the next-generation WAP, use TCP/IP and other native Internet protocols. Once you have TCP/IP access, of course, you will be able to bypass any kind of gateway that exists between the phone and the server, although for performance reasons you might want to retain it when you do not need end-to-end security. This is because it contains a mobile profile of TCP, which is optimized for the narrowband network.

The MIDP APIs are logically composed of high-level and low-level APIs.

The high-level APIs are designed for applications where software portability across a range of handsets is desired. These APIs use a high level of abstraction. The tradeoff is that the high-level APIs limit the amount of control the developer has over the look and feel of the user interface. The underlying implementation of the user interface APIs, which is accomplished by the handset manufacturer, is responsible for adapting the human interface to the hardware and native user interface style of the device.

The low-level API provides very little abstraction and requires a bit more design work to remain portable. It is designed for applications that need precise placement and control of graphic elements as well as access to low-level input events. The low-level API enables the application to access special, device-specific features. Games are a typical application of this API.

The high-level portion of the user interface API is screen-based. That is, the API is designed so that interaction with the user is based around a succession of screens, each of which presents a reasonable amount of data to the user. Commands are presented to the user on a per-screen basis. They enable the application to determine which screen to display next, what computation to perform, what request to make of a network service, and so on.

A screen is an object that encapsulates device-specific graphics rendering user input. Screens can only scroll vertically; there is no horizontal scrolling. The screen handles all events that occur as the user navigates the screen. Only high-level events are passed on to the application. Only one screen can be visible at a time, and the user can only interact with the items on that screen.

For this reason, you cannot make screens too complicated. Use as few user interface components as possible, and make sure they are related. It makes it easier for the user if he or she only has to perform one task per screen.

Applications are composed of a set of screens through which the user steps as he or she completes the tasks (which the application consists of). The screens need not be organized linearly; they can branch out, and it is possible to jump between them (and make conditional jumps, as well).

The MID Profile has four types of screens: List, Alert, Text Box, and Form. The List screen provides three variations: an implicit (menu) list; a single-choice list; and a multiple-choice list. The implicit list can be used to display a list of menu commands inside the application.

The Alert and Text Box screens contain alert and text, but alerts are simpler than in other environments and will use the entire screen. Text boxes generally will not support fancy text formatting or font styles and enable only basic text-editing capabilities. Form screens contain an arbitrary mix of items including images, read-only text, editable text, date and time fields, gauges, and choice groups.

The form screen is where you can actually develop something. There are theoretically no limits on what a form can include. The implementation, not the application, handles layout, traversal, and scrolling. There are quite a few primitives that you can use for development, such as text boxes, multiple-choice menus, and so on. There are many good books about Java, so I do not need to list them here.

None of the components contained on the form screen are able to scroll independently. The entire contents scroll together vertically (there is no horizontal scrolling). When a form is presented on the display, the user can interact with it and with its items indefinitely.

Interaction can be made by using the pointer on the screen or using editable data. Devices following the MIDP profile are assumed to have either a one-handed ITU-T phone keypad or a two-handed QWERTY-style keyboard. Beyond that, devices can have any number of programmable (soft) buttons or no programmable buttons at all. The manufacturer ultimately maps these commands onto the device.

Developers can create portable applications, but they lose control over where each specific command will map on the device.

MIDP also contains a database in the device that can be controlled with three primitives in the API. The RecordStore primitive consists of a collection of records that remain persistent across multiple invocations of the MIDlet. RecordComparator defines a comparator that compares two records in a RecordStore to see whether they match or what their relative sort order is. RecordFilter defines a filter that is used to extract sets of records that match a criterion.

Java is available for mobile phones, and NTT DoCoMo in Japan (as well as its competitor, Jphone) includes it in all new mobile phones (albeit in different versions). In terms of location-based applications, though, there are basically two ways where Java comes in handy: as an animation system for the display of maps (where you can add an animated line to a downloaded map, for instance), and as a system to program a dedicated client, which is independent of the browser and does more than change the map when the position changes. The challenge in developing applications of this kind lies in two different areas: you have to create applications that work in the minimal environment of the mobile phone, and it is a user interface challenge. When you move beyond the static display, you get an entirely new set of problems—as we saw in Chapter 9, "Maps as User Interfaces."

The other type of application is an active application that runs in the background of your PDA (mobile phones are hardly up to this point yet) and that conducts an action based on the position, such as checking your gasoline level and beeping you when you are coming within 500 meters of a filling station run by the company where you have a discount. It could actually run in the application server as well and be triggered when you pass a boundary, but then it would be harder for it to communicate with the gasoline meter, of course. This type of application belongs to the future (although it is not science fiction; similar applications have been demonstrated, although not dependent on position).

It is still not entirely clear how Java in the mobile phone will play out (although play is just what it is being used for). More and more manufacturers are implementing it in their phones, though, and just as it has developed into a complement to the Web browser on the fixed Internet, it is likely that it will become a very important execution environment in the mobile phone. But there is still room for developers—and new applications.

Car Navigation and In-Car Telematics

In Japan, car navigation devices are extremely popular. Because there are no street names, roads are almost never straight for more than 50 meters and because drivers are rarely familiar with the layout of cities other than their own, car navigation systems have become a hot seller.

Car navigation systems are often standalone systems with the map database on a CD-ROM or DVD, and the in-car GPS receiver is used to center the map around the car's position. Car navigation systems also usually have functions to calculate routes and sometimes contain databases of points of interest.

What they do not have, however, is communication with the world outside. If you want to update the information in a car navigation system, you have to change the stored maps (for instance, by changing the CD-ROM). Most car navigation systems do not have any communication with the outside.

Given the popularity of car navigation systems in Japan, however, NTT DoCoMo has formed an alliance with car navigation companies to connect car navigation systems with its mobile phones to update them with traffic information and other information that is dynamic.

In the United States, General Motors has been successful with the OnStar system, which actually is nothing else than a call center where the user calls automatically when the OnStar button is pressed. The car is positioned automatically (using GPS), however.

OnStar is one example of telematics in the car industry, but in reality it is not different from other location-based services. One problem with the term is that it has come to mean so many things because it has been hijacked by a number of different organizations.

Telematics, in the current definition of the automotive industry, encompasses a collection of information services delivered to the vehicle by a wireless link. These include emergency and breakdown calls, assistance in locating points of interest and services, remote diagnostics, door unlocking, stolen vehicle location and tracking, and concierge services.

Much telematics functionality, originally developed for the automobile, has extended to mobile people in general, and unless it is used specifically for functions in the car, there is no reason your phone should be restricted to in-car use. The same terminal can be used outside the car as well as in it. A case in point is Japanese car navigation system manufacturers, who now build their systems as removable, general-purpose DVD players.

Some specialized applications for location-based services already exist, and it is usually the car that is the platform. Often, the purpose is to track the vehicle—for instance, for value transports. The communication with the vehicle in that case normally uses SMS, and there is no user interface (indeed, in many cases, the positioning equipment is hidden away or secured so it can not be removed by possible robbers).

When there can be interactions between the car and the communications device, however, interesting things start to happen. Ericsson and Saab have demonstrated control of the engine in the car by using WAP telephones to reset the number of allowed engine revolutions, for instance. In other words, you could buy a "service contract" for your car, and when you need more engine power (for instance, when you are moving to a different home), you buy more power (at the expense, probably, of a shorter engine life). The car is an awesome platform compared to wearable and even portable environments with a continuous power supply of 24 volts (soon to be upgraded to 48) and no weight and space restrictions to speak about.

Location-based services for cars are still in their infancy, however. Nobody quite knows how telematics will play out. DaimlerChrysler seems to be concentrating on entertainment in the car, rather than utility services. Right now, it is possible to imagine what could be done when the car and its navigation system can receive personalized updates about conditions, but so far, alternative technologies such as broadcast and signage have proven more cost efficient.

CHAPTER 12

Privacy and Location

The most common definition of informational privacy in current use is the one by Alan Westin, "Privacy is the claim of individuals, groups and institutions to determine for themselves, when, how and to what extent information about them is communicated to others." This can be interpreted in very different ways, depending on the local legal system, the practice of it, and your own preconceptions. In most Western countries, it is illegal to use data for purposes other than those for which it was collected; in any case, you will seriously damage the user's trust in you (and your future business) if you try to use data for purposes other than those approved by the user. Personalization, especially combined with location data and logging by the service provider, has the potential to be extremely contentious. The line is fine, and users are willing to give up quite a great deal of privacy in the interest of convenience, provided that they can trust you.

But some people have a hard time feeling trust. When traffic cameras became allowed in Sweden, the marketing manager of the paper company I was working at suddenly became very interested in his privacy.

"What if someone finds out that you were not in the place you said you would be? And even worse, if they found out whom you were with? And used that to blackmail you?"

He had, of course, told his wife that he golfed, but what he really did was go to see his mistress.

Immoral though the anecdote may be, it points to a very real risk: What if someone finds out your position and misuses it? The legal systems in most countries forbid the collection of this type of information without a court injunction, but legal interception of information is permitted in most countries, something that has happened with far less debate than when legislation about the interception of phone calls was passed. Criminals in the United States have already gotten wise and are using pagers (which can not be traced) instead of cell phones. And, as a user, you get the same types of problems that you get with your credit card: You can be traced. Also, as the system is constructed, you cannot request a location anonymously. If you are using GPS, you can request positioned information without giving out any identifiable information, but that does not work everywhere. And if you are sending your position along with your telephone number, then you have already identified yourself and told the service where you are located.

There is also the case in which you give away your identity or information that can be used to identify you, at one point, where this information can later be used to tie you to a position—for example, if your device capabilities, your choice of operator and some personal settings may be very typical for you alone. If you use one service where you give away your name, and all this information is logged, it can be reused to identify you when using another service in which you don't give away your identity, but only your position. It would be like using your company's envelope and a very typical handwriting when sending the same person bills and anonymous loveletters.

If there is an identifier, it becomes relatively simple to mine the log files for your position and combine that with other information. If there is enough information about you, a lot of personal information can be deduced, and you will not have privacy anymore.

Government and Standardization Initiatives

Privacy is actually part of the specifications for the third generation of mobile telephony systems, because the ITU has included a section in the recommendations for the IMT-2000 mobile system (in Service Recommendation ITU- R M. 816, (§ 8.2.2)) that says "In order to protect the privacy of the IMT- 2000 user, access to location information must be restricted to specific applications authorised by the IMT- 2000 user and the administration concerned." There is, however, no such provision for the 2G mobile telephone systems such as GSM, CDMA, and so on; and there is no way of enforcing such a directive technically.

The *Internet Societal Task Force* (ISTF) has tried to define some of the important issues when it comes to privacy. It basically comes down to three things: who collects the information, what information is collected, and how is it being collected. There is also a third sensitive question: Where does it go?

Then, there is the question of how you give your consent (if you are able to do so). Existing protocols give three general possibilities:

- Out of band—no protocol action at all
- Prior consent—action completed by some protocol mechanism that has a given lifetime (which could be infinite)
- Consent at the point of use

The European Union has also formulated a set of rules to safeguard personal privacy in location-based services. They overlap the principles of the ISTF to a significant degree. The first principle is to *obtain the subscriber's explicit consent*. This could, as said before, be done out of band; you do not need to have the explicit agreement for every time you do a positioning, and if you do, it could be done in terms and conditions for the use of a service (by clicking on a link, you accept that your position will be used). In such cases, the P3P technology from the W3C could be used to ensure that the policy of the service is one that is acceptable to you.

This makes the second principle of the European Union necessary: *To provide complete information about the use and storage of data*. If that information is expressed in a formal language, such as the P3P policy language, the description can be read automatically and the approval given automatically. However, an important point is that such a policy be followed. *Only use data for the purpose for which it was collected* is important in building trust for the service, because if it is discovered that the data was misused (however great the short-term gain may be), the user is very unlikely to use the service in the future. The same goes for the principle to *erase personal data after use or make it anonymous*, and to *give the user the possibility to restrict or prevent transmission of personal data*, which would require something like the P3P system as it is applied to another profile technology, CC/PP. Finally, the last principle, *do not transfer the data to a third party without the user's consent*, is strictly speaking unnecessary if you are keeping within the legal system, at least in some countries; it may actually be illegal to sell data that contains a personal component.

One major problem that is implicitly addressed only in all these initiatives, however, is who owns the location information. If it is the property of the user, he or she is at liberty to use it as desired, including putting restrictions on its use, and the network operator has to follow those restrictions. But the alternative, that the information is the property of the network, is also possible. After all, the position information is generated as a part of the normal operation of the network. Why should this not be the property of the network in the same way as the frequency usage information?

This, of course, applies only to situations in which the data is generated by the network. If the user is creating the information himself, for example through the use of a GPS receiver, it will be different. But, what if the information could not be created without the assistance of the network, for instance the transmission of an almanac? The situation is not at all as clearcut as it may sound. The only certain case is where there is no involvement of another party, for example when the information is created using geocoding. Even then it is not so clearcut. You may think that your IP address is your property, for example. But all IP addresses are handed out in groups—called Class A, Class B, and Class C addresses—by a central authority, on delegation from the Internet Central Authority for Names and Numbers (ICANN). In theory, they could be revoked (and there has been discussion about whether this should not be done in some cases and redistributed; the Massachusetts Institute of Technology, for instance, has more IP addresses than China).

Privacy becomes extra sensitive when it is associated with personal information. And, as you have seen in the previous chapters of this book, personalization and location information go hand in hand. The more personal information you gather, the more personal information you have, and the more sensitive the information becomes—and the greater the potential for misuse.

The Organization for Economic Cooperation and Development (OECD) has also created a set of principles for fair information practices, which are a common set of privacy principles that satisfy data protection requirements. Data security can satisfy the requirements, but not in itself. No matter how secure the data is, there is nothing you can do if it is compromised at the receiving end.

The practices are:

- Collection Limitation
- Purpose Specification
- Data Quality
- Use Limitation
- Security safeguards (i.e., confidentiality)
- Openness
- Individual Participation
- Accountability

All these principles—the ISTF, the EU, and the OECD—overlap. However, they do not have the force of law everywhere (which, by the way, introduces a location-dependent problem: If you are roaming from one country in which the privacy protection is contractual into another in which there is a legal requirement on privacy protection, how do you make sure that the requirements are satisfied?).

It is important to understand that the process is always initiated by users granting other users or applications permission to locate them, never by users requesting permission to locate other people. Such requests will, of course, be matched with the database of which users have allowed themselves to be located, and the users who do not want to do so, will not appear in the location system. Depending on the contractual relationship and the system used, users can access their approved permission list at any time via Web, WAP, and SMS interface to review and update their permissions for other users or applications to locate them. Further, operators can set the service to initiate a periodical MT-SMS message to users with their permission details (this can be cancelled by users).

The European Commission is working to integrate positioning in the data protection law, which regulates privacy of information. In the United States, the Wireless Communication and Public Safety Act of 1999 (section 222(h)(1)) added location to the definition of customer proprietary network information. This means that the operator has to have approval from the customer before releasing this information to third parties.

The law says that the operator should get explicit consent to use the information ("opt-in"). This can be done electronically or by using a written or oral agreement. However, the U.S. Court of Appeals for the Tenth Circuit found that the opt-in principle violated the First and Fifth amendment of the U.S. Constitution. Neither did the court think that the Federal Communications Commission demonstrated clearly that the method improved the privacy protection; and it should have investigated an opt-out approach, in which the approval can be inferred from the contract between the customer and the operator, and the user has to state explicitly that he does not want to release his information.

It is, however, required to release the location information for emergency purposes, which is the foundation for the E911 directive. That includes the name and address of the user. And, of course, law enforcement authorities have the ability—with a warrant—to access the information. However, the American law is unclear. In cases of services based on telephony, some kind of approval from the user is necessary, but this can be seen as given when a certain number (for example, 911) is called. In services based on the World Wide Web (or other information service), however, there may not be any need for a consent at all. And, whichever is true, aggregated location information (in which the user identity has been removed) can be used without restrictions.

This means that it would not be possible for a company like DoubleClick to aggregate positioned information, retaining user information, with information about which advertisements users had clicked on in the fixed Web. But after you have started aggregating information, it is possible to derive patterns that can be used to personalize information, without knowing the identity of the user. This is an area in which legislation has not caught up, neither in Europe nor in the United States.

Fair Privacy Practices: The FTC Rules

Inform customers about the collection practices of the site

Give the customer choice regarding any uses of the information

Allow for access to the data so that customers can ensure it is correct and correct it if not

Maintain the data securely

Enforce and audit the policies

As a matter of fact, the Federal Trade Commission in the United States, which has stated a number of rules that apply to providers of financial services, has defined a set of rules for fair privacy practices. They are supposed to inform the user about the collection practices of the site, give the user a choice regarding how the information should be used, let the user see and correct the data, maintain the data securely, and enforce and audit the policies.

The CTIA, another industry group in the United States, also has adopted a set of rules. These rules are based on U.S. law and have been made as a recommendation to group's members. They also stress the notice to the user, the consent from the user, the security and integrity of information, and in an additional point, they declare that rules should be technology neutral, so they can be applied to any location technology, including those which work in fixed systems.

The Wireless Advertising Association (WAA), potentially one of the biggest bugbears in mobile privacy, has adopted a policy to make sure that they do not infringe on the users' privacy. According to the rules, *users should be notified about what rules and policies apply.* There needs to be information about exactly which information is being collected, how it is stored, if collection is optional or mandatory for the individual, if third parties will be allowed to use the information, if it is indeed collected through a third-party organization, a statement that declares the advertiser's commitment to data security, and how

Location Privacy Principles of the CTIA

Notice

Consent

Security and Integrity of Information

Technology Neutral Rules

they attempt to safeguard data quality and give users access to the information collected about them.

Users also must be *given the ability to choose how the information is used* and the ability to opt out (through a confirmed opt-in, actually). The WAA also declared that it will *make every effort to store the data securely, and also accurately.* Members of the WAA should not spam users—that is, they should not send wireless push advertising or content without a confirmed opt-in. Nor should they transfer the data to third parties without the explicit consent of the user. The WAA is trying hard to distance itself from the "spammers" of the Internet, but unfortunately those companies are not members of the WAA and not likely to ever be.

The Magic Services Forum, who develops the Magic API for location services, have included a privacy model in their work (although it is somewhere in the future, related to the current specifications). They state that users must understand the model and have the means to control access to location information, and they must believe that they have ultimate and absolute control. Without this understanding, users may consider mobile applications inherently invasive of their privacy, and the development of a broad market could be delayed as a consequence. It is interesting to note that one of the leaders of the forum is Microsoft, who certainly have had their fingers burned in this context.

The Magic Services Forum intends their privacy model to define the following things:

- The types of location information that may be published
- The roles that different parties can play in scenarios involving published location information
- A logical model for publishing and viewing location information
- The mechanisms for publishing and viewing location information under the logical model, including the role of networks and proxies.

The WLIA Rules

The Wireless Location Industry Alliance (WLIA) has published a set of rules for its members to adhere to, and there are other organizations that have done similar things. The WLIA guidelines prescribe the existence of documentation on the Web site for how privacy-sensitive information should be used and to make sure that there is an individual who is responsible for them in the company.

They also state that if the use of personal data is not an intrinsic feature of the application, then the service provider should not use it unless the user gives his or her OK. The service provider will also only use the data for purposes for which it was obtained with the knowledge of the user. Other uses require express permission.

The way it works is that the location information service will have an agreement with the network operator, and he or she will have agreements with the users. In other words, you cannot determine which services can position you—it is all or nothing. To do it a different way would require the creation of some kind of blacklist, and it would mean that you would have to encrypt and/or sign requests and responses (to make sure that they are not intercepted by some malicious third party who wants to insert coupons for Café Veloce instead of Mosburger). It becomes several magnitudes more complicated.

If you own the subscription, however, you can decide whether it should be positioned or not. If you do not, then the owner of the subscription can. In other words, there is a potential risk for anyone who is using a corporate mobile telephone. Has your company allowed positioning of the phone? Are they maybe positioning it continuously? Does that mean that if you go to a hotel with your secretary instead of having a 2 PM meeting, you will get a call from human resources when you get back (because, of course, they will know when you got back)?

The IETF Geopriv Working Group

The IETF has set up a working group to address the questions of mobility and privacy. It might seem a bit strange that an organization that does not standardize anything else in the area of mobility, or the area of privacy performs this task, but if nothing else, it will be unbiased toward development concerns.

The working group has published a requirements document that describes how it perceives that privacy management should be applied to location-related objects, but because it starts defining that it will work with a transport protocol and data structure that are not supported by any existing protocols, it might be difficult for them to have an impact.

The group is looking at creating a policy-based system, much like P3P. It requires the protocol to provide a mandatory-to-implement, optional-to-use policy enforcement point mechanism and an optional policy decision-point mechanism. The protocols should specify how servers could obtain a policy from a policy storage facility if the policy decision point mechanism is implemented.

They will also specify a policy information base and provide a mechanism to restrict the information in reports by accuracy of location, frequency of report generation, and representation format.

Because the group is relatively new, it is possible that it might change its requirements and possibly align them with P3P, which represents existing implementations and deployed mechanisms.

W3C P3P

The Web had not had time to grow very large when concerns were first raised by parents. There was pornography and morally offensive material available to all. This was when the pornography industry had just discovered the new medium, and there was a rush of people who sought to publish their opinions, as well—some of which were considered offensive by others.

The reaction came mainly from the Christian right in the United States, but even they realized that censorship is not possible in a global medium, especially when it would infringe upon the laws of other countries. The debate raged on for some time, and the W3C, which was then fairly recently formed, decided to create a filtering tool so concerned parents could deselect sites that had been marked undesirable. The marking itself was done by a third party, the Platform for Internet Content Selection (PICS), which is used by several of the tools that enable parental control on the Internet today.

The work that led to PICS also led to other work concerning filtering, and the discussion of the collection of information by service providers, information that could potentially be used for violations of privacy, led to the creation of a system that was intended to allow the users to create policies for how the information they gave out should be used. This became the Platform for Privacy Preferences Project (P3P).

The P3P has some major disadvantages, but it is the only game in town, and because the work started in 1997, it has gained widespread acceptance, if not implementation. This includes the mobile field, in which a workshop held in 2000 jointly by the WAP Forum and W3C pointed out some clear points that anyone interested in mobile privacy needed to think about.

Privacy in Practice: Operators and User Profiles

There are two main problems with privacy and position. The first is that you do not know with whom you are making a contract. The second is that you do not know what that contract contains. Then, there is a third big problem: How do you allow or disallow positioning?

When you are connecting to a Web server, you do not have a clue about who owns it—and that person or company might not at all be the same as the ones who run it. You can assume that the Disney corporation owns disney.com and that the material will be family friendly and that playboy.com is owned by the Playboy corporation and that the material will be raunchier. But what about disney.st or playboy.st? Or wireless-information.net? There is nothing that tells you who owns these sites.

When location-dependent information sites become common, you will be giving out a lot more information than you do when you connect to a Web page. If your computer has stored a cookie for a service, it can use that for a database lookup and put the "Welcome, Johan! We have recommendations for you (if you are not Johan, click here)" text on the page and change it depending on the name. This scenario is innocent enough, but you will notice that sites like Amazon.com also keep your credit card details on file (very convenient if you intend to buy again, of course).

As it is, a Web browser cannot give out much more information than the e-mail address you have stored in it, the brand and model of browser you are using, and any cookies that it has stored that relate to the service (and a few more system-related things). That is scary enough—someone can get your e-mail address when you look at their Web site? But it becomes far worse when you start using position-dependent services. Now, the service provider can determine that you look for Chinese restaurants in Shinagawa on Tuesdays and for Soba restaurants in Shibuya on Wednesdays. He might want to sell that information to restaurant owners so they can put out extra big signs as you are coming. Or worse, he might sell an advertisement that sends you advertising for Soba restaurants in Shibuya on Wednesdays. If you happen to be somewhere else, it becomes a nuisance.

But the real danger comes when this information is combined with other information. Say that he sells your whereabouts to the credit card company. They will then be able to determine that you request information about restaurants every day but that you only pay with your credit card on Saturdays. Perhaps they will give you a discount if you start paying with your credit card in soba restaurants on Wednesdays?

This scenario would be highly contentious in some countries, like Germany, which had a very bad experience with the registration of personal information some 60 years ago (and again 30 years later). If this information is available, someone is liable to get it—and if you know who goes to a Communist party meeting, you might draw the conclusion that they are engaged in anti-American activities. Or, if you know who goes to an abortion clinic, you can target them with personal visits to try to make them repent (or worse).

Once you open the Pandora's box of releasing this kind of information into a database somewhere, it is liable to be misused—just like credit card numbers are used by hackers to buy things that their owners never had a clue about. The only security provided in the application servers I have looked at for this book is that provided by their DBMSs, which often is good enough but can be hacked if you are not careful because it is both well-known and well-documented.

When you are connecting to a Web site today, you are implicitly making a contract with the site owner that he or she can use the information you have set in your Web browser. If you have disabled cookies, he or she cannot use those; if you have not typed in an e-mail address, he or she cannot use it. He or she might ask for it, but then you have to give your explicit consent for him or her to get it. P3P makes this contract explicit.

The contract with a position-dependent service is different. There is no way you can decline positioning; as a matter of fact, the case can be made that you, by accessing a site you know will use position information to customize the information returned to you, have accepted that this action is performed.

The only way to make this situation work is for you to be able to check the location service's credentials to be sure that it does not give out or sell your information. A system for this process, P3P, was developed by the W3C in response to worries from governments and others. This system enables you to retrieve a policy file in a structured format from the Web server, which declares how the information will be used. Your browser (or a separate application) can then read it, and you can decline or accept to give out the personal information on those terms (many Web servers already have the same type of declarations, but they are usually hard to find, and the format is totally arbitrary).

The W3C and the WAP Forum held a workshop in 2000, discussing privacy for mobile systems. Although they did not come up with any solutions, the workshop agreed on a number of points. The first, and most important point, was that *the user must be in control of personal information*. Granularity of control, however, is a crucial—and difficult—issue. The second point was that *there is a need for privacy tools and privacy architectures*. Tools, however, need to be built, as they do not yet exist. There is a variety of possible tools, as well. And, it is not just about tools; it is also a matter of designing the underlying information architectures in a way that allows users to control access to specific fields and types of information, so that data does not leak out in passing intermediaries, such as gateways and proxies.

At the same time, this should not be a uniquely wireless experience. *Keeping the user experience consistent* over the wireline and wireless environments is important, so that the user understands that his or her privacy preferences are being respected, whatever the environment. This is part of a different view on privacy—not just as a legal requirement, but as a business opportunity. Privacy should not be seen as a technology or legal problem alone; rather, it is a societal issue, and technology and laws must work together to ensure the privacy of the user. Otherwise, the high level of trust that is required in the mobile environment will not be kept, and the opportunity of mobile network operators to become trusted agents for transactions will disappear.

However, different requirements exist in different environments. The mobile environment—at least as of today—is special, because the small CPU, small memory, and low bandwidth of the user means that solutions which work well in a PC environment do not automatically work for a mobile device. Different business models also entail different privacy protection mechanisms.

There is, however, a second party in all location-dependent service transactions: the network (or service bureau) that gives out the position information. You get the data from the service provider, but he or she gets the position from the network operator. That means that your subscription must allow the network to position you, either implicitly or explicitly. This task could be done electronically, but it is probably simpler to do by asking you to sign a new version of your paper agreement with the network operator. If you have given your permission, he or she knows that he or she is not out of line in giving out your position (giving out a position can be prohibited by setting the privacy flag in the HLR of the network, for instance). Of course, any settings can be overridden by law enforcement with the proper credentials (such as a court order).

You might be surprised at how much information Web sites have about you. But that is nothing against the mobile telephone network operator, who also has a way of identifying you every time you make a request to be positioned. In other words, the burden of guaranteeing the privacy of the user largely falls on the operator.

A typical subscriber personalization profile includes the following:

- General information: name, nickname, login ID, device, address, and so on
- Buddy list: friends, family, and so on—people to whom the user grants access to locate him/her
- Favorite locations: home, office, club, friends' homes, and so on—part of the platform-unique shortcuts mechanism that saves time to access favorite, commonly used locations and address tasks
- Favorite routes: from home to office and vice-versa, similar in concept to the favorite locations mentioned previously—enabling users to query routine driving directions (changing by traffic conditions) via simple shortcuts

- Permission profile to locate other: by days, hours, area, accuracy, and so on
- Permission to be located by others: by days, hours, area, accuracy, and so on

Several of these make it necessary for the users to give permission to locate each other. In other words, there is a potential for the user's privacy to be violated. In general, whether the user wants to allow this action is a personal matter, and the privacy restrictions (if any) should be part of the contract between the user and the service provider.

It is important to understand that the process is always initiated by users granting other users or applications permission to locate them, never by users requesting permission to locate other people. Such requests will, of course, be matched with the database of the users who have allowed themselves to be located, and the users who do not want to do so will not appear in the location system. Depending on the contractual relationship and the system used, users can access their approved permission list at any time via Web, WAP, and SMS interface to review and update their permission for other users or applications to locate them. Furthermore, operators can optionally set the service to initiate a periodical MT-SMS message to users with their permission details (this action can be canceled by the user's personalization).

Typically, the provisioning database is also interfaced with the personalization system, whether based on the telecoms' subscriber profile data or a portal system database. Personalizing data is very useful in minimizing the difficulties in using a service (the more tasks the user has to perform, the less likely it is that he or she will use the service). We discussed this topic in more detail in Chapter 6, "Providing Databases and Doing Searches," when we talked about user interfaces. Just keep in mind that the database system does not exist in isolation; rather, it is part of the environment that produces the user's experience. This statement also goes for how easy it is to update the information.

Options for Location Privacy

There are three distinct options for location privacy:

- **Location OFF (disabled, not allowed to any user or application)**
- **Location ON (enabled for users and applications on the user's predefined permissions)**
- **Upon Approval (asking the user's approval for each request, on demand)**

Some operators have already tried to safeguard users' privacy. NTT DoCoMo, in its iMode system, has a feature that hashes the MSISDN when it is sent outside the DoCoMo site. The hash value is kept, and the value can be deciphered when a request comes from a Web site because all accesses are made through their gateway. In the iArea service, this situation means that the request for position information is never done with the user's MSISDN from the Web site but always with the hashed number (and it is even more complex because you have to install a special program in your application server to participate in the service).

One possibility is for the industry itself to decide on the practices it wants to apply (and as I previously discussed, there is no lack of practices to apply). Self-determination is in many ways more important than privacy as such. The important matter is not to store or process any data at all, but to do it to the extent that the end user decided was reasonable and acceptable.

Creating fair location information practices for the industry and getting it accepted by all parties involved will be an enormous effort, but it would be easier on the industry than having legislation forced upon it. Users, it seems, are willing to trust service providers if they feel that the information given to them is truthful about how the information will be used. Another way of ensuring trust is to give the user a choice not to participate and to ensure the user that the personal information is safeguarded (for example, that the service provider has a believable system to safeguard against someone stealing the information).

Users may also have unrealistic expectations, although it is probable that European users are better educated in the systems' technical workings than users in Japan (because the network operators in Europe are not as good at selling consumer services but have been trying to sell the systems as technologies). But, the industry has also consistently published stories that set expectations very high—although after the World Trade Center incident, these may have been tempered because cellphones seem to have been frequently used in calling for help, even if no help could be given.

The privacy practices are mainly safeguarded through the data management. It is there that the user controls over collection and use of personal data are safeguarded, as well as provide a secure data repository, authenticated data access, confidentiality in the transmission of data, and the anonymization (or de-identification) of information. If these practices are built into the system, there needs to be some way to audit that they actually work, and there has to be uniform practices between all the different involved parties, so that the operator and the location-dependent service provider does not have different practices—and systems—for gathering and providing the location-sensitive and private information. There must be a way for the different parties involved to share and communicate privacy policies, so that one party does not have a less restrictive policy than that which is presented to the user, who is the ultimate arbiter for whether the information should be given out or not.

The only way to make absolutely sure your position is totally untraceable is not to use a mobile phone at all—not even for phone calls. If you want to make sure that you cannot be tracked in other ways, use cash instead of credit cards and walk instead of driving (because many modern cars will have built-in mobile phones or other transceivers).

APPENDIX A

Who Does What in Location-Dependent Standards?

There are a number of groups that create standards in the location-dependent area (some would say way too many).

Standards used to be produced by formal standards bodies with representatives from countries that voted on the documents when they had progressed to a finished status. ITU, the leading standards body in the telecommunications field, is even a United Nations body.

Not any more. Standards are now developed by industry consortia who seek to obtain a consensus among the most important players about the recommendations that they issue. The documents produced do not have any formal standing; the W3C specifications are recommendations to its members, nothing more.

The important thing, of course, is to get involvement from the important players. A consortium such as the Location Interoperability Forum will get its specifications implemented merely because its members are the companies that produce the products that will use them.

All the different consortia and associations use different processes to produce their documents. The only thing that they have in common is that they are based on peer review—that other knowledgeable persons can read and comment on the specifications. The problem, of course, is that there are too many specifications and too few reviewers.

Wireless Community Standards

The actual standards for the wireless networks are produced by several bodies. The standard for IMT-2000 is being developed by a consortium of standards bodies, the 3rd Generation Partnership Project, where standards bodies from Europe, the United States, Japan, China, and Korea participate.

The standardization work for GSM was done by the *European Telecommunications Standard Institute* (ETSI). Originally called Groupe Speciale Mobile, it changed its name to Global System for Mobile when it became obvious what impact a standard for mobile telephony would have.

The standardization effort for third-generation mobile systems started at about the same time in three different places: ETSI, which was working with results from research projects sponsored by the European Commission to create a system for mobile communication called *Universal Mobile Telecommunications System* (UTMS); the Japanese *Association of Radio Industry and Broadcasters* (ARIB); and ITU, which was developing a specification for a system with the unwieldy name *Future Land Mobile Portable Telecommunications System* (the acronym, FLMPTS, has the distinction of being unpronounceable in all human languages).

But working on the same thing in three different places is not very economical, so the three decided to join forces and created the 3rd Generation Partnership Project, together with other standards bodies and some companies.

The location-dependent work had been driven by the subcommittee T1P1 of the American T1 Standards Committee under its collaboration agreement with ETSI. Technical Committee SMG has been working on the standardization of *Location Services* (LCS) for GSM. This work was headquartered in the United States because an FCC mandate to provide location information for emergency calls had made this issue critical for PCS-1900 operators. The scope of the LCS standardization, however, also addressed providing the capabilities for value-added location services.

GSM already has some rudimentary location capability. The system knows the cell site/sector where an active GSM user is currently located. For some location services, that granularity is sufficient. By combining this information with timing advance measurements, it is possible to constrain the user's location even further. These are valid positioning methods but are inadequate for emergency services and for many of the proposed value-added services, such as locating the nearest pizza restaurant. To satisfy the tighter accuracy requirements, three positioning methods were standardized by T1P1. The methods standardized by T1P1 for GSM are *Time of Arrival* (TOA), *Enhanced Observed Time Difference* (E-OTD), and *Assisted GPS* (GPS). Each method has advantages and drawbacks, so no single method satisfies all requirements.

TOA requires that multiple base stations listen to handover access bursts and triangulate the position of the mobile. TOA has the advantage of working with existing GSM mobiles, but it has the disadvantage of requiring the greatest investment in supporting infrastructure.

In E-OTD, the handset listens to bursts from multiple base stations and measures the observed time difference. These measurements are used to triangulate the position of the mobile. This process requires handset modifications but requires less positioning infrastructure support than TOA.

GPS relies on mobiles having an integrated GPS receiver. Assistance data is transmitted from the network in order to expedite the GPS signal search and possibly improve sensitivity. This method is potentially the most accurate; however, GPS is limited by the weak penetration of the satellite signal.

Computing the position of the mobile station is only one aspect of LCS. Others include interfacing with location applications, authorizing services, ensuring user privacy, and so on.

New network nodes have been defined and are shown in the figure. The *Gateway Mobile Location Center* (GMLC) is the interface point for location applications. The *Serving Mobile Location Center* (SMLC) is responsible for requesting that a mobile be positioned. It might, in some cases, maintain information on network topology to allow efficient positioning of the mobile. Additional functionality to support location services will be integrated into existing nodes (MSC/VLR, BSC, HLR, SCP, and so on).

All the positioning methods rely on very precise timing measurements. The *Location Measurement Units* (LMUs) provide reference information that is combined with the received measurement data or broadcast to assist in the calculations. There are two types of LMUs depending on whether they are signaled: the A-bis interface or the over-the-air interface.

The network architecture is designed to enable easy migration to UMTS; however, the T1P1 work does not currently support either GPRS or UMTS. The completion of the LCS standards has taken more than two years. An early version of LCS was approved at SMG#28 (February 1999), which specified the TOA positioning method and an NSS-based architecture. This specification was adequate to meet the needs of PCS-1900 operators in complying with the FCC mandate. The T1P1 mandate was to address the needs of the GSM community, however, and not just the needs of the PCS-1900 operators. Two new positioning methods were added, and the network architecture was revised to enable easy migration to GPRS and UMTS.

In the microlocation area, IEEE-sponsored bodies stand out. The 802.11 standardization committee for wireless LANs is working on technologies that take the LAN in campus areas and make it wireless. The Bluetooth *Special Interest Group* (SIG) has been set up to develop standards for the Bluetooth short-range wireless network.

The ITU was formed May 17, 1865, after $2\frac{1}{2}$ months of arduous negotiations, when the first International Telegraph Convention was signed by the 20 participating countries and the International Telegraph Union was set up to enable subsequent amendments to this initial agreement to be agreed upon. This action marked the birth of the ITU. Today, it is a specialized agency of the United Nations and standardizes telecommunications between countries and the use of radio frequencies both on land and in satellites.

Geospatial Community Standards

In the geospatial community, there are three important international organizations leading the development of industry standards and specifications: ISO/TC211, ISO SQL3/MM, and OGC. The ISO/TC211 workgroup is a formal international standards body creating an entire family of geospatial standards ranging from exchange formats to metadata to spatial data models. The ISO/SQL3/MM workgroup, another formal international standards body, is working on multimedia and spatial extensions to the SQL3 dialect of the SQL standard. The OGC, on the other hand, is a consortium of industry, government, and academic organizations from around the world, working to develop software specifications (APIs) that enable location-based technologies to interoperate. OGC's Interoperability Program offers a unique, fast-track approach to developing interface specifications and protocols for location-based application services. It has started an intitative to integrate mobile services into its GIS (Geographical Information Systems) testbed. The *Open Location Services* (OpenLS) initiative is intended to engage the wireless community in taking location services to the next level.

Standards for the Prime Meridian, time keeping, and so on are set by the International Organization for Standardization (the abbreviation ISO is correct, and the organization itself claims that it is a word in its own right)—a worldwide federation of national standards bodies from some 140 countries (one from each country).

The ISO is a non-governmental organization established in 1947. Its mission is to promote the development of standardization and related activities in the world. The work results in international agreements that are published as international standards.

The *International Earth Rotation Service* (IERS) maintains the *International Terrestrial Reference Frame* (ITRF). The IERS was established in 1987 by the International Astronomical Union and the International Union of Geodesy and Geophysics, and it began operation on January 1, 1988.

The primary objectives of the IERS are to serve the astronomical, geodetic, and geophysical communities.

Internet and Web Community Foundations

Most of the Internet and Web technology standards today are oriented toward enabling interoperability through common encodings for data exchange (for example, HTTP and XML) and wire transport protocols (such as HTTP/HTTPS and IIOP) for connectivity.

Of particular interest is the newly created *Location Interoperability Forum* (LIF), a joint initiative of Motorola, Ericsson, and Nokia whose mission is to define simple and secure access methods that enable wireless devices to access location information from the wireless networks and to promote standards-based location determination methods.

The IETF develops standards for the Internet.This group is an informal, ad-hoc body that nevertheless has developed the standards upon which today's Internet runs (and possibly those of tomorrow, as well). IETF is strongest in the network layer and IP, and its associated applications and protocols are without a doubt its largest success. The IETF only has one working group in the location-dependent field, however: the GEO-PRIV working group.

The W3C has some 500-plus members and was set up by Tim Berners-Lee, the inventor of the World Wide Web, to make sure that implementations stayed interoperable and the system continued to develop into a great collaboration zone for users of computers as well as other devices. The member companies come from all different industries and consist of both small and large companies. Its biggest success is XML, of course discounting the monumental success that it has had with HTML; but now the move is continuing in the direction of Web services.

The WAP Forum was formed in 1997 by Ericsson, Nokia, Motorola, and a small company then called Unwired Planet to develop technologies for Web browsing on mobile phones. Initially drawing large interest, it has been hampered by the fact that it does not release documents for open review until they are finished and that the membership often thinks in old telecommunications terms. WAP 2.0 represents a big step forward, however—integrating Web technologies and producing a testing framework (ensuring that they work together in a wireless environment).

The *Object Management Group* (OMG) was formed in April 1989 by 11 companies and has now grown to 650. It rapidly drew a large interest for being the first industry consortium in the computer field to produce open standards and has developed the Common Object Request Broker framework, the Unified Modeling Language, and several other standards. The work is done by member representatives, but the consortium maintains an administrative staff. It not only develops base technologies, but also technologies for special areas (such as healthcare).

The Parlay group was founded in 1998 and straddles the wireless core network and the Internet by creating APIs. It has some 50 members, mostly from the telecommunications industry.

Java is not an international standard. It has been on the verge of being submitted to international standards organizations several times, but Sun Microsystems, the company that owns the technology, has always retracted the proposals. The Java community uses the *Java Community Process* (JCP) to incubate standards. JCP is a formal process defined and managed by Sun Microsystems for the development and revision of Java technology specifications in cooperation with the international Java community.

Java standards are developed collaboratively, but they use the JCP to get reviews of the development. The process also includes an approval stage by the members of the working groups.

APPENDIX B

XML: An Introduction

With XML, companies and organizations on the Web have a common markup language that can be extended to support their own needs. A book publishing company might use XML to mark up books, for example. A mobile phone manufacturer might use XML to mark up information about the phones. *Extensible Markup Language* (XML) is a universal data format and a rule set for defining markup languages (called XML applications). In a markup language, you do not actually work with the data itself; rather, you create elements that have contents (which are the actual data) and work with them. The structures of the documents are expressed through the nesting of the elements. The content of the elements can be almost anything because this content does not affect the workings of the XML application itself. For instance, *Synchronized Multimedia Integration Language* (SMIL) is an XML application that is used to handle multimedia data (in other words, synchronize and integrate presentations of video, audio, text, and images).

XML can be used to mark up almost any information. Today, XML is used to describe the primitives of communication protocols, documents, multimedia presentations, images, and the state of a computer system, user interfaces, and much more. In the context of location-dependent services, there is GML, the Geographic Markup Language, defined by the Open GIS Consortium. There is no central registry, according to the principle of the Web and the Internet that there should not be a central point of control. So, there is no one central place where information about XML applications can be found. Without a central authority that controls the birth of new XML applications, there is a risk that two or more applications will use the same element name to mean different things. To avoid naming conflicts, XML supports a simple namespace mechanism. More information about this topic is available in the section about namespaces in this chapter.

Almost every data model that can be expressed in XML can be represented as a tree. An XML document contains a tree of elements. Each element has an element type name and zero or more attributes, and each attribute consists of a name and a value. Let's look at an example phone description:

```
<phone>
     <manufacturer>Ericsson</manufacturer>
<model>R320</model>
     <network>GSM 900</network>
</phone>
```

Elements have relationships with each other that will be familiar to anyone who has worked with object-oriented programming. The manufacturer, model, and network elements are called children of the phone element, which then becomes the parent of these elements. In terms of the tree, as the tree branches out the leaves and twigs become children of the branches. Elements that are on the same level in the documents are called siblings. In this example, the phone element is also the root of the XML document.

The start and end tags are together called an element, and elements can have attributes, as well, as in the following example:

```
<phone type="mobile">
     <manufacturer type="name" identifier="brand">Ericsson
</manufacturer>
<model type="name" model_reference="manufacturer_reference">R320</model>
     <network type="string" network_type="ITU_definition">GSM 900
</network>
</phone>
```

An XML document can be seen as a tree, where the root is the top element and the branches are the other elements nested in it. In an XML document, the information tree consists of a set of nodes of different types. The most common types of nodes are elements and attributes. Each element has a name and can contain a set of other element nodes or plain text. An element consists of two tags: a start tag and an end tag. The difference from HTML is that there are no end tags in many HTML elements (those that were introduced before HTML 4). <p> is perfectly valid HTML markup, but in an XML-based markup language, you have to use both the start and end tags to describe the element. A paragraph in XHTML, the XML version of HTML, has to be written as <p></p>. Otherwise, the computer has no way of knowing where the paragraph ends.

An element can have attributes, which are written inside the start tag <p class="paragraph"></p>. Each attribute has a name and a value. Usually, the element name gives a hint about what type the element is and says something about what kind of data is inside the content of the element, and the attribute values says something about the element itself. Programs that will affect the document in some way (for instance, an XSLT transformation) can use this information. Attributes can be used to enhance the data, like in the previous example where they declare the type of the content of the different elements. There can be multiple attributes of one element, and the content of the element and the values of the attributes are to a very large degree interchangeable. The names of the elements and the attributes, what data types they can contain, and the structure into which they can be combined (in other words, which elements can be parents and which can be children) are together called a document type and are defined in a *Document Type Definition* (DTD). The DTD is actually not an XML document, however, because the mechanism was inherited from SGML. This situation can cause problems if you are trying to find consistent ways of dealing with data sets, because you cannot use the same tools to work with DTDs as with the XML documents themselves. This situation is one reason why the W3C created XML schemas. Another reason was to create better descriptions of the data (because the DTD enables a very limited set of data types, for instance). The schema language of XML Schema can also be used to describe names and structures—in essence, declaring what element structures are allowed.

Elements can be nested in other elements, creating a structure within the document. This structure is mandatory for an XML document that has to have a root element in which the other elements are nested. A document cannot be in XML without having this structure.

When a program or a human being checks whether the names and the structure of a document matches that in the DTD, that process is called validation. An application that checks whether a particular document breaks any of the rules in the DTD and the XML specification is called a validating XML parser. An application that does not validate the document is, not surprisingly, called a non-validating XML parser. Both types of parsers check that the document follows the basic rules that make the XML document different from plain text and other data formats (the number of < must be equal to the number of > and so forth).

A well-formed document satisfies all of the requirements of the XML format (but does not have to be valid). A document that is not well formed is not an XML document. An XML document does not actually have to be valid to be used, but it must be well formed. Otherwise, the parser will stop and will throw an error.

There are two ways to declare the elements and attributes of an XML document. The first is the DTD, which is used to associate an XML document with a document type. This function is performed at the start of the document. A document is an instance of a document type, which is defined with a DTD. This relation is similar to the relationship between an interface and an object:

```
<!DOCTYPE html PUBLIC "-//W3C//DTD XHTML Basic 1.0//EN"
"xhtml-basic10.dtd">
<html>
(-)
</html>
```

The string after the word PUBLIC is a global identifier that identifies the DTD. It can be used by an application that has the DTD built in and that does not have to read the DTD from the specified file; in this example, xhtml-basic10.dtd.

When you create an XML application, you create an information model that accurately describes your information set instead of using an all-purpose model or one developed for some other purpose (the problem that has plagued the *artificial intelligence* [AI] industry). This model is expressed in the DTD or the XML schema—a document that describes the elements in an XML language. In the DTD or the XML schema, you specify a set of elements that will contain your information, declare which rules they have to follow, and give them names that are globally unique (which is possible because you are using the global URI naming system). XML makes sure that all new languages follow the same basic rules (if this process sounds confusing, you can play football, rugby, and soccer on the same field, but the rules for how and when you can use the field are the same, and you have teams that use a ball and that score goals in all of the games). Belonging to a family of common rules enables you to transform one XML language into another (by using XSLT, which is another standard in the XML family), and it enables you to write software that can work with the markup without having to be rewritten for each new markup language with which you want to work. It is impossible to say how many different types of XML applications there are, because there is no central registry (indeed, one of the central ideas for XML is that there should not be a central registry, although several organizations have undertaken to register XML applications in their domains).

It is not necessary to have a DTD in order to write XML documents. And, whether it is necessary to validate the document, if you have a DTD, depends on the situation in which the document is used. The following example shows a DTD for the XML document in the phone example that we have been using:

```
<!ELEMENT phone (model, network)>
<!ELEMENT model (#PCDATA)>
<!ELEMENT network (#PCDATA)>
<!ELEMENT manufacturer (#PCDATA)>
```

How to create a good DTD deserves its own book. So, we will not go further into any DTD adventures. Ultimately, transformations of XML documents can be seen as transformations between DTDs (because you are replacing one markup structure with another). For practical purposes, however, we do not need to go that deep into the specifics of DTDs. It is enough that you know that it exists.

Because the DTD is an SGML document, it cannot be handled with XML tools. But the W3C has created another format, XML Schema, which can be used to describe the document format that you are using. It fulfills the same functions as the DTD, but it is an XML format (and somewhat modernized). Because XML Schemas are XML documents, they can be transformed by using XSLT—for instance, into DTDs.

The character set in the element names, attribute names, and attribute values of an XML document is always Unicode or a subset of Unicode, such as UTF-16 (the data itself can be binary). The first octets of the document always indicate the character encoding. An XML document does not rely on an external type system to indicate the character encoding.

Unicode is a character-encoding system that was originally developed in cooperation with a number of international standards organizations but is now developed and maintained by a consortium of organizations and companies (the Unicode consortium). It is a 16-bit format, which means that there is space for almost all characters that are used in writing systems today (almost, because Chinese actually has an enormous number of characters if you start looking at the unusual ones). The character encoding that was originally used on the Web, 7-bit ASCII, does not have the space to encode more than the 26 letters of the Latin alphabet in lower and upper case, plus some special characters. This situation causes a problem for languages that use accented characters and umlauts (such as the ö, ä, and å, which are Swedish characters), and it is an enormous headache for the Japanese, who have several mutually incompatible encodings in 7-bit ASCII of their rich character set (Japanese actually uses three different sets of characters in writing, plus the letters of the Latin alphabet. But that merits its own book, and we will not go into it now).

As the Web (actually, the MIME types that are used to carry the information of the Web) has developed, character encodings in other formats have been brought in to enable non-English languages to use the system (first, Latin-1, which is an 8-bit encoding of the Latin alphabet; then UTF-8 and UTF-16, which both have been defined as subsets of Unicode).

Because Unicode does actually contain most of the characters that are in use today, it is possible to have element names in XML that are not in English or that even use Latin letters. In other words, it is quite possible to have a Japanese document that has Japanese element names, for instance. If documents are to be used by an international audience, however, the fallback language is English, and that is also what (in reality) is used in most element names. If the document is not to be read by humans, however, any bit string will do. Computers cannot perceive meaning.

Different DTDs can be used to create different XML applications, such as XHTML, WML, and XSLT.

XML, HTML, and SGML

Like HTML, XML makes use of tags (words bracketed by < and >) and attributes (of the form *name="value"*), but while HTML specifies what each tag and attribute means (and often, how the text between them will look in a browser), XML uses the tags only to delimit pieces of data and leaves the interpretation of the data completely to the application that reads it. The XML specification specifies neither semantics nor a tag set. In other words, if you see <p> in an XML file, do not assume that it is a paragraph. In fact, XML is really a meta-language for describing markup languages. In other words, XML provides a facility to define tags and the structural relationships between them. This information goes into the DTD or the XML schema.

Because there is no predefined tag set, there cannot be any preconceived semantics. All of the semantics of an XML document will either be defined by the applications that process them or by style sheets. The *Resource Description Framework* (RDF) is an application that adds semantics to XML (actually, RDF Schema adds the semantics). The development of XML started in 1996, and it has been a W3C recommendation since February 1998. SGML was developed in the early 1980s and has been an *International Standards Organization* (ISO) standard since 1986. XML is defined as an application profile of the *Standard Generalized Markup Language* (SGML) defined by ISO 8879, but it is also defined as an application of itself. DTDs express the XML application as an application of SGML, and XML schemas as an application of XML. XML is, roughly speaking, a restricted form of SGML. The designers of XML simply took the best parts of SGML, guided by the experience with HTML, and produced something that is no less powerful than SGML.

XML is different from HTML in many more ways than being an application profile of SGML. It uses Unicode, which is a 16-bit format for representing almost all characters that are being used all over the world (it was really intended to be used for all characters, but there are some that are not mapped into the character set, mostly in Japanese and traditional Chinese writing—although the most frequently used characters are covered). HTML up to version 4.0, on the other hand, used 7-bit ASCII as its least common denominator. The reason why HTML used 7-bit ASCII is the same as for mail systems, where some older mail servers are not equipped to handle modern character sets. 7-bit ASCII misses many characters that are important to people outside the United States, however, such as inflections, accents, and umlauts, and it is very complicated to represent Chinese and Japanese characters. This situation is one reason for using Unicode: it is possible to parse all XML content by using the same parsers, regardless of the language in which the content is written. URIs, however, which identify the resources, still have to be 7-bit ASCII on the insistence of the *Internet Engineering Task Force* (IETF). And the URI encoding excludes some characters, as well.

Writing XML is also different from HTML in that it requires that the elements—the combination of start and end markup tags—be closed. Open-ended tags such as <P> are not allowed; rather, they have to be closed, such as <p> . . . </p> (and element names must be in lower case, according to the XML convention—although technically, they could as well be upper case). Elements can either be structured with a start and end tag (<tag> content </tag>) or as an empty element with the end tag included in the start tag (<tag/>). The first type is elements with content, and the second is elements with no content (used for elements with attributes only). Figure 2.1 shows an XSLT document in an XML editor.

In XML, as in HTML 4.0, it is possible to have attributes on markup. Attributes are placed inside the start tag, so <start beginning="now"> means that the element start has an attribute name of *beginning* and an attribute value that is *now*. The attribute value must be in quotes. Attributes, though, have a much larger role to play in XML than in HTML. They also play a very large role in RDF.

Element names describe what the element is about while attributes provide further information (which can be used in the processing of the content or the application). Attribute values give you control over the element, but the element drives the application. You can, for instance, define an alternate representation, when it should be used, and in what ways (you recognize the alt element from HTML). This function can be used to facilitate transformations, to control how the content is applied, and to do many other things. Note, however, that in applications such as *Wireless Markup Language* (WML), the interpretation in the device is very restricted, and you cannot use attributes as you like.

XML is case-sensitive, so Creator, creator, CREATOR, and cREATOR are interpreted as four totally different elements. Putting element names in lower case is the existing best practice (so you would use creator, not Creator), but the practice that has developed is also to use capitalization in elements where it increases readability (documentCreator, not documentcreator, for instance). This usage is called the interCap convention. The important thing is to watch this capitalization very carefully, because you cannot take upper-case element names and render them in lower case automatically.

HTML browsers normally render a document line by line as it is received at the client. The XML model is different. When a document arrives at a client, it is processed through a number of steps. First, the character data is decoded from the binary encoding that is used over the network, creating a stream of Unicode characters. The document is then parsed; that is, the XML processor steps through the document and identifies the elements that it contains and determines how they should be handled.

Unlike HTML documents, XML documents do not have to be structured in the order that they should be displayed or processed (you might not want them to be displayed at all). That they are ordered as they should be processed, instead of as they should be rendered, results in a tree structure that can be manipulated by programs and scripts via the *Document Object Model* (DOM) of the W3C, which essentially is an *Application Programming Interface* (API) to the data in the document. There have been discussions about APIs for RDF, but so far, there do not seem to be any winners.

To further confuse things, an XML document actually has two different object structures. Each XML document has both a logical and a physical structure. Physically, the document is composed of units called entities. An entity can refer to other entities in order to cause their inclusion in the document (an inline reference). A document begins in a root or document entity. Logically, the document is composed of declarations, elements, comments, character references, and processing instructions—all of which are indicated in the document by explicit markup. The logical and physical structures must nest properly within each other.

A document has to contain one or more elements. There must be one element, called the root or document element, that does not appear in the content of any other element. The document entity serves as the root of the entity tree and as a starting point for an XML processor. For all other elements, if the start tag is in the content of another element, the end tag is in the content of the same element. More simply stated, the elements, delimited by start and end tags, nest properly within each other.

XHTML

Most of the documents on the Web today are HTML documents. HTML is not XML. Instead, the documents are SGML documents, which might look like XML from a distance but are more complicated (some would even say more powerful). In reality, very few people understand all of the ins and outs of SGML, and to become useful, it had to be simplified. When Tim Berners-Lee started creating the World Wide Web, he decided that HTML should be an application of SGML because he intended it to have the structured features of SGML.

Once XML had become a W3C recommendation, work began to create a new HTML based on XML instead of on SGML. HTML had developed quite extensively since the first version was presented in 1991, and HTML 4.0 represents a full-blown markup language with elements and attributes that enable content management through the markup.

The new version of HTML, a rendering of HTML 4.01 in XML, is called XHTML and is defined in the W3C Recommendation XHTML 1.0. There are no new functions in XHTML compared to HTML 4.0, but all elements, attributes, and everything in XHTML 1.0 are defined in HTML 4.01 (which, in turn, merely contains corrections to HTML 4.0 and does not have any new functions). The expectation is that XML will be adopted as the universal data format on the Web, and because HTML is the most common type of data, XHTML fits better into this architecture than HTML. Languages that use XML as the data format are said to be applications of XML. So, XHTML is an application of XML. Figure 2.3 illustrates the relationships between the different modules of XHTML.

Many document editing programs still produce HTML, but if the HTML is correctly written, it is relatively easy to transform it into XHTML. You cannot perform this action by using XSLT, however, because HTML is not an XML application. But if you write your documents in XHTML, you can transform them into different HTML variants (for instance, Compact HTML, which is used in the Japanese iMode phones) by using XSLT.

Other XML Applications

Today, there are hundreds of XML applications used on the Web. Some of the most important, apart from WML and XHTML, are SMIL, the Synchronized Multimedia Integration Language (which is used to define how multimedia streams should interrelate); Xforms (the language that the W3C has defined as a replacement for the current forms mechanism on the Web); SVG (Scalable Vector Graphics, the format defined by the W3C to represent images as vectors); RDF (the Resource Description Framework); and SyncML (the Synchronization Markup Language).

It is also used for protocols (WAP Push, RSVP, and SOAP) to synchronize multimedia objects (SMIL) and the next generation of forms on the Web (Xform). Several markup languages can be mixed together in one document. This function is a feature of XML, but it is more or less useful depending on what the markup is intended to do. XHTML, for instance, is actually intended to function as a host language for other languages.

Namespaces

When you mix elements from several markup languages in a document, you need some way of keeping them apart. Otherwise, how would you know that the <a></a> element that you are using is a WML element and not an element from XHTML (because the same element name is used in both languages)? The way of separating them is to use XML namespaces.

An XSLT style sheet might contain elements and attributes from many different XML applications. A style sheet that transforms WML documents into XHTML documents will contain elements from three different XML applications: WML, XHTML, and XSLT. When elements and attributes from different XML applications are present in the same style sheet, there is a risk that an element or an attribute exists in more than one application but with a different meaning. If this situation happens, the XSLT processor cannot tell the difference between the names.

Applications use the element type name to determine how to process the element. In a distributed environment like the Web, names must be globally unique, or otherwise one name might accidentally be used for different purposes. For example, originally the element type VAR was used in WML to bind a value to a variable name, but in HTML an element type with the same name was used to describe that the following text was the name of a variable. Because the WML specification had not been published yet, it was possible to change the name to SETVAR. In practice, it is not always possible to check that a particular name is not used somewhere else for a different purpose. For this reason, namespaces are a part of most computer languages.

Namespaces are defined in the W3C recommendation Namespaces in XML (www.w3.org/TR/WD-xml-names). In a document, they are given by using the colon-delimited prefixes. The prefix is significant when comparing element names within a document; therefore, xsl:template and template are different. The prefix can vary between documents, however. The important thing is the association of a prefix string with a URI. That is the function of the xmlns: attribute in the style sheets. For instance, the namespace declaration "xmlns: xsl="http://www.w3.org/TR/WD-xsl"" associates the namespace prefix xsl with the URI that follows it: "http://www.w3.org/TR/WD-xsl". Because the prefix is arbitrary, it could be xmlns:flasklock=http://www.w3.org/TR/WD-xsl instead.

The names in an XML document can be associated with a URI that is globally unique. The name that is used inside the document is called a local name, and the name plus the associated URI is called a qualified name. That URIs are unique, by virtue of their association with the DNS, means that the qualified name will be globally unique and that the elements will also be globally unique. The URI is actually only used to identify the element names, however. There is no requirement that there should be a description, schema, DTD, or something else behind it.

XML namespaces are used to resolve naming conflicts between XML applications. In an XSLT style sheet, every element and attribute belongs to an XML namespace. A namespace is a set of names that has a unique identifier. The identifier can be a URL. All element and attribute names in XHTML belong to the namespace identified by the http://www.w3.org/1999/xhtml URL. All element and attribute names in XSLT belong to the http://www.w3.org/1999/XSL/ Transform URL. By comparing both the names and the respective namespaces, the XSLT processor can distinguish between two elements that have the same name but that belong to two different namespaces. It is, however, up to the one that is responsible for the namespace to make sure that there are no names that conflict in the same namespace.

The standard for XML namespaces is published in a separate specification, not part of the core XML specification, at http://www.w3.org/TR/REC-xml-names/. Since its publication in January 1999, most new XML applications use XML namespaces.

Qualified Names

A qualified name is an element or attribute name that can be associated with an XML namespace. In an XSLT style sheet, all element and attribute names are qualified names.

A qualified name consists of two parts. The local part is the name used inside the document and distinguishes the element or attribute name inside its namespace. Names such as style sheet, template, value-of, and copy-of are all local names inside the XSLT namespace.

The namespace name is the second part of the qualified name. It distinguishes the name globally and is typically a URL because URLs are globally unique. The namespace name for XSLT is http://www.w3.org/1999/XSL/Transform. It is part of every element and attribute name in XSLT.

So, the fully qualified name of the <stylesheet> element in XSLT is "stylesheet" plus the http://www.w3.org/1999/XSL/Transform URL. The fully qualified name of the <template> element is "template" plus the http://www.w3.org/1999/XSL/ Transform URL. The fully qualified name of the <a> element in XHTML is "a" plus the http://www.w3.org/1999/xhtml URL (and so on).

Qualified names are essential for interoperability on the Web. XML documents—and an XSLT style sheet is an XML document—are shared by many users who use different XML applications defined by different people, organizations, and companies. It is unavoidable that the same name will be used in different XML applications. By qualifying names with a unique URI, conflicts will be avoided.

Declaring and Using Namespaces

Every element and attribute name in an XSLT style sheet is a qualified name. Also, many other objects of an XSLT style sheet are identified by using qualified names (for example, variables, named templates, and attribute sets).

The xmlns and xmlns: Attributes

Before a qualified name can be used in an XML document, the namespace must be declared. To achieve this task, use a special attribute whose name starts with either xmlns: or xmlns.

The XML specification reserves names that start with the letters "xml" (both lower- and uppercase).

The xmlns: attribute declares a namespace prefix and a namespace name. Elements and attribute names that use the prefix in their name get associated with the namespace. In the following XSLT style sheet, all element names get associated with the XSLT namespace name:

```
<xsl:stylesheet xmlns:xsl="http://www.w3.org/1999/XSL/Transform">
     <xsl:template match="/">
          <xsl:copy-of select="." />
     </xsl:template>
</xsl:stylesheet>
```

The xmlns attribute declares a default namespace. Every element that does not have a prefix is bound to the namespace. In the following example, all elements are bound to the XHTML namespace:

```
<html xmlns="http://www.w3.org/1999/xhtml">
     <head>
          <title>An XHTML document</title>
     </head>
     <body>
     ...
     </body>
</html>
```

The default namespace does not apply to attributes. It is a common mistake to think that it does. An attribute belongs to the element on which it is declared, and that element can be in any namespace declared in the document. An attribute that has a prefix, however, belongs to the namespace to which the prefix is bound.

Many XSLT attributes can contain references to element or attribute names. The references are qualified names. In the following example, the names in the match and select attributes refer to the elements in the XHTML namespace:

```
<xsl:stylesheet
     xmlns:xsl="http://www.w3.org/1999/XSL/Transform"
     xmlns:html="http://www.w3.org/1999/xhtml">
     <xsl:template match="html:ol">
          <xsl:value-of select="html:li" />
     </xsl:template>
</xsl:stylesheet>
```

The default namespace does not apply to names in attributes values. In the following modified example, the names in the match and select attributes refer to the <ol> and <li> elements in no particular namespace:

```
<xsl:stylesheet
     xmlns:xsl="http://www.w3.org/1999/XSL/Transform"
     xmlns="http://www.w3.org/1999/xhtml">
     <xsl:template match="ol">
          <xsl:value-of select="li" />
     </xsl:template>
</xsl:stylesheet>
```

The default namespace does not apply to names in attribute values.

The commonly used attributes xml:lang and xml:space are in the XML namespace. It is an implied namespace that is never explicitly declared in the document. The namespace is controlled by the W3C, the organization that specifies XML. More names can be added to the namespace in the future.

The scope of a namespace declaration is the element on which it is declared and all descendant elements until it is overridden by another declaration that declares either the same namespace prefix (or, in the case of the default namespace, a new namespace) as the default.

In the following example, each template has a different default namespace:

```
<xsl:stylesheet
     xmlns:xsl="http://www.w3.org/1999/XSL/Transform">
     <xsl:template
          match="html-version"
          xmlns="http://www.w3.org/1999/xhtml">
          <body>
               <xsl:value-of select="." />
          </body>
     </xsl:template>
     <xsl:template
          match="wml-version"
          xmlns="http://www.wapforum.org/2001/wml">
          <card>
               <xsl:value-of select="." />
          </card>
     </xsl:template>
</xsl:stylesheet>
```

For readability, it is common to put all namespace declarations in the beginning of the document. Also, it is common practice to always use the same namespace prefix for a namespace name. For XSLT, we always use xsl as the namespace prefix. Sometimes XSLT elements are referred to with the xsl prefix, as in the <xsl:stylesheet> element and the <xsl:value-of> element. The namespace prefix, however, does nothing more than associate the name with a namespace declaration. The prefix must be declared in each document where it is used.

Transformations Across Namespaces

You will need to transform XML documents from one namespace into another (for example, a WML document into an XHTML document, a POIX document into an NVML document, or a document in your own private namespace into the XHTML or WML namespaces). All you need to do is declare the namespaces you want to use in the beginning of the style sheet.

The following example transforms a WML <card> element into an XHTML <body> element:

```
<xsl:stylesheet
     xmlns:xsl="http://www.w3.org/1999/XSL/Transform"
     xmlns:wml="http://www.wapforum.org/2001/wml"
     xmlns="http://www.w3.org/1999/xhtml">
     <xsl:template match="wml:card" >
          <body>
               <xsl:apply-templates />
          </body>
     </xsl:template>
</xsl:stylesheet>
```

The following example illustrates how XML namespaces can be used to embed an XHTML anchor element inside of our now familiar phone description:

```
<?xml version="1.0" encoding="UTF-8"?>
<?xml-stylesheet type="text/css" href="phone.css" ?>
<phone xmlns ="http://www.ericsson.com/phones"
  xmlns:html="http://www.w3.org/1999/xhtml">
<manufacturer>
<html:a href="http://www.ericsson.com/">
Ericsson
</html:a>
</manufacturer>
<model>R320</model>
     <network>GSM 900</network>
</phone>
```

The prefix is used as the link between the local name and the namespace name. There is also a default namespace that does not use any prefixes. In this example, names in the XHTML namespace use html as the prefix while all of the names in our own invented "phone description language" are in the default namespace and do not use any prefix. Without the prefix, it would have been necessary to include the namespace name (the URI) in front of every local name, which would have been very inconvenient.

Hybrid Document Types

The Web browser Internet Explorer 5.5 presents documents that contain a mix of HTML and SMIL. Two different document types exist: HTML, which is familiar to everyone when describing documents, and another, SMIL, which is less familiar and describes how multimedia objects are synchronized. The document types are mixed in order to create new document types. This class of document types has been given a special name: hybrid document types. The motivation for creating new document types from existing ones is the classic argument of reuse.

The basis for all hybrid document types is a core set of document types that represent functions that a large number of content authors need. Here are some examples:

Create presentations with different kinds of multimedia objects (for example, images, video, and sound) that are synchronized with each other and with events from the user interface. This task can be done with the *Synchronized Multimedia Markup Language* (SMIL).

Create maps and simple logos with vector graphics that are defined directly in the markup. This task can be done with the *Scalable Vector Graphics* (SVG) image format.

In a structured way, include information about the document; for example, the name of the author, copyright information, and so on. This task can be done with the *Resource Definition Format* (RDF).

Control presentation and layout of the markup. This task can be done with style sheets; either with *Cascading Style Sheets* (CSS) or *Extensible Style Sheets* (XSL).

Present a document as a set of cards and maintain state between cards and documents. This task can be done with the *Wireless Markup Language* (WML).

Describe text structures such as paragraphs and headers and include links to other documents. This task can be done with *Hypertext Markup Language* (HTML).

It is possible to mix and match markup in different XML formats by using namespaces in order to create truly multimedial or multimodal presentations.

One document instance can be presented in different ways (so-called presentation instances). The step from document to presentation is controlled with a style sheet. Presentation in different formats is not the entire solution, however. Often, the markup needs to be transformed into a different markup language. For instance, most WAP phones do not have the memory capacity to receive even a modestly long XHTML document. If it is transformed to WML instead, the card and deck mechanisms can be used to make sure that the document does not overflow the memory. The walk from one document type to another is where transformation sheets fit into the equation. It can also be used to do some primitive filtering.

Validity and Well-Formedness

If a document is not well formed, it is not XML, and the XML processor cannot handle it. This knowledge is very important when you write WML and XHTML: Documents that are not well formed will not be allowed. Well-formedness is, in turn, a subset of validity. A document can do without both the processing instruction at the start and the DTD, but it is then neither well formed nor valid. XML documents must be well formed, which is a minimum conformance level for the WML parser, for instance. They can also be valid. The validity check is basically a check of the document elements, their placement, and ordering against the DTD of the document. There is no automated way to check the content of a document (and indeed, it would be quite dangerous for our right to express ourselves if this situation were possible).

XML elements are defined in the DTD. Unlike SGML, an XML document is not required to have a DTD. Neither the XML declaration nor the DTD are required for an XML document to be well formed, which is all that is needed for a document to be read by the XML processor and rendered on the screen by the browser. An XML document always begins with a processing instruction, which is a way to define a document as an XML document and declare in which version of XML it is authored.

A well-formed XML document contains start tags, end tags, and content. A start tag contains a descriptive name (the "element name") surrounded by angle brackets: <french_army>. An end tag looks similar to a start tag except that it has a slash before the element name: </french_army>. Content is everything between the start tag and the end tag.

Another aspect of well-formedness is that document elements must be properly nested. You cannot have one element inside another. The sequence <french_army><prussian_army> Battle of Waterloo </french_army></prussian_army> is incorrect because the <prussian_army> element is not nested in the <french_army> element. If the processing is to work, you have to take great care to make sure that all of the tags you use in the document nest properly inside each other—including the root of the document, <html></html> in classic HTML.

A well-formed XML document must have a single root element that contains the rest of the document. This root element cannot exist anywhere else inside the document. If you are used to writing HTML, you will recognize that the HTML element is the root element because it surrounds the HEAD and BODY elements.

At the top of the document, you can place an XML Declaration to indicate the version of XML you are using and which character encoding the document has. XML documents start with the XML declaration <?xml ...?>. This declaration tells which version of XML is used and identifies that XML is being used. Apart from the xml version declaration, this declaration can also contain the character encoding, which is important if the document is to be rendered correctly. Here is an example:

```
<?xml version="1.0" encoding="UTF-8"?>
```

The XML declaration looks like a processing instruction, which is a way to provide information to an application. Processing instructions are not part of the document (in the same way as a comment), but the XML processor is required to pass them on to the application. Processing instructions contain the name of the instruction, which is how it is identified by the application. They also contain the data that are the parameters for the name. The application should process only the instructions that are directed at itself and leave all others aside. Any data after the name is optional, because what it is and how it is treated is up to the application. There are also processing instructions that have names that are prefaced with xml: and those are reserved for future XML standardization.

Therefore, you should also include any comments enclosed by <!-- and -->. Comments can contain any data except the two minus signs (the literal string --). Also, comments are not part of the textual content of the XML document, which means that an XML processor will not pass the content of a comment on to an application (unless there are other instructions saying so). In other words, you cannot include comments inside a document unless you are certain that the document will be used only in the current form or unless the comments do not matter. In that case, it might be a bit silly to include them anyway.

To be well formed, a document has to obey the syntax of XML. If the document cannot be parsed because it misuses markup characters, it cannot be well formed. That also applies if it does not follow the grammar for XML documents. Some types of markup, like parameter entity references, are allowed only in specific places and circumstances. If the document has them in other places, it is not well formed (even if it is well formed from all other points of view). In other words, you have to be careful about where you use parameter entity references, which are used for variable substitution in WML (among other things).

If the replacement text for parameter entities referenced inside a markup declaration is not correct, for instance, consisting only of a part of the markup declaration, the document cannot be well formed. All entities except & (the ampersand &);, < (<);, > (>);, &apos (');, and " ("); must be declared (and remember, those were declared in the XML specification). The document is also not well formed if an attribute appears more than once on the same start tag. And, because string attribute values are restricted to internal elements, they cannot contain references to external entities. The content flow also cannot reference binary entities (which are allowed only in an attribute that is declared as ENTITY or ENTITIES). Although the binary entity might contain information that can be used by the application, it will not be possible for other applications to understand it. Finally, no parameters or text entities are allowed to be recursive (directly or indirectly).

Strange although it might sound, a document can be well formed and invalid. A document is valid only if it contains a proper document type declaration (either in the document or referenced) and if the document obeys the constraints of the declaration (for instance, that element sequences and nesting of elements are correct and that required attributes are provided and attribute values are of the correct type). If you want the XML parser to ensure that all of your XML documents adhere to the same structure, make sure that your XML documents are valid. Valid XML requires the DTD or XML Schema that specifies the structure of the document in a very unambiguous, machine-readable way. You do not have to write this yourself; it can be referenced. When you are writing in a predefined format like WML or XHTML, you can just reference the DTD (or XML Schema).

To determine whether a document is valid, the XML processor has to read the entire DTD (both internal and external subsets). The big hurdle is whether a document is well formed. That it is valid does not become a requirement until it is transformed into something else, in which case the schema of the document has to be transformed (which puts very different requirements on what is needed in the document—in essence, all that is covered by the DTD).

XSLT: Rules Declarations for Transformations

The biggest change to the computing industry that XML brings is that it enables the programmer (or more properly, the system designer—who can be a programmer, a Web site designer, or someone else) to declare the rules for how information should be processed and handled. The XML application itself describes the rules for how information should be structured and how the markup should be constructed. The representation of the information is shaped by the use of a markup vocabulary determined by the designer, which also declares the data type of the markup vocabulary (note that the markup is separated from the actual content of the document). This situation enables tools to constrain the creation of an instance of information and enable users to validate a properly created instance of information against a set of constraints. The rule set for how to handle information is described in XSLT.

The generic XML processor has no idea what is meant by the XML, and the XML markup does not (usually) include formatting information. The information in an XML document might not be in the form desired to present it, but this information has to be described somewhere else.

An XML document is just an instance of well-formed XML. The two terms *document* and *instance* could be used interchangeably. A document is an object collection (of instances of the objects defined in the DTD or XML schema), where the content is held in elements (which are described in the schema or DTD).

The XML family of standards is starting to grow quite large, and in this book we will concentrate on four of them: Xpath, XSLT, and style sheets (XSL, which is an XML standard; and CSS, which is not an XML document type but still is used with XML documents). There are a few supporting standards that we also have mentioned already, such as XML namespaces and URI:s. XSLT is (for purposes of transformation) the most important. The other two are Xpath, which is used to define patterns (or parts of an XML document), and XSL Formatting Objects, which defines how XML documents should be displayed (to a viewer or to a listener). To confuse things, the W3C has another display language as well, called *Cascading Style Sheets* (CSS). It is not an XML-based language, but it can still interact with XML documents.

We declare our choice of an associated style sheet for an XML instance by embedding the construct described in the Stylesheet Association Recommendation. Recipients of the document and applications that process it can choose to respect or ignore this choice, but the declaration indicates that we have tied some process (typically, rendering) to our data, which specifies how to consume or work with our information. In the case of rendering, this process will normally be a CSS style sheet.

With XML, we can use any elements (combinations of starttags and endtags) we want. We can write documents by using our own element names—names that are meaningful where we intend to use them and that offer us greater control not only over presentation but also over the way in which the document will be processed. But this freedom comes at a price: XML tag names have no predefined semantics. An <h1> might just as legitimately identify a tall hedge as a first-level heading. Is <img> an image or an imaginary number? Who knows. You have to look in the DTD or XML schema to find out.

The characteristics of an XML element are declared in the XML schema (or DTD), but the presentation semantics are not part of the element's characteristics. If the content of the element <nuts> always had to be displayed as 10-point Times Roman, it would be impossible to import it into a database and process it automatically, to say nothing of importing it into a voice browser and rendering it as speech. Style sheets create a mechanism for how to present content, and different style sheets can be applied to the same content (depending on the circumstances).

In simplest terms, a style sheet contains instructions that tell a processor (such as a Web browser, print composition engine, or document reader) how to translate the logical structure of a source document into a presentational structure.

Style sheets typically contain instructions such as the following:

- Display hypertext links in blue.
- Start chapters on a new, left-hand page.
- Number figures sequentially throughout the document.
- Speak emphasized text in a slightly louder voice.

Many style-sheet languages augment the presentation of elements that have a built-in semantic meaning. For example, a Microsoft Word paragraph style can change the presentation of a paragraph, but even without the style, Word knows that the object in question is a paragraph.

The challenge for XSL, the XML Style Language that was the immediate origin of XSLT, is slightly greater. It is used for presentation of entire documents or document sets, transforming them to the relevant format in the process. Because there is no underlying semantic for an XML element, XSL must specify how each element should be presented and what the element is. For this reason, XSL defines not only a language for expressing style sheets but also a vocabulary of "formatting objects" that have the necessary base semantics.

The definition from the specification is actually as follows: "An XSL style sheet specifies the presentation of a class of XML documents by describing how an instance of the class is transformed into an XML document that uses the formatting vocabulary." In other words, a style sheet tells a processor how to convert logical structures (the source XML document represented as a tree) into a presentational structure (the result tree). An XSL style sheet is, in itself, an XML document, and during processing it is handled by the XML processor before it is handed to the XSL processor.

The key to understanding how you can transform one markup language into another—for instance, GML into SVG—is to understand XML documents in terms of tree structures. When you are creating a transformation sheet, you are creating an instruction for how a source tree should be transformed into a result tree. You are not manipulating the file itself, however. The original is still unchanged after the transformation.

In a WML document, for instance, the card element is nested in the deck element. It can be seen as a tree, where the deck is the root element and the card is a branch. Multiple cards will mean multiple branches. A branch is, in object-oriented terminology, a child of the root element. The root will then, of course, be the parent. As an aside, the tree can only branch further and branches cannot cross. An instance of an element in a tree cannot be the child of two parents (there can be other instances with the same element name, of course). If you have an element, which is the child of two (or more) parents, you no longer have a tree but a graph. And, to express graphs in XML, you need the *Resource Description Format* (RDF).

Contrasted with a file format where information identification relies on some proprietary hidden format, predetermined ordering, or some kind of explicit labeling, the tree-like hierarchical storage structure infers relationships by the scope of values encompassing the scopes of other values. The addressing of elements is done by using the Xpath language.

Although trees shape a number of areas of XML, both logically (markup) and physically (entities such as files or other resources), they are not the only means by which relationships are specified. For example, an information object (such as an element) can arbitrarily point to other information elsewhere by using *Universal Resource Identifiers* (URIs).

Because an XML application has to be well formed, applications also provide a language for specifying how a system can constrain the allowed logical hierarchy of information structures. Well-formed XML does not only dictate a certain syntax and use of characters in a certain way, but it also creates an implicit document model that is defined by the way elements are nested. There is no need to declare this model separately, because the syntax rules governing well-formedness guarantee the information to be seen as a hierarchy. The hierarchy also translates into a tree structure, which can be described in object-oriented terms. As with all hierarchies, there are family tree-like relationships of parent, child, and sibling constructs relative to each element.

Take, for instance, this simple XML example:

```
<?xml version="1.0"?>
 <purchase id="p001">
   <customer db="cust123"/>
   <product db="prod345">
        <amount>23.45</amount>
   </product>
 </purchase>
```

If you were to draw this code as a tree, it would look something like the following:

```
purchase -- customer
      |
product-amount
```

The hierarchy is implicit in the nesting of the elements. The customer element is a child of the document element, which is named purchase, and so is the product element. Amount, in turn, is the child of the product element. The elements nest in each other from the root out, and the element that does not nest in the others is the leaf node—the final tip of the tree.

The formal model does not affect the structural model of the document, and it does not really influence the interpretation of content. It is useful for an XML processor in handling the information in handling the document or to other types of systems that work with the XML document. This statement is true regardless of whether the model is expressed as a DTD or as a schema. References to those can be used for validation, that the information content conforms not only to the lexical rules for XML (well-formedness) but also that the syntax rules that the model dictates (validity). There is also other information that can be derived from the document model (such as the data type of an element or that it is required).

XML semantics, while one of the foundations of XSLT, is actually a gray area. The document model is only one of many components that are used to describe the semantics of the information found in the document. Although well-formed documents do not have to have a formal document model (because there does not have to be a schema or a DTD), the names of the elements and attributes themselves will give hints to the associated semantics. There is currently no way of expressing these formally, however. You have to describe them in comments to the schema.

The XML 1.0 Recommendation only describes what an XML processor acting on an XML stream should do, including how it should identify the data and provide it to the application that is using the processor. Because there are no formalized semantic description facilities in XML, any XML that is used is not tied to any one particular concept or application. One way of looking at it is that the only purpose of XML is to unambiguously identify and deliver constituent components of data. Nothing is imposed by any process when you are creating a new XML vocabulary. Applications using XML processors to access XML information must be instructed how to interpret and implement the semantics because there is nothing to stop several applications from using the same names.

The first working draft of XSLT from the W3C was produced in August 1998, and in November it became a formal recommendation of that organization. Its companion recommendation Xpath is about the same age. XML by itself does not actually do anything. XSLT makes it possible to match the output form of one process to the input form of another process or application.

Xpath was designed to be used together with XSLT. It is actually used with several other XML languages, as well. Xpath is a language to address parts of XML documents in XSLT. Xpath is also used in the XPointer W3C recommendation and in the emerging XML Query Language.

It is important when we think about styling information to remember that two distinct processes are involved, not just one. First, we must transform the information from the organization when it was created into the organization needed for consumption. Second, when rendering we must express, whatever the target medium, the aspects of the appearance of the reorganized information.

XSLT can be used to create output markup in any format that can be structured as characters. This format can, for instance, mean HTML but not binary formats. XML is based on Unicode, and that means it encompasses practically all character formats in existence. XSLT can also remove elements as well as add completely new elements. Elements can be rearranged and sorted, and elements can be tested against preconditions in order to make decisions about which elements to display.

The XSLT 1.0 Recommendation describes a transformation instruction vocabulary of constructs that can be expressed in an XML model of elements and attributes. One way of describing XSLT is "transformation by example," as opposed to other techniques such as "transformation by program logic." We tell an XSLT processor what we want as an end result, rather than describing how to do the changes, and it is the responsibility of the processor itself to do the actual work.

The XSLT Recommendation contains a vocabulary for specifying templates that function as "examples of the result." Based on how we instruct the XSLT processor to access the source of the data being transformed, the processor will incrementally build the result by adding the filled-in templates. It takes the source tree, applies the rules in the transformation sheet, and creates the result tree.

XSLT is similar to other forms of content transformation in that it deals with the documents as trees of abstract nodes. XSLT, or rather Xpath, enables you to identify structures and to make as many passes as required over them, modifying the structures in the source information (or rather, the markup of the information). The information being transformed can be traversed in any order needed and as many times as required to produce the desired result. The algorithms are declared in the XSLT code and are handled by the XSLT processor. It only works with the markup as abstract nodes, not with the source data or with the semantics of the markup.

The XSLT processor handles the mechanics of the operations. High-level functions such as sorting and counting are available when required as functions in the language. The XSLT processor handles low-level functions such as memory-management, node manipulation, how nodes are node traversed and created, and garbage collection. In other words, the XSLT programmer does not have to think about the mechanics of the operations. Nor do you have to consider the mechanics of the presentation, which can be left to the browser and the style sheet processor.

Writing a style sheet is actually a way of using markup to declare the behavior of the XSLT processor, much like HTML is used to declare the behavior of the Web browser to paint information on the screen. One effect is that XSLT might be more accessible to non-programmers (although this is doubtful, because programmers tend to underestimate the complexity of what they do).

While XSLT is declarative, there are procedural constructs as well. You can use it to write just about any program you can imagine. XSLT is (in theory) "Turing complete." It is possible to implement any algorithm, however complex it is. But this implementation will come at the price of considerable verbosity—the more complex the algorithm, the more verbose the XSLT code. Implementing business rules and semantic processing in XSLT is exactly what it is intended for (although it has its constraints).

XSLT does not actually transform the original document. It works with a copy of the source tree, which is transformed into the result tree. Or, if you will, the XSLT transformation sheet is an instruction for how to turn the source document into the result document (although it does not work with the actual source document, but only with the representation in memory). It only affects the parts that are to be transformed. The parts of the source document that are not addressed in the transformation sheet are unchanged and are passed on directly to the result tree. The fact that it works with a "virtual" document also means that it is possible for an XSLT transformation sheet to work with documents that are dynamically composed; for instance, created by another transformation sheet.

A transformation sheet consists of a set of templates (the instructions to the processor) that are matched to the source document (by addressing them by using Xpath). When a match is found, the matching part of the source document will be transformed to the result document.

One way of parameterizing an element is by its place in the tree structure. Another way is by giving it attributes. The attributes of an XML element are part of the element, not the content. Content in XML elements can be anything, and it can basically be binary—it does not concern the element structure. In other words, if a document has the same structure but the content is different, it can be processed as part of the presentation and the content will be the same as for the instance of the document. For instance, pages presenting flight information resulting from a request containing a flight number can be presented in WML on a WAP phone but in XHTML in a browser. The content will still be the same.

The most common usage of XSLT is probably producing human-readable forms of XML source documents. XSLT can be used to produce plain ASCII text through HTML, XML, and WML as well as formatted print documents in PDF and Postscript. One interesting use is to transform descriptions of geographical features, for instance in GML, into maps in a vector format, such as SVG.

XSLT is a general-purpose transformation language for XML-based content. It works fine for manipulating basic markup. It can be used to transform the content, but in XML, content can be binary, so if the content is not characters, it can be passed through untouched.

On the other hand, XSLT cannot be used to access system resources. There is no way to call subroutines that are written in other languages (although there is an extension mechanism that enables the processor to handle code from other declarative languages). It cannot deal with binary files (although the element content of the XML element can be binary). It is also an interpreted language, which is slower than compiled processing (although there are compilers available to speed up processing).

Because an XSLT style sheet is an XML document, it is no harder than creating any other XML document. Any text editor will suffice, although it is easier to use an XML editor (because it will give you a warning if you are trying to create a document that is not well-formed).

The following is a simple example that runs in most XSLT processors. There is no source document, or if you prefer, the source document is in the style sheet. It will generate a piece of content. It is embedded in HTML tags, but it is actually an XSL document that will generate the text “Welcome. This is an XSLT example” on the screen.

```
<HTML xsl:version="1.0" xmlns:xsl="http://www.w3.org/1999/XSL/
Transform">
  <HEAD>
    <TITLE>Welcome</TITLE>
  </HEAD>
  <BODY>
<P>This is an XSLT example</P>
  </BODY>
</HTML>
```

The namespace xsl:version=“1.0” xmlns:xsl=“http://www.w3.org/1999/XSL/ Transform” tells the processor that this document is a style sheet document. The processor will then look for elements in that namespace (in other words, <xsl:....>) and do something with them. Because there are no such elements, nothing happens.

This rule is actually the first rule of XSLT processors: If there is no XSLT, do nothing. The second rule for processors is that anything that is not in the XSLT namespace is passed straight through to the output and is included in the result tree. We are actually cheating a bit in this case.

If we instead create a source file like the one below (and call it example1.xml) and apply a style sheet to it, the result will be different. Here is example1.xml:

```
<?xml version="1.0" ?>
<example1>
<h1>The title goes here</h1>
<p>And this is the first paragraph</p>
</example1>
and the style sheet is:
<?xml vers ion="1.0"?>
```

This style sheet will generate some content and put it together with the content from the source document into an HTML document:

```
<xsl:stylesheet version="1.0"
                xmlns:xsl="http://www.w3.org/1999/XSL/Transform">
  <xsl:template match="/example1">
    <html>
      <head><title>Test Document</title></head>
      <body>
        <xsl:apply-templates/>
        <i>More content generated by the style sheet</i>
      </body>
    </html>
  </xsl:template>
<xsl:template match="head">
  <h1><xsl:apply-templates/></h1>
</xsl:template>
<xsl:template match="para">
  <p><xsl:apply-templates/></p>
</xsl:template>
</xsl:stylesheet>
```

Now, this coding is already a bit verbose, as you can see. Each iteration through the source document has to be described separately, which means that when you want to change <head> to <h1>, you have to write three lines of code and another three lines of code for changing <para> to <p>. This situation would be bad if it were not for the fact that the same template could be applied to every instance of <head> or <para>. The example will produce the following document:

```
<html>
<head><title>Test Document</title></head>
<body>
<h1>The title goes here</h1>
<p>And this is the first paragraph</p>
<i>More content generated by the style sheet</i>
```

```
</body>
    </html>
```

You can also embed the XSLT style sheet in the XML document. To perform this action, you have to use the xml-stylesheet processing instruction for that document. You also have to have a DTD that defines the xsl:stylesheet element as having an id attribute of type ID—otherwise, the href pseudo-attribute in the xml-stylesheet processing instruction won't be able to find the style sheet.

You also need a template matching xsl:stylesheet that does nothing so that the stylesheet is ignored when it runs. Otherwise, it will try to run on itself.

Here is an example of an XML document with an embedded style sheet:

```
<?xml version="1.0"?>
<?xml-stylesheet type="text/xml" href="#stylesheet"?>
<!DOCTYPE doc [
<!ATTLIST xsl:stylesheet
  id     ID     #REQUIRED>
]>
<doc>
<xsl:stylesheet id="stylesheet"
                version="1.0"
                xmlns:xsl="http://www.w3.org/1999/XSL/Transform">
  <!-- xsl:import elements to include content -->
  <xsl:template match="xsl:stylesheet" />
  <!- the rest of the stylesheet -->
</xsl:stylesheet>
<!- the rest of the XML document -->
</doc>
```

Because there are no reserved element names in XSL or in any other XML document (except for those containing the letters X, M, and L and the prefix xmlns:), it is necessary to use some other mechanism to distinguish between elements that have XSL semantics and other elements. Namespaces were designed to solve this problem.

In this case, the XSLT style sheet is an XML fragment, not a document in itself. Fragments, however, preserve the context of the document, so the style sheet is still XSLT even while embedded in some other type of document. The style sheet is contained within a style sheet element, and its content is template elements. (Style sheets can contain elements in addition to the template, but most style sheets consist of mostly templates.)

The XSL processing model is built on that of XML. An XSLT processor takes over after the XML processor, or parser, which reads the byte sequences of the XML document from the input stream, resolves any entities or inline references (for example, fetches any included document fragments), and passes any well-formed XML structures that it finds to the application that called it. The XML processor is a program that is required to handle the processing, but it does not execute instructions in the same way as the XSLT processor (or a Java virtual machine) does. It only checks the well-formedness of the document and resolves the references and entities. The input to the XML processor can be either as static entities in the notation described in the XML recommendation (that is, files using tags following the XML definition of "well-formed"), or more generally, as "documents" constructed through some other method (for example, presented by some other application such as a pre-built DOM tree or fired as a series of SAX events). Popular XML processor implementations produce either a stream of SAX events such as "begin element foo, begin CDATA, CDATA bar, end element foo . . ." or a DOM object. In other words, the XSLT processor is dependent on the XML processor. It will take the stream (or DOM objects, but most XSLT processors use streams) and uses that information in a node tree (the style sheet tree) to create a new node tree (the result tree), possibly using several other node trees as input (the source tree or trees).

Most XSLT processors accept SAX events generated by some XML processors as input, and some accept DOM objects that might be generated by other XML parsers or created from scratch. The SAX events or DOM objects are used as the basis for the node trees that follow the Xpath/XSLT data model.

Some XSLT processors are bundled with XML processors and are provided as standalone applications. Others are toolkits that application developers can integrate into their own software. In this book, we will assume that you are using one of the standalone XSLT processors, and we will not separate the XML and XSLT processing. XSLT processors are sometimes built so they can accept input in different forms. Because the XML notation is normative but an XML data model or "Infoset" is not, however, the usual case is for an XSLT processor to be combined with an XML processor to accept XML files by using the XML notation as input.

The XSLT processing can be done in the browser but also in the server. It has turned out to be done more economically in the server, especially for formats that are to be displayed in devices that are less capable in terms of display and processor (such as a mobile phone).

Conceptually, the XSL processor begins at the root node in the source tree and processes it by finding the template in the style sheet that describes how that element should be displayed. Each node is then processed in turn until there are no more nodes left to be processed. (In fact, it is a little more complicated than this situation, because each template can specify which nodes to process—so some nodes might be processed more than once and some might not be processed at all. We will examine this topic later.)

MSXML, for instance, provides both an XML parser and an XSL processor in one Windows DLL. IE 5.0 is an application that utilizes MSXML to handle the processing of XML documents that have been associated with a style sheet by means of a processing instruction in the XML. The version of MSXML that ships with IE 5.0 does not fully implement XSLT but is instead based on an older working draft of the XSL specification, before XSL was split into XSLT, XSLFO, and Xpath. This version is the source of much confusion among people who are new to XSLT.

Contrary to the case of browsers, where designers often feel that it is necessary to create a specific version of the document for each browser, the programmer can create an XSLT transformation sheet that will work on all processors by avoiding the differences.

The XML processor reads the XML input. The XSLT processor performs the actual XSL transformations. It will have to use an XML processor to read the source XML and the

XSLT. An XML processor (often called a parser, but called a processor in the XML recommendation) reads a source XML file and identifies the syntactic units (such as elements, attributes, and text content).

An XSLT processor takes a style sheet and applies it to the tree representation of a source XML document (produced by an XML parser) and generates a tree representation of an output XML document.

Because XSLT transformations operate on the tree-like document data model described in the Xpath specification, inputs and outputs to the XSLT processor will be a representation of a tree. These trees often start and end life as documents in an XML notation. In other words, it might look like we are taking the document and changing the tags, but that is not how it works. XML tags (elements, actually) are only a way of representing underlying data structures. So, XSLT actually works with the underlying tree structure through the XML elements. You must realize this fact, because many misconceptions about how XSLT style sheets are best written come from looking at them as a way of switching around tags.

The product of the XSLT processing is a "result tree." If the result tree is composed of XSL formatting objects, then it describes how to present the source document. There is nothing, however, that says that the result tree has to be composed of XSL formatting objects. It can be composed of any elements, and it does not have to be XML, either. When HTML is used in the result tree, XSL will transform an XML source document into an HTML document. There are some restrictions, however: empty elements, for instance, will use the XML syntax for empty elements.

While checking source documents for validity can be very useful for diagnostic purposes, all of the hierarchical relationships of content are based on what is found inside the specific document that is being input into the processing (the instance), not based on the document data model. An XSLT style sheet is independent of any DTD or other explicit schema that defines the abstract model of the instance. In other words, XSLT can process well-formed XML that does not have an explicit data model (for example, a DTD or schema).

If there is a DTD or an XML Schema describing the document, however, certain information such as attribute types and defaulted values can be used to improve the processing. Without this information, the processor can still perform style sheet processing as long as the absence of the information does not influence the desired results.

Because we do not depend on a specific DTD for the current document, we can design a single style sheet that can process different (but similar) documents. For instance, if you have a damage report for cars, which will include certain elements when the car has a diesel engine and others when it has a gasoline engine, the same style sheet can be used to process them. When the models are very similar, much of the style sheet operates the same way each time and the rest of the style sheet only processes that which it finds in the sources.

The corollary is that one single source file can be processed with multiple style sheets for different purposes. In other words, it is possible to process a source file with a style sheet designed for an entirely different vocabulary. The results will probably be very inappropriate, but there is nothing inherent to an instance that ties it to a single style sheet or a set of style sheets. Style sheets can actually be used to validate input as well as present output, but this topic is beyond the scope of this book.

A style sheet can be the synthesis of the starting style sheet and a number of supplemental files that are included or imported by the main file. There are restrictions, however. It is not possible to build style sheets that are dynamically composed, for instance.

APPENDIX C

WAP 2: How It Works

Web services that are adapted to the wireless environment have been a tremendous success, especially in Japan where more than 25 million people use iMode. The architecture of iMode has limitations, however, which will be overcome with the next generation of WAP—an architecture where NTT DoCoMo has been participating in the development.

Most Web services cannot be used directly in a mobile terminal because they have been designed for larger screens and for a use that is incompatible with mobile terminals.

WAP and iMode are not compatible. The markup in iMode is based on an older version of HTML, WAP on the new XML data format. The protocols used are totally different, with iMode using optimized versions of the protocols used on the Web—something that does not fulfill the design constraints for WAP.

To make life easier for developers, DoCoMo has been working in the WAP Forum to develop the new version of WAP, which will contain the best features from both iMode and WAP while maintaining compatibility with future standards and the installed WAP base and using features from the next generation of the World Wide Web. The result, which DoCoMo has declared it will use in coming versions of iMode, is WAP NG (or WAP Release 2001, to use the formal name). In the second-generation WAP architecture, more of the Web technologies are adopted directly (as they are) rather than adapted and then extended with WAP-specific functions. These functions enable service providers and developers of systems, content, and devices to provide users with greater added value.

The WAP 1.x architecture consisted of the origin server, gateway, and user-terminal environment. The server could be a WAP or HTTP server; the gateway translated the protocol layer and application information. By contrast, the second-generation WAP architecture consists of four conceptual components, namely:

- Application environment
- Protocol framework
- Security services
- Service discovery

The WAP Location Framework, which I discussed in Chapter 6, "Providing Databases and Doing Searches," fits into the application environment but leverages several different aspects of the other components. How the system looks is easier to understand if you look at Figure C.1.

The second-generation WAP architecture does not have strict divisions between the server, gateway, and user-terminal environment. Also, there is no longer any intermingling between transport and service. Instead, functions—which are accessed via the Internet—can be outsourced to capability servers in a WAP network that implements support for, say:

- A *wireless telephony application* (WTA)
- A *public-key infrastructure* (PKI) portal
- A provisioning server
- A *user-agent profile* (UAPROF) repository

Communication from WAP clients can take place directly with the origin server (in other words, the Web server that is located at the URI that they are requesting), but it will most likely take place through a proxy, which is a new role for the WAP gateway. Proxies have been around almost since the beginning of the Web (the idea is far older) and are being established as one of the main points of control (for example, through firewalls) and as central points for resource interconnection. WAP clients support a proxy-selection mech-

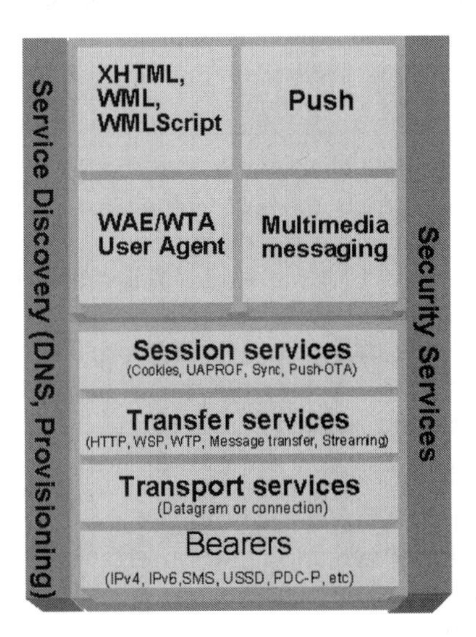

Figure C.1 The second-generation WAP architecture.

anism that enables them to choose the most appropriate proxy for a specific task. This functionality extends the current Internet proxy model.

The second-generation WAP protocol is compatible with WAP 1.x, but it relies more extensively on standard Web protocols (such as HTTP) and formats (such as extensible hypertext markup language, namely XHTML). WAP 2.x also clearly separates the bearer (for example, GPRS, IMT-2000), transport (such as WDP, TCP), session layer (for example, Cookies, CC/PP), and applications.

Most protocol services in the WAP 1.x suite are also available in new Web protocols. But the WAP push service cannot be realized through existing Web protocols without significant changes to the existing Web architecture. Both the WAP 1.x stack and Internet protocols (such as hypertext and multimedia transfer services) can provide some services, but only WAP is capable of providing others, such as the WAP push service.

The architecture enables components to interact. Developers can select modules from different components and create new user services. Conceivably, a minimal device can be created or developed by selecting components with the smallest footprints. In practice, devices and proxies will most likely implement either a dual stack or only the Internet stack. Backward compatibility is achieved by translating XHTML to WML in the proxy.

The application environment component enables the following services:

- *WAP application environment* (WAE); in other words, the browser, calendar agent, and other user agents
- User agent profile
- Multimedia messaging and other data formats
- Push service

The application environment provides the user interface and other functions that display content. Because it is a flexible environment, modules can be added on an ad-hoc basis (optional) or through the WAP Forum specification development process.

The WAP application environment resides in the mobile terminal. It contains a subset of XHTML and the CSS language (for content formatting). It also contains user agents for WTA and programming interfaces for use in mobile devices. The WAE also supports ECMAscript (formerly Javascript). WML and WMLscript execute in the WAP application environment.

Using cascading style sheets, an author can define how each element in a document is to be displayed. This feature gives authors greater control compared to when the display is specified inside the markup. A style sheet need only be downloaded once from the network server. After that, it can be retrieved from a local cache.

CSS can adapt automatically to the capabilities declared by a device's user-agent profile. This feature is particularly important because display capabilities vary significantly among devices. A format that looks good on one device might look different on another. The user-agent profile ensures that the device gets the most appropriate style sheet. Also, because style sheets separate display from content, authors can use the same WML document for many different devices with significantly different display capabilities.

WAP 1.x versions contained the vCal and vCard data types, which are not part of the browsing environment. The Internet Mail Consortium standardized vCal and vCard as structured data types for displaying contact and calendar information. iCal, which was developed from vCal, is used in products such as Microsoft Outlook and Lotus Organizer. It is also used in Ericsson's AirCalendar, which enables users to synchronize the electronic calendars they use in the fixed environment with the calendars on their mobile terminals. WAP also accommodates other data types, such as audio and video.

Many of the functions of the new WAP environment are available in existing Internet architectures, but the second-generation WAP application environment contains two modules that were developed in WAP 1.x, which contain functions that are not available in other systems. *Multimedia messaging* (MMS) was one of the highlights of this year's GSM-UMTS Forum, and Push is not possible on the Web when using standard HTTP. The MMS is an e-mail-like mechanism for the transmission of multimedia messages (electronic postcards with sound), which are expected to become very popular applications (especially on third-generation networks). It contains a multimedia transport mechanism for asynchronous message transport (messages are encapsulated for transmission between multimedia and WAP servers in a WAP-specific protocol). The data-transport mechanisms also include IETF streaming formats.

Service providers use push services to send information to users (who need not initiate any action). As simple as it might sound, the push-service architecture has been a major item on the WAP Forum's agenda. In the second-generation WAP architecture, the push service has been divided into the user-agent module and the session-layer module. The push OTA session service enables the establishment of push sessions across communication links that might not be persistent and in instances when addresses are dynamically assigned.

The WAP application environment relies on a protocol framework component that enables the functions needed to provide the services described previously. The protocol framework consists of four modular layers, which can be combined:

- The session service layer
- The transfer service layer
- The transport service layer
- The bearer service layer

In traditional Internet environments, the protocol framework solely provides transport services for applications, such as hypermedia transport (HTTP), streaming (RSVP and RTP Internet protocols), and message transport (standard Internet protocols, such as SMTP). In the WAP architecture, a logical layer has been added: the session services layer.

In the WAP 2.x architecture, the session service layer, which resides between the transport layer and the application environment, brings several new services to applications. Sessions do not exist in HTTP, but cookies can provide sessions. Cookies, which are database markers included in the request and looked up on the server side to identify the user, are part of the WAP 2.x architecture. They enable the reuse of mechanisms that already exist in the Internet and solely give an indication about the relationship between a single server and the user agent. Cookies cannot be used as a general information source.

The session services layer also includes a technology for reporting to the server information on terminal capabilities and on the terminal application environment. This information is used to optimize the display format.

Synchronization is another new service in WAP. The WAP Forum has been working with SyncML (another industry consortium) to create a language for data synchronization. The synchronization of data that has been updated in mobile and fixed environments can be a thorny issue. Users retrieve data from the network and store it on a mobile device, which they use to access and manipulate the local copy of the data. Periodically, users reconnect to the network to send changes to the networked data repository. Users also have the opportunity to learn about updates made to the networked data while their terminal was offline. Occasionally, users need to resolve conflicts between their local updates and the networked data. This reconciliation operation (during which updates are exchanged and conflicts are resolved) is known as data synchronization. The data synchronization protocol synchronizes networked data with that on many different devices, including handheld computers, mobile phones, automotive computers, and desktop PCs.

There is no binding between the transport of data and the session on the Web. The data transport is transparent to the session. Once a hypertext transport transaction is finished, the state it created disappears. In the WAP 1.x stack, the *wireless session protocol* (WSP) and *wireless transaction protocol* (WTP) can be used in combination to create and maintain a state, and through it, sessions. This has several advantages (for example, to enable push). By including HTTP as a transport method, the WAP Forum now enables both stateful and stateless transports. Session services provide a "memory" of previous transactions (which does not exist in HTTP because it is a stateless protocol) that enables the retention of terminal characteristics and makes for faster initialization of complex transports (such as data streaming).

The transport services in the next generation of WAP are either datagrams (connectionless service) or connections. Datagrams are more efficient for services that are not dependent on a persistent connection. In WAP 2.x, the datagrams comply with either the *user datagram protocol* (UDP), which is used on the Internet, or the wireless datagram protocol, which was defined for the WAP 1.x architecture.

The connection-oriented aspects of the new architecture are handled by the *transmission control protocol* (TCP). TCP, however, does not work well over mobile networks, so the WAP Forum is discussing an optimized mobile profile to enable the mobile terminal to function optimally over the mobile network with its special characteristics.

In WAP 2.x, the bearer services—the networks for which there are bindings—have been extended considerably. They now include the mobile radio bearers used to transport WAP (such as SMS, FLEX, USSD, and GUTS) as well as IP version 4 (IPv4) and IP version 6 (IPv6).

WAP can be transported over different networks, and mapping can be handled directly from the WAP stack to several bearer services. WAP 1.x contains several modules for bearer networks, some of which (broadcast networks, for example) could not be handled by using TCP transport. In WAP 2.x, bearers can be managed by the IP stack or directly by the WAP datagram or connection service, which uses the Internet's *transmission control protocol* (TCP).

The security services component is positioned orthogonally to data-transfer and data-use services within the protocol framework component. Security on the Internet is currently a hot issue, and the WAP architecture has received a lot of criticism. The telecommunications industry has been a leader in the security area for a long time, and this experience has been transferred into the WAP 2.x architecture.

WAP security services span all layers of the WAP architecture, thus creating opportunities for users to set up extremely secure environments by combining application-layer, transfer-layer, transport-layer, and bearer-layer security—the possibilities are endless. Security services include the following:

- Mechanisms for signing and encrypting data as a WMLScript crypto library
- Authentication services
- An identification service that uses the *wireless identity module* (WIM)
- A PKI system (Public-Key Infrastructure)
- *Transport layer security* (TLS, previously called SSL)
- WTLS, the WAP 1.x-adapted version of TLS

The service-discovery component is another orthogonal component in the WAP 2.x architecture that embraces what is available on the Internet and extends it by adding mobile-specific components. One example, the service lookup protocol, uses the existing DNS from the Internet. Terminal functionality is extended through the *extended functionality interface* (EFI), which enables a WAP device to have external entities attached to it (thermometers, pressure gauges, and so on).

Provisioning, which is another telecommunications-specific protocol, is translated into WAP. Devices can thus be provided with all the parameters they need to function through the network.

Navigation discovery enables a client to discover services in the network as part of its navigation—for example, a client might need to find a proxy in order to download data.

APPENDIX D

What Is CC/PP?

CC/PP is, simply put, a data structure that enables you to send device capabilities and other situational parameters from a client to a server. It can be included in a separate header in HTTP, and there are transports defined for other protocols. What happens when it gets to the server, and how it is created, is not discussed in the specification (but this document attempts to provide some guidance). CC/PP is organized into components. They are not fixed, and anyone can create components as well as vocabularies.

CC/PP contains a vocabulary for the handling of structural items, such as how to append a proxy description to the description of a client (if, for instance, the proxy can provide services for the client—for instance, a transcoding proxy can add the languages it can transcode into to the languages that the client can accept).

The WAP Forum defined the term *user-agent profile* (UAPROF) for use within the *composite capabilities and preferences profile* (CC/PP), which is used to describe the capabilities of the user terminal application environment. The user-agent profile is a data format. A specific set of properties and values describes each terminal. The WAP Forum standardized the property names and values as part of the UAPROF vocabulary.

CC/PP is defined in an XML framework—called the *resource description framework* (RDF)—which enables users to connect a property to an object (the CC/PP is an application of the RDF). The resource description framework can be used for annotations, metadata, and profiles that describe users or their terminals. By knowing the information display capabilities of a terminal, the server can create a display that is optimized for that terminal. Including profile information with the request minimizes the number of transactions needed to optimize the information, and it can be cached in a proxy or retrieved from a repository that the device manufacturer maintains. This feature minimizes information transmitted over the air and speeds up information access. Designers can create pages or page templates to be used with database servers (such as the

Ericsson WAP application server), displaying them in formats that are adapted to user devices.

Using CC/PP is actually very simple. Based on a description of a device (or a requirement for a device) that is part of the request for a resource, you output a document that is formatted according to some heuristics that are contingent on that description and deliver it to the user. There are a number of vocabularies that can be used to describe device capabilities, user preferences, and other information related to the user's delivery context. CC/PP provides a way of expressing such vocabularies in the W3C RDF and describes how they could be organized to be transported and handled by user agents (such as browsers) in an efficient manner.

Descriptions of single features in a description of a terminal or the user's preferences for how that terminal should be used cannot in themselves be used to infringe on a user's privacy. When a profile contains a collection of such properties, however, it can be used to personalize information; the closer the personalization, the bigger the risk that the user can be identified from his specific use of the terminal and the bigger the risk of misuse from a privacy standpoint. There is also a possible risk that a user can be identified as having certain abilities (for example, if he or she constantly requests text in double-sized fonts, it is likely that he or she has poor eyesight), and this situation might have the potential for misuse. Generally speaking, a user can consider some or all of the data in a CC/PP profile as private and might wish to have control over who receives that information and when. Origin servers customizing the content should therefore communicate to the user or user agent their privacy practices with regard to the use of CC/PP data so that the user can make a decision on whether or not to share that data with the server.

P3P is a way for an origin server to express the privacy policy it adheres to for the user and/or his or her terminal. While the fit between P3P and CC/PP is not perfect, and while the intent and implementation of the two are different, they can be used together to enhance the privacy protection of the user.

In general, user preferences and device capabilities need to be protected from malicious use but there is no trust management framework for CC/PP so far. Without trust management, privacy-sensitive information opens to attacks by malicious servers or content providers.

This situation does not mean, however, to create new technologies in terms of network security, authentication, message validation, personal privacy protection, and cryptography. We intend to employ the existing technologies in terms of trust management while considering how to apply such technologies to the use cases of CC/PP.

There is currently no CC/PP protocol defined as part of the standard. But there are two main ways of transporting CC/PP over HTTP: using the CC/PP Exchange Protocol or using the UAProf W-HTTP Protocol.

CC/PP Exchange Protocol

The CC/PP Exchange Protocol was presented as a W3C Note on June 24, 1999. It uses the HTTP Extension Framework (RFC2396), which is a major problem since it is not likely to be implemented. However, the concepts described in CC/PP-Ex have been reused in the WAP W-HTTP implementation, so it is interesting to discuss how it is constructed. The intent was to provide a framework that was both possible to map into HTTP headers and that can handle defaults as URIs. The idea was to minimize data transfer over the air, a goal that was accomplished as demonstrated by Kiniko Yasuda. CCPP-ex uses two headers, one for the defaults and one for the updates (profile-diff:s), which are separated by using MD5 hashes. A third header carries warning information. The protocol is documented in the W3C Note "CC/PP exchange protocol based on HTTP Extension Framework."

The CC/PP framework is a mechanism for describing the capabilities and preferences associated with users and user agents accessing the World Wide Web. Information about user agents includes the hardware platform, system software, applications, and user preferences. The user agent capabilities and preferences can be thought of as metadata, or properties and descriptions of the user agent's hardware and software. The CC/PP descriptions are intended to provide the information necessary to adapt the content and the content delivery mechanisms to best fit the capabilities and preferences of the user and the agents.

The major disadvantage of this format is that it is verbose. Some networks are very slow, and this method would be a moderately expensive way to handle metadata. There are several optimizations possible to help deal with network performance issues. One strategy is to use a compressed form of XML, and a complementary strategy is to use references (URIs).

Instead of enumerating each set of attributes, a reference can be used to name a collection of attributes such as the hardware platform defaults. This feature has the advantage of enabling the separate fetching and caching of functional subsets.

Another problem is to propagate changes to the current CC/PP descriptions to an origin server, a gateway, or a proxy. One solution is to transmit the entire CC/PP descriptions with each change. This method is not ideal for slow networks. An alternative is to send only the changes.

The CC/PP exchange protocol does not depend on the profile format that it conveys. Therefore, another profile format besides the CC/PP description format could be applied to the CC/PP exchange protocol.

For example, a user agent issues a request with URIs that address the profile information, and if the user agent changes the value of an attribute, such as turning sound off, only that change is sent together with the URIs. When an origin server receives the request, the origin server inquires of CC/PP repositories the CC/PP descriptions using the list of URIs. Then the origin server creates a tailored content using the fully enumerated CC/PP descriptions. The origin server might not obtain the fully enumerated CC/PP descriptions when any one of the CC/PP repositories is not available. In this case, it depends on the implementation whether the origin server should respond to the request with a tailored content, a non-tailored content, or an error. In any case, the origin server should inform the user agent of this fact. A warning mechanism has been introduced for this purpose.

It is likely that an origin server, a gateway, or a proxy will be concerned with different device capabilities or user preferences. For example, the origin server might have the responsibility to select content according to the user's preferred language, while the proxy might have the responsibility to transform the encoding format of the content. Therefore, gateways or proxies might not forward all profile information to an origin server.

The CC/PP exchange protocol might convey natural language codes within header field-values. Therefore, internationalization issues must be considered. The internationalization policy of the CC/PP exchange protocol is based on RFC2277.

Considering how to maintain a session like RTSP [RFC2326] is worthwhile from the point of view of minimizing transactions (in other words, the session mechanism could permit the client to avoid resending the elements of the CC/ PP descriptions that have not changed since the last time the information was transmitted). A session mechanism would reduce cache efficiency, however, and would require maintaining states between a user agent and an origin server.

An extension declaration is used to indicate that an extension has been applied to a message and possibly to reserve a part of the header namespace identified by a header field prefix. The HTTP Extension Framework introduces two types of extension declaration strength: mandatory and optional, and two types of extension declaration scope: hop-by-hop and end-to-end. Which type of the extension declaration strengths and/or which type of the extension declaration scopes should be used depends on what tasks the user agent needs to perform.

The strength of the extension declaration should be mandatory if the user agent needs to obtain an error response when a server (an origin server, a gateway, or a proxy) does not comply with the CC/PP exchange protocol. The strength of the extension declaration should be optional if the user agent needs to obtain the non-tailored content when a server does not comply with the CC/PP exchange protocol.

The scope of the extension declaration should be hop-by-hop if the user agent has an a priori knowledge that the first hop proxy complies with the CC/PP exchange protocol. The scope of the extension declaration should be end-to-end if the user agent has an a priori knowledge that the first hop proxy does not comply with the CC/PP exchange protocol or the user agent does not use a proxy.

The Profile header field is a request-header field that conveys a list of references that address CC/PP descriptions.

The grammar for the Profile header field is as follows:

HEADER FIELD **GRAMMAR**

```
Profile             = profile-field-name ":" 1#reference
profile-field-name  = "Profile"
reference           = <"> ( absoluteURI | profile-diff-name ) <">
profile-diff-name   = profile-diff-number "-" profile-diff-digest
profile-diff-number = 1#DIGIT
```

```
profile-diff-digest = sp; < MD5 message digest encoded by base64 >
DIGIT               = <any US-ASCII digit "0".."9">
```

The Profile header field-value is a list of references. Each reference in the Profile header field represents the corresponding entity of the CC/PP description. A reference is either an absolute URI or a profile-diff-name. An entity of a CC/PP description which is represented by an absolute URI exists outside of the request, and an entity of a CC/PP description which is represented by a profile-diff-name exists inside the request (in other words, in the Profile-Diff header field).

The profile-diff-name in the Profile header field addresses a CC/PP description in the corresponding Profile-Diff header within the same request. When the Profile header field includes a profile-diff-name, the corresponding Profile-Diff header *must* be included within the same request. The main reason why the profile-diff-name is introduced is to specify the priority of each CC/PP description in the Profile header field-value. The priority is indicated by the order of references (in other words, absolute URI or profile-diff-name) in the Profile header field-value. The latest reference in the Profile header field-value has the highest priority. Therefore, a CC/PP description, which is represented by the latest reference, can override CC/PP descriptions that are represented by the precedent references. This behavior is the default in the absence of schema rules.

All profile information could be represented by absolute URIs in the Profile header. In this case, the Profile-Diff header field does not have to be added to the request. On the other hand, only one Profile-Diff header can contain all profile information. In this case, the Profile header includes only the profile-diff-name, which indicates the Profile-Diff header.

W-HTTP

The WAP Forum UAProf group has defined a transport for CC/PP over W-HTTP (Wireless Profiled HTTP). It can be found in Section 9 of the UAProf specification. In the case where the mobile terminal supports wireless profiled HTTP, the profile is transported using metadata defined by this specification. The CC/PP Framework remains unaltered. The defined mechanism provides a functional equivalent for the CC/PP exchange protocol, but the definition of the syntax and semantics of the transport remains in this specification.

The following extension headers are defined to transport CC/PP in W-HTTP. The defined extension headers are considered to be end-to-end headers. The x-wap-profile header is a general header field that *must* contain the following: a URI referencing the CC/PP profile or a reference to a profile difference, transported by using the x-wap-profile-diff; or a combination of multiple instances of these two types of data. This data can be generated by the mobile terminal or attached by an intermediary point in a request to an origin server.

In the case of Push, this header can be generated as a response to a request. In the case of Push, this data can be cached; however, this header must be present in any request or response when UAProf is used.

The x-wap-profile header may contain references to instances of the x-wap-profile-diff header (defined in the following section). Each reference contains two parts, the sequence number and the profile-digest. The sequence number is used to determine the order of how the x-wap-profile-diff headers should be applied, and the digest is used to validate that the profile-desc in the x-wap-profile-diff header value is correct.

The x-wap-profile-diff header is a general header and *can* be generated by the mobile terminal or an intermediate proxy to enhance or alter the CPI. There might be multiple profile differences; each profile difference must also have a reference in the x-wap-profile header (which indicates the order in which differences should be applied). This header contains two parts, a sequence identifier and the entity (which represents the part of the CC/PP description that is being enhanced). This header *might* be present in a request or response. In the case of Push, this data can be cached.

The x-wap-profile-warning header is a general header. Essentially, it is the same as the X-WAP-PROFILE-WARNING header in the CC/PP Exchange Protocol. Its presence indicates the level to which the response has been tailored in relation to profile data that has been supplied in the request. This header *might* be present in a request or response. The warning codes that are defined fall into the following categories:

- 1xx—reserved
- 100—reserved
- 2xx—indicates whether the content has been adapted depending on the profile
- 5xx—indicates the server is incapable of processing CPI.

The x-wap-profile-warning can have the following values:

- 200 Not applied

  This value *must* be included if the content has not been tailored and is sent in a representation that is the only representation available in the server.

- 201 Content selection applied

  This value *must* be included if the included content has been selected from one of the representations available.

- 202 Content generation applied

  This value *must* be included if the content has been tailored or generated as a result of applying the included profile.

- 203 Transformation applied

  This value *must* be added by an intermediate proxy if it applies any transformation changing the content-coding based on the CPI data.

- 500 Not Supported

  This value indicates that the entity sending this warning code does not support UAProf.

In this transport variant, the headers and their values are not compressed for over the air transmission. It is recommended that the hardware and software manufacturer only use the x-wap-profile header to indicate an absolute URI as the reference for CPI for the mobile terminal when transmitted over the air. This specification does not preclude the use of x-wap-profile-diff in this case, however.

If the x-wap-profile-diff header is included, the profile-diff-seq *must* match a sequence number in the x-wap-profile header; otherwise, the associated profile-diff-desc *must not* be processed. The reference to each x-wap-profile-diff header contains two parts: a sequence number that governs the order in which the x-wap-profile-diff headers are processed and an MD5 message digest that is used to validate the profile-desc, which is contained in the x-wap-profile-diff. If neither the sequence nor the MD5 validation match, the particular profile-diff must be ignored.

If the x-wap-profile-diff header is added by an intermediate proxy, it *must not* alter the existing sequence of x-wap-profile-diff headers; the proxy must append using the next available sequence number in numeric order. The digest associated with an x-wap-profile-diff *must* be generated by applying the MD5 message digest algorithm [RFC1321] and Base64 algorithm, section 6.8 in the MIME specification [RFC2045], to the corresponding profile-desc part of the header field-value.

The MD5 algorithm takes as input a message of arbitrary length and produces as output a 128-bit "fingerprint" or "message digest" of the input. The Base64 algorithm takes as input arbitrary binary data and produces as output printable encoding data of the input.

The profile information referred to in the x-wap-profile and x-wap-profile-diff header does not supersede HTTP request or response header information.

The x-wap-profile-warning header *must not* be used for cache control purposes. If a server wishes to indicate a caching dependency based on these headers, then it should use the Vary header as defined in section 14.44 of the HTTP 1.1 specification.

CC/PP Vocabularies

The CC/PP Structure and Vocabularies Working Draft describes how such vocabularies can be created and communicated to the origin server using the CC/PP framework. CC/PP in itself does not describe a vocabulary (although a simple demonstration vocabulary is provided in the Structure and Vocabularies working draft). This document describes a number of existing vocabularies, and how they can be conveyed as part of a CC/PP profile. It also discusses the harmonization that has taken place with other groups. The CC/PP specification contains a few vocabulary items. We should stress, however, that to the extent that they are not structural, they are there for illustration only. Other groups have been encouraged to create their own schemas for vocabularies, and have indeed done so, as the FIPA and UAPROF examples demonstrate.

In principle, there is no limit on the number of vocabularies that can be created or used. In fact, RDF and XML namespaces enable independent creation of interoperable vocabularies. Clearly, there can be no mandate that a specific set of vocabularies should be used, nor can there be a mandate that a specific set of attributes within a vocabulary should be used within the context of an application. Schema interoperability, enabled by RDF and the use of XML namespaces, enables the selection of any number of attributes within the context of a user agent profile, because it becomes possible to select any element that is present in a schema using the XML Namespaces mechanism.

A vocabulary is analogous to a dictionary. It identifies all the possible attributes in a schema, which is similar to a database schema. A profile is an instance of the vocabulary. Different devices and user agents may refer to the same schema and support the same vocabulary but communicate different profiles to the origin servers. Any vocabularies can be used in CC/PP, provided there is an RDF Schema for them, because this requirement is issued by the RDF processor.

Writing vocabularies that conform to CC/PP is relatively easy. They have to be created as an RDF Schema, made available at a URI, and be usable in a component in the CC/PP framework. Vocabularies can be included in CC/PP using XML namespaces. If a vocabulary is not written as an RDF Schema, and does not conform to the CC/PP component structure, it will not be possible to use it in CC/PP.

Adapting Content Using CC/PP

Using technologies developed by the W3C, there are two basic ways of creating adapted content: Client-based and server-based. These are similar to the HTTP agent-driven and server-driven forms of content negotiation but differ in that the client-driven approach is assumed to take place entirely in the client. Note that the server-driven approach as described here can also take place in the client, provided it has an XSLT processor. For optimization reasons, however (for example, in narrow-band channels such as the wireless environment), it is more efficient to send only the adapted information. There may also be security reasons (in other words, not giving out too much content). The simplest way is to create a simple heuristic to tie the HTTP content negotiation to parameters from the profile. It is also possible to generate a style sheet that is adapted to the device dynamically. The working group has not developed these heuristics, since they will depend on the manufacturers, the server owners, and the content provider's preferences. The rule sets will have to be defined separately, preferrably in an RDF rules language. The CC/PP working group decided not to tackle feature-independent profile matching. It is my hope and expectation that the ongoing DAML+OIL work will provide mechanisms that enable us to incorporate equivalent functionality into CC/PP, provided that vocabulary development is performed in a disciplined manner (as noted previously).

CC/PP covers an important area not addressed by CONNEG; namely, defaults, proxy behavior descriptions, and distributed profile descriptions. Reviewing the CC/PP design with some benefit of hindsight reinforces the view I previously held that such features should be incorporated at a layer that encapsulates any CONNEG-style feature expression. Specifically, I think that CONNEG-style feature expressions should be contained within a CC/PP-style component, and that feature matching will take place only at the component level. A simple higher-level rule, such as requiring a match for every specified component, could define whether or not two complete profiles match or not. (Any asymmetry between client- and server-side features could be accommodated at this level.) This topic is for future development, if required. As the user interaction changes, so will the need for presentation. As the author, you will no longer be able to write content that assumes a single presentation model, because it might be presented in a totally different way from what you expected. In the abstract, the interaction of the user with the device consists of two functions: presentation of data and input of data. Presentation, in turn, breaks down into navigation and style.

The input models of devices can be abstracted to a textual input format, regardless of whether input is received from a keyboard, a keypad, voice recognition, character recognition (for example, pen-based input), or other means. As yet, there are few non-textual, non-verbal input devices (such as cameras) using image recognition. A special problem, which can be abstracted at a higher level, is presented by devices that use shortcut keys, macro keys, soft keys, or other devices to represent a textual input as a shortcut. Menus can be seen as a special case of such devices.

The styling of presentation of textual information today varies from no style at all (for example, on pagers) to highly styled, graphical presentations (for instance, the Web as it is usually presented). Abstracting the style to a higher level, instead of including it in the content, enables the insertion of different styles, using mechanisms such as CSS. Using CSS, style information can be handled quite independently of the device.

When receiving information using a mobile or handheld device, however, traditional presentation formats tend to become irrelevant. Table layouts in fixed pixel widths adapted to a 21-inch screen become unusable on a device that has a screen that is one-inch square. Instead of assisting it, the formatting tends to obscure the presentation of the information. Disabling style sheets is a simplistic way of achieving a presentation that is usable on a small-screen device (anecdotal evidence seems to point to screens having to be quarter-VGA before styling becomes relevant). Applying different styling methods, such as aural styles, is another way of representing the content.

Style is not the most difficult part, however. The navigation is the other aspect of the presentation and the hardest to abstract from the device. Currently, navigation through Web content is subsumed in the "desktop" paradigm of the personal computer. There is no way of structuring content in ways other than an infinite scroll (possibly as shorter sections of the infinite scroll), because the page model is seen as a property of the presentation in CSS.

In devices with very small screens, however, as well as in voice-menu systems, alternative paradigms are emerging for navigation of content. In small-screen devices, the "deck of cards" paradigm of WML is emerging; in voice menus, an entire body of work exists to structure menus for natural interaction.

Modeling all possible future presentation formats is not feasible, however. There are two possible ways to meet this goal: either restrict presentation to a single set of devices or model the navigation of your content in a device-independent way.

Modeling the navigation in a device-independent way also lets you retain control over how it will be navigated on different devices. Navigation is actually a larger part of the user experience than the presentation, although these tend to be mixed up in popular discussion.

In the W3C CSS paradigm, the user has the ability to override the style presented by the service. This feature is intended to enable users who cannot receive the style selected; for example, users using assistive technologies to receive the content to take part in the presentation. In other words, the designer can never be assured that content is presented in the way it was designed on a device. On the other hand, the Web has never been pixel-perfect, and presentation has always varied somewhat between devices and implementations.

In traditional HTML presentations, content is not structured for other navigation formats at all. In this paper, we will discuss possible solutions to enable the navigation of different presentation paradigms in the same document as well as the ability to insert different styles.

Applying these technologies also enables the author to retain control over the navigation of the content (given that the presentation takes place on a device implementing CC/PP), as well as retaining some control over the styling. We contend that the most important part, however, is not the style but the navigation. Users actually tend to ignore style information, frequently switching off images to enhance transmission speed and ignoring advertisements.

Different styles require different formatting of the content presentation, but different navigation paradigms require different formatting of the content itself. This task can be accomplished in several different ways. One is to enable a customization engine to interact with the content at the client; another is to format the content according to the navigation paradigm in the origin server. Both ways imply that there will be a way for the content adaption process to conduct the required filtering and formatting.

In the case where the adaptation of content takes place at the client, the entire content set has to be transmitted to the client and the parts of it that are not required have to be discarded (alternatively, a set of programs could be transmitted which generates the content from data available at the client; this statement is speculative, however). The adaption can be based on the CC/PP presented by the client. This method is indeed one of the use cases for CC/PP.

In the case where adaptation is done by the server, the content can be adapted using a transformation process; for instance, XSLT—applying different transformation sheets to the content depending on which CC/PP is received.

For the author, controlling the transformation then comes down to two aspects: controlling the selection of the transformation sheets, and the transformation of the content itself.

Control over which transformation sheets are applied to the document depending on which CC/PP is received can be done using the document profile, which in essence is a reverse of the client profile, using the CC/PP framework. The author can determine which transformation sheet should be applied when a certain profile, or certain parameters of a profile, is used. This would entail creating a vocabulary for the selection process. Creating such a vocabulary could be part of the effort of the HTML working group to create a vocabulary describing XHTML modules, or a separate effort (for instance, spearheaded by a commercial vendor).

To change the navigational paradigm for a document, however, it is not enough to change the formatting of the content. In the case of a small-screen device, the information will need to be filtered, if the user will not have to wade through enormous amounts of irrelevant information (if you think that this is not an issue, consider that a normal WAP telephone can take just about half a line of the text presented on a 15-inch screen). It goes without saying that browsing becomes extremely tiresome in this environment. Given that most users do not read the text (but instead scan it, according to Jacob Nielsen), the salient features will have to be highlighted.

Determining which features of the content are salient is the heart of the matter. To accomplish this task, knowledge about the user's goal for the interaction is required. In certain cases, the author can contribute to the setting of these goals; for instance, in services where the interaction is tightly controlled (in forms). In other cases, where the goals for the interaction can be more ambiguous, this situation becomes a problem.

As the content is filtered, navigational aids need to be inserted. Traditionally, navigational tools on the Web include menus, navigation bars, and pointers (most frequently represented as icons). The Web also has a navigation paradigm built into the browser.

Enabling other navigation paradigms is possible if the content is formatted differently. For instance, the same content can be navigated with the card-and-deck paradigm of WML if it is reorganized. Content can be reformatted as WML if it is marked up as WML in advance. While it is possible to mark up content both as WML and HTML, for instance, it is not a pretty sight. And while the HTML browser will ignore the WML tags, the WML browser will not ignore the HTML. Then, there is the question of filtering out content that cannot be displayed on the WML device. Rather, the content should be filtered into the target format from the start. This situation implies that the base markup could either contain multiple formats, or that the XSLT transformation sheets should contain transformation rules to enable the conversion of the generic element set to HTML, WML, VoiceML, or whatever the markup format should be. The selection of the transformation sheet can be done in several different ways, all implementation-dependent.

In the CC/PP working group, we are satisfying ourselves with allowing for a selection to take place based on the received CC/PP, but we do not attempt to mandate the mechanism by which this event takes place. (It could be a Perl program that enabled the selection from a transformation sheet based on the received screen size and browser software, or it could be a rule base which acted on the profile and the document profile to compose a transformation sheet. It could be HTTP content negotiation. Determining the mechanism is out of the scope of our group.)

To enable the transformation sheet to filter and format the content appropriately for the navigation model of the device at hand, the content or the transformation sheet must contain a representation of the navigation model. This situation implies that a document that is device-independent must contain all possible navigational models (the style is divorced from the content because it is contained in a style sheet) or a subset of all possible navigational models for which the content is enabled.

Using XSLT to Generate an Appropriate Presentation

There are a few prerequisites for content that is to be transformed in the way described in the previous section. One is that a transformation mechanism exists. For content that is written in XML, this is possible to achieve using XSLT; for content in HTML, no simple way exists to transform it into other formats, even if it follows the HTML 4.0 specification. There are too many parameters that can be interpreted differently. Then, there is the high percentage of erroneously encoded HTML pages (up to 60%, according to some estimates). In summary, HTML content is not usable in transformations that enable the control over content in the manner that an author should expect. Separating the content from the navigation using transformation sheets as filters is a possibility that enables the same kind of device independence as for style; however, this is a chimera. There seems to be no complete device independency, because preliminary investigations demonstrate that you have to retain at least some concept of the elements in the generic document to be able to filter them properly (which also seems to be true for style—in narrating the content as the generic format, you have to have some awareness of the way the style will represent it to ensure that it is not misrepresented).

So far, we have identified four different models for creating content that is independent of the display device.

The first includes all possible markup in the document. Such a document will not display on some types of terminals without filtering, since the diverse kinds of markup will be rejected. It may be used as a generic document for filtering out the appropriate representations, however.

```
Example: (...) <body> <card> <p>Some text here</p> </card> (...)
```

The second situation assumes that the different types of markup are clearly separated. Because the names overlap for certain elements and attributes without their functions being the same, however, a variation of this using namespaces is possibly a better solution:

```
(...) <html:body> <wml:card> <html:p><wml:p>Some text
here</wml:p></html:p> </wml:card> (...)
```

A third way is to use a navigation modeling language like XDNL. This action essentially enables the creation of a leaf structure of nodes within the document. It does not in itself, however, enable the filtering out and reordering of sections or chunks of a document, which might be necessary to prepare the document for a different navigational model.

Another way of handling the same problem is to use Xpointer, as described in the W3C Note "Annotation of Web content for transcoding." The advantage of this method is that it enables the markup to point into the document; essentially, a WML markup could shadow the HTML markup. It is possible to create a markup (or at least navigational markup) that is separated from the document itself. The document need not change. It requires a separate mechanism to apply to the transformation process, however.

Which of these models will be selected will most likely depend on the preferences and predilections of the author, his or her technical environment, the end device to be primarily used, and so on. It is quite possible that other models are also used.

The important thing is not which model is used, but that the transformation system understands which model is used in the document so it either can apply the appropriate mechanisms or defer the transformation (if it does not have the appropriate mechanism implemented). This negotiation needs to take place between the transcoding entity and the origin server. In the CC/PP working group, we have tried to separate these logical functions and define ways to enable CC/PP to be used in this negotiation, too; one crucial part here is the use of document profiles, which enables the author to define his or her preferences for the transformation mechanisms to be used and declare the encoding used in the document.

To transform a document using XSLT, you need a transformation sheet that is specific to the transformation you want to achieve. Therefore, we will not give any examples in this document. It is relatively easy to use variables from an external document, however, such as the CC/PP profile, to select the transformations that will be done. Technically, this action might be easier to do by having three style sheets: one that selects the elements, one that comprises all of the transformations, and another that is the result of the application of the first style sheet to the second (and which will be the style sheet that is actually used in transforming the document).

There are currently seven implementations of CC/PP that have been reported to the working group:

NAME OF IMPLEMENTATION	ORIGINATOR	URI	REPORTED TO WORKING GROUP	OPEN SOURCE
Musashi	Ericsson Wasalab	www.w3.org/Mobile/CCPP/implday/#demo1	Nov. 15, 2000	No
WAP Application Server	Ericsson	www.w3.org/Mobile/CCPP/implday/#demo1	Nov. 15, 2000	No
Panda/Sasa	Kiniko Yasuda, Keio University	yax.tom.sfc.keio.ac.jp/panda/slidemaker/0011ccpp/Overview.html	Nov. 15, 2001	Yes
SBC/TRI Reference implementation	SBC/TRI	www.w3.org/Mobile/CCPP/implday/#demo1	Nov. 15, 2000	No
Information Architects	Chris Woodrow, Information Architects	www.w3.org/Mobile/CCPP/implday/#demo1	Nov. 15, 2000	No
W3C	Jigsaw Team	www.w3.org/Jigsaw/	Nov. 15, 2000	Yes

NAME OF IMPLEMENTATION	ORIGINATOR	URI	REPORTED TO WORKING GROUP	OPEN SOURCE
University of Wales	Stuart Lewis	www.ccpp.co.uk/, http://users.aber.ac.uk/sdl/ccpp/cs39030.html	July 26, 2001	Maybe
DELI	Mark Butler, Hewlett Packard Laboratories	www-uk.hpl.hp.com/people/marbut/	Nov. 2, 2001	Yes

Glosssary

3G Third generation. The generic name for a new breed of mobile network that will support a number of data services, for instance, streaming data, and not just voice communications. It is called the third generation, since it is the generation after the analog and the current digital systems, such as GSM.

3GPP The Third Generation Partnership Project. A body set up by ARIB, CWTS, ETSI, T1, TTA, and TTC, to coordinate the development of globally accepted standards for the third generation of mobile phone systems (after analog and the current digital systems, such as GSM).

3-sector-site A site consisting of three sector cells served by one mast with three directed base stations.

802.11 An IEEE standard for wireless local area networks.

absolute positioning GPS positioning using data from only one GPS receiver.

Accuracy The difference between the actual position of the target MS and the position estimate (in other words, as provided by the position-determining entity). Also applies to the quality of data and the number of errors contained in a dataset or map. Accuracy, or error, is distinguished from precision, which concerns the level of measurement or detail of data in a database.

Actual Position The actual coordinates of the target mobile system based on a geographic measure.

Admin System The Administration System enables the operator to configure and provision location-based services.

AEIR Administrative Equipment Identity Register.

Agent A kind of intermediary service that acts on behalf of another service (service provider or requester) according to rules established upon its invocation. Also known as an "intelligent agent."

A-GPS (Assisted Global Positioning System) GPS solution enhanced by assistance data provided from the network.

Air interface The way radio waves are radiated from the antenna of a transmission system (i.e., the frequency, the wavelength, the encoding, and so on).

A-KEY Authorization KEY.

ALI Automatic Location Identification.

Almanac Long-term model of the GPS satellite trajectories as well as the ephemeris data for the satellites, updated together with the ephemeris tables as new data comes in, maintained by the GPS receiver.

Altitude The geodetic position of the target mobile station in terms of distance above or below the WGS-84 ellipsoid surface. Most often expressed in meters above mean sea level.

AMPS Advanced Mobile Phone Service/System. The original standard for analogue cellular mobile telephone systems still used in North America.

AN/LI Auto Location Number / Location Interface.

Anchor MPC The mobile positioning center that is associated with the Anchor MSC for the Target MS.

ANI Automatic Number Identification.

ANSI American National Standards Institute. ANSI standards have been established for many elements of computer systems to aid research and development. The existence of standards enables designers to develop general solutions to common problems.

Ante Meridian (A.M.) Time occurring before noon.

Anywhere fix The ability of a GPS receiver to start position calculations without being given an approximate location and approximate time.

AOA Angle of Arrival.

API Application Programming Interface. A set of commands (interface definitions) that a programmer can use to access functions. Gives software developers a unified way of addressing functionality on dissimilar systems. An API is typically a library of functions or subroutines that give application programmers access to the functionality available in a resource such as an operating system, imaging system, graphics device, and so on.

APLMN Associated PLMN.

Applet A small application, with limited functionality, designed to operate in a componentware and/or middleware environment. Typically written in Java and executed within the scope of the browser.

Application The use of capabilities, including hardware, software, and data, provided by an information system that is specific to the satisfaction of a set of user requirements.

Application developer A software programmer who creates applications, usually by integrating a variety of pre-existing elements such as application programming interfaces and software and hardware platforms.

Application platform The collection of hardware and software components that provide the infrastructure services used by application programs. APIs make the specific characteristics of the platform transparent to the application.

Application portability The ability to move software among computers without rewriting it. This feature may be provided in three ways: as source code portability, pseudocode portability, or binary code portability.

Application Server The server, on which a given resource resides or is to be created; often referred to as Web server or an HTTP server.

Application software The computing elements supporting users' particular needs. Includes data, documentation, and training, as well as programs.

Architecture An abstract technical description of a system or collection of systems. Architectural frameworks identify key interfaces and services, and provide a context for identifying and resolving policy, management, and strategic technical issues. Architecture constrains implementation by focusing on interfaces, but does not dictate design or specific technical solutions. Conceptually based, architecture does not contain the level of detail needed for construction.

ARIB Association of Radio Industries and Businesses. Japanese telecommunications standardization body.

ARPU Average Revenue per User—key measure of a mobile operator's financial perfomance.

ASCII American Standard Code for Information Interchange. A character encoding format, now superceded by Unicode.

Assisted GPS A method which enables GPS positioning by using an almanac and ephemeris data, either having all the functionality in the handset or transmitting it via the network.

ATD Absolute Time Difference.

ATDM Asynchronous Time Division Multiplexing.

Attribute data Descriptive information about features or elements of a database. For a database feature like census tract, attributes might include many demographic facts including total population, average income, and age. In statistical parlance, an attribute is a "variable," whereas the database feature represents an "observation" of the variable.

AUC (Authentication Centre) Functional unit of the GSM network.

Authentication The process of verifying the identity and legitimacy of a person, object, or system.

AVL Automatic Vehicle Location.

B2B, B2C Business-to-business, business to consumer—two main models of e-commerce activity.

Bandwidth The range of frequencies in a signal. Often misused to mean transmission capacity.

Base information definition An information model that provides a common, logically consistent definition of the elements from the Open Geodata Model used within an OpenGIS application environment.

Base maps Maps that provide the background upon which thematic data is overlaid and analyzed. As inputs into a GIS, the term base map is usually applied to those sources of information about relatively permanent, sometimes timeless, features including topography, pedology (soil data), geology, cadastral divisions, and political divisions. Within a GIS database, such information might become part of a land base to which other information is indexed and referenced.

Base standard An approved international standard, technical report, CCITT recommendation, or national standard.

Base station Fixed transceiver in the mobile network, connecting the mobile network to the telephone (or other) network.

Billing Mediation Billing mediation allows the operator's billing system to receive data from the Location-Based Services solution to be able to bill for the location-based services.

Biometrics Identification of a person by a physical or behavioral characteristic (such as the way they sign their name, their fingerprint, or the marks on the iris of their eye).

BLOB (Binary Large Object) A term for large binary data sets, such as images, stored as a whole in a database.

Bluetooth A short-range, low-effect, spread-spectrum, frequency-hopping radio technology, originally intended as a cable replacement. Connects devices to each other for ad hoc information exchange.

Bounding rectangle A set of minimum and maximum coordinates for each of two horizontal dimensions. Bounding rectangles are typically used to specify a retrieval region or to index spatial information.

Broker A kind of intermediary service whose responsibility is only to bring other services together (typically a service requester and a service provider) and has no responsibility for satisfactory completion of the "contract" established between the requester and the provider.

BSC Base Station Controller. Determines the position (and several other things) of the specified mobile station. Functional unit in the GSM network.

BSSAP-LE BSS Application Part LCS Extension for Lb, Lp, and Ls Interface.

BSSLAP BSS LCS Assistance Protocol.

BTS Base Transceiver Station. The correct term for base station (because it is a two-way radio transmission station). Functional unit in the GSM network.

burst The information transmitted during one timeslot in a TDMA system.

C/A code The standard (Course/Acquisition) GPS code. A sequence of 1023 pseudo-random, binary, biphase modulations on the GPS carrier at a chip rate of 1.023 MHz. Also known as the "civilian code."

CA Certification Authority: a body that is capable of certifying the identity of one or more parties to an exchange or transaction.

cache Memory where frequently accessed data can be stored for rapid access, or the process of storing data in the cache.

CAD (Computer Aided Design) Computer systems used when drawing blueprints, sketches and sometimes maps.

Cadastral survey The means by which private and public land is defined, divided, traced, and recorded. The term derives from the French *cadastre*, a register of the survey of lands, and is, in effect, the public record of the extent, value, and ownership of land for purposes of taxation. Cartesian coordinates are a system of positional reference in which location is measured along two or three orthogonal (perpendicular) axes. Every location can be defined uniquely by its X, Y, and Z coordinates. Locations in the coordinate system can be established by using any unit of measurement, such as meters, feet, or miles.

CAI Common Air Interface.

CAMEL Customized Applications for Mobile Network Enhanced Logic. CAMEL specifies how features normally associated with IN (Intelligent Networks) can be integrated into a GSM network. The greatest benefit CAMEL provides is to enable information on the caller's location to be passed from the network to an Internet Web site.

Carrier frequency The frequency of the unmodulated fundamental output of a radio transmitter.

Carrier phase GPS GPS measurements based on the L1 or L2 carrier signal.

Carrier signal A signal that can be varied from a known reference by modulation.

Carrier-aided tracking A signal-processing strategy that uses the GPS carrier signal to achieve an exact lock on the pseudo-random code.

Cartesian coordinates Coordinates that differ from latitude-longitude coordinates in that the latter comprise a spherical (rather than planar) reference system.

Catalog A collection of entries, each of which describes and points to a feature collection. Catalogs include indexed listings of feature collections, their contents, their coverages, and other metadata. Registers the existence, location, and description of feature collections held by an Information Community. Catalogs provide the capability to add and delete entries. At a minimum, the Catalog will include the name for the feature collection and the locational handle that specifies where this data can be found. The means by which an Information Community advertises its holdings to members of the Information Community and to the rest of the world, each catalog is unique to its Information Community.

CBC Cell Broadcast Center.

CC/PP Composite Capabilities/Preferences Profile. Abbreviation for a standard for transmitting characteristics of a terminal and a user's situation to a server, which is responsible for content adaption.

CDG CDMA Development Group; the industry body representing the CDMA industry.

CDMA Code Division Multiple Access. A means of splitting radio channels depending on codes rather than splitting into time slots and frequencies (as GSM does). CDMA digital mobile systems have been championed in the United States and other countries such as Korea by Qualcomm (as cdmaOne) and Samsung.

CDMA 2000 A 3G version of cdmaOne, one of three technologies accepted by the ITU for IMT-2000. In the run up to 3G, there are a number of cdmaOne variants (including 1xEV).

CDR Call Detail Records.

CDR Customer Data Record.

Cell (GSM) The radio coverage area of a single base transceiver station in a cellular system.

Cell ID Each cell (the radio coverage area of a base station) in a mobile network has a unique identity. The cell ID might be that identity, or an alias, that is broadcast to all mobile stations using the cell.

Cellular A cellular system is built up with many low-power cells to simultaneously achieve a good coverage and a high capacity in a region. Frequencies can be reused in non-neighboring cells.

CEN Centre Européen pour la Normalisation (European Standards Centre).

Centroid The center of an area, region, or polygon. In the case of irregularly shaped polygons, the centroid is derived mathematically and is weighted to approximate a sort of "center of gravity." Centroids are important in GIS because these discrete X-Y locations are often used to index or reference the polygon within which they are located. Sometimes attribute information is "attached," "hung," or "hooked" to the centroid location.

CGI (cell) Cell Global Identity. Identification code for a cell in a cellular network.

CGI+TA Cell (Global Identity and Timing Advance) Network-based positioning solution. Enhancement of the CGI positioning solution, by determining the distance to the BTS.

Channel A channel of a GPS receiver consists of the circuitry necessary to receive the signal from a single GPS satellite.

Character encoding A specific method for assigning codes corresponding to character glyphs and representing those codes in a binary data stream.

Chip The transition time for individual bits in the pseudo-random sequence. Also, an integrated circuit.

cHTML Compact HTML, the markup language employed by NTT DoCoMo for sites on its i-mode service. A subset of HTML; will be replaced by XHTML.

CHV Card Holder Verification.

CI Cell Identity.

Circuit-switched A kind of network technology where each call is allocated a (radio) channel that remains open for the duration of the call.

CL Current Location.

CLI Call Line Identity/Identification.

Client The entity that desired to learn the location of the target; one end of the protocol. Components, within a client-server context, that require a service. Even though clients might also provide service to higher higher-order clients in the client-server model, the fundamental aspect of clients in all of the following discussion is that they are the requesters of some service.

Clock bias The difference between the clock's indicated time and true universal time.

Code phase GPS GPS measurements based on the pseudo random code (C/A or P) as opposed to the carrier of that code.

Common Object Request Broker Architecture (CORBA) The basic distributed object scheme developed by the *Object Management Group* (OMG). *Object Request Brokers* (ORBs) help clients find programs on servers.

Communications Service Interface (CSI) The interface by which an application platform accesses external entities that provide data transport services. The service provided is data transport among application platforms.

Computer environment The general term describing the people, hardware, software, and databases comprising a single computer system or several network-connected computer systems, and the associated standards.

Confidence The likelihood by which the position of a target mobile station is known to be within the shape description (expressed as a percentage).

Confirmed Opt-In The process of verifying a consumer's permission each time the service is provided. For example, in order to ensure that Push Messaging is not accidentally sent to the subscriber's wireless mobile device, an advertiser sends a message to the subscriber to which he or she must positively reply in order to confirm permission to start receiving Push Messaging.

Connectivity A topological property relating to how geographical features are attached to one another functionally, spatially, or logically. In a water distribution system, connectivity would refer to the way pipes, valves, and reservoirs are attached—implying that water could be "traced" from its source in the network, from connection to connection, to any given final point. Functional, spatial, and logical connectivity are examples of relationships that can be represented and analyzed in a GIS database.

Consumer Defined as an individual person who is identifiable by use of personally identifiable data.

Contact Card A smart card with a visible chip and gold contacts.

Contactless card A smart card with no visible contacts; uses the passive radio to transfer data (usually up to one meter).

Content type The type of media to be retrieved. This can be, for example, cartographic display, 3D display, text-to-speech, text display, HTML, audio, or video.

Control segment A worldwide network of GPS monitor and control stations that ensure the accuracy of satellite positions and their clocks.

Conversion The process of transferring data derived from existing records and maps to a digital database. Conversion is a major input problem and can consume the greatest share of time in a GIS project.

COO Cell Of Origin.

Coordinates A type of external reference defined by numerical values that define a location in a cartographic grid system or other spatial reference system.

CORBA See "Common Object Request Broker Architecture."

CORS Continuous Operating Reference Station.

CRM Customer relationship management.

CSS Cascading style sheet; the W3C standard for describing the formatting of a document. Used in WAP 2 and on the Web.

CTI Computer Telephony Integration.

Current Position After a position determination has successfully provided a position estimate and its associated time stamp, the position estimate is referred to as the "current position" at that point in time.

CVM card A verification method for identifying that the person presenting the card is genuine.

CWTS China Wireless Telecommunication Standard; a Chinese telecommunications standardization body.

Cycle slip A discontinuity in the measured carrier beat phase resulting from a temporary loss of lock in the carrier tracking loop of a GPS receiver.

D-AMPS Digital AMPS. An old term for TDMA (IS-136) networks.

Data Access Servers Data access servers are components, within a client-server context, that provide data access in response to a specific client request. Although data access servers can also be clients in the client-service provider mode, the fundamental aspect of data access servers in this discussion is that they are suppliers of access to data.

Data message A message included in the GPS signal that reports the satellite's location, clock corrections, and health. Included is rough information on the other satellites in the constellation.

DBMS Database management system. DBMS sometimes refers to the software that contains and organizes the data, and sometimes refers to an organizational plan for the use of information within a single project, or within one unit or the whole of an organization.

DCS1800 Digital Cellular System at 1800MHz, now GSM 1800.

DDI Direct Dialling In.

de facto standard A standard that has been informally adopted, often because a particular vendor was first to market with a product that became widely adopted. MS-DOS and Microsoft Windows are examples. Standards from bodies such as the W3C have no legal standing, but are accepted by its members. They are also de facto standards.

de jure standard An official standard created in a formal judicial process, such as the *International Organization for Standards Technical Committee 211* (ISO TC/211). De jure standards are produced by international standards bodies and become law in many countries when published.

Decryption The process of making encrypted data readable (reversing the process of encoding it).

DECT Digital European Cordless Telephony System, a standard for short-range telecommunications over wireless. Mostly used for indoor, local communications.

Deferred Position Request A position request where the position estimate response is not required immediately.

Deferred Query Service In the WAP Location Framework, it allows an application to send a query to the WAP Location Query Functionality and to get back the location of a WAP client with (possibly multiple) deferred responses. Deferred response means that there is no immediate response, for instance, sending a request and getting responses every two minutes. Tracking applications (for example, for fleet management) that want to track devices periodically are use cases for this type of query as well.

DEM Digital elevation model, a data exchange format developed by the United States Geological Survey for geographical and topographical data.

DES Data Encryption Standard (or Data Encryption Algorithm); the most widely used method for "symmetric" encryption (i.e., using the same key for encryption and decryption). The main source is ANSI X3.92.

Device A network entity (e.g., terminal, gateway, or server) that is capable of sending and/or receiving packets of information and that has a unique device address. A device can act either as a client or a server within a given context or across multiple contexts. For example, a device can serve a number of clients (as a server) while being a client to another server.

DGPS Differential GPS.

Differential GPS A method to increase GPS accuracy during selective availability by using a reference signal from a known position. It requires that at least two GPS receivers be used to compensate for signal and satellite errors, increasing the accuracy.

Differential positioning Accurate measurement of the relative positions of two receivers tracking the same GPS signals.

Digital orthoimages Orthorectified images produced by using photogrammetric techniques to orthorectify scans of aerial photos and paper maps.

Digital Signature An encrypted field, normally encrypted by using the sender's private key, which is attached to a message to prove its source and integrity (more secure than handwritten ones, or at least harder to forge).

Digitize The process of converting information into the digital codes stored and processed by computers. In geographic applications, digitizing usually means tracing map features into a computer by using a digitizing tablet, graphics tablet, mouse, or keyboard cursor.

Dilution of Precision The multiplicative factor that modifies ranging error in GPS. It is caused solely by the geometry between the user and his or her set of satellites and is known as DOP or GDOP.

Dithering The introduction of digital noise. This process is what the U.S. Department of Defense used to add inaccuracy to GPS signals to induce Selective Availability.

DLG Digital line graph, a form of digital map developed by the United States Geological Survey. DLGs supply users with the digital version of information printed on USGS topographical quadrangle maps.

DNS Domain name service. A system to bind names to Internet addresses, deployed on the Internet to make addressing human-readable.

DOM (Document Object Model) A language and platform independent interface making it possible to access elements inside an XML document from scripts and programs.

Doppler shift The apparent change in the frequency of a signal caused by the relative motion of the transmitter and receiver.

Doppler-aiding A signal processing strategy that uses a measured doppler shift to help the receiver smoothly track the GPS signal. Allows more precise velocity and position measurement.

Downlink The transmission from the base station to the mobile station.

DSL Digital Subscriber Line.

DTD Document Type Definition. An SGML definition of the elements, attributes, variables, and so on that can occur within the type of documents defined in the DTD. Conceptually, if not in practice, replaced by XML Schema.

DTM Digital terrain model, a method of transforming elevation data into a contoured surface of a three-dimensional display.

DTMF Dual Tone Multifrequency.

DXF Drawing interchange format, a file exchange format developed by Autodesk, Inc. for its AutoCAD drafting software. DXF files are ASCII records of all objects in a drawing file. DXF is used by GIS systems for exchanging map files.

Dynamic segmentation Points along a line that vary in value, for example, pavement thickness along a road center line.

E911 Enhanced 911.

EA Evaluation Authority.

Earth Model An approach to abstracting the Earth. The data model for the Earth.

Eastings The base meridians longitudes in sectors in the UTM coordinate system. Eastings are in meters with respect to a central meridian drawn through the center of each grid zone (and given an arbitrary easting of 500,000 meters).

E-cash Electronic money, often held on a smart card.

ECC Elliptic Curve Cryptography.

ECEF Earth Centered Earth Fixed. Three-dimensional coordinate system centered in the center of the earth.

E-CID Enhanced Cell ID.

ECMAscript Formerly Javascript, ECMAscript is a language used to create simple programs for use on Web pages.

E-commerce Electronic commerce, traditionally conducted over the Internet, (now extending to other devices such as mobile phones). See also m-commerce.

ECR Enhanced Call Routing.

EDGE Enhanced Data for Global Evolution (or Evolved Data for GSM Evolution). A high speed version of GPRS that works on both GSM and TDMA networks. Sometimes referred to as 2.5 5-G (generation) technology. In theory, its top speed should be to 384 kbit/s—the minimum requirement for a 3G network.

EFI Extended functionality interface, a WAP function to connect external devices (e.g., sensors) to a mobile terminal.

E-GPRS Enhanced GPRS; original name for EDGE.

EIR Equipment Identity Register. Functional unit of the GSM network.

EIRP Equipment Identity Register Procedure.

Ellipse A two-dimensional shape which is defined by its radii being of different length around its circumference (a circle has the same radius all around).

Ellipsoid A three-dimensional shape like a flattened sphere, with the ellipse as its cross section, instead of a circle (which forms the cross section of a sphere).

Ellipsoid point A point on a surface of an ellipsoid. In other words, it can be used to refer to a point on the surface of the Earth (or close to it).

Elliptic arc A shape defined as an area along a sector of an ellipse with a fixed length from the circumference of the ellipsoid, which does not comprise the entire radius.

Emergency Services Routing Digits (ESRD) Number in the national Numbering Plan that can be used to identify an emergency services provider and its associated Location Services client. The ESRD also identifies the base station, cell site, or sector from which an emergency call originates. ESRD and ESRK can be of two different types: NA for North American and EU for European, plus Other for other expected formats.

Emergency Services Routing Key (ESRK) Number in the national Numbering Plan that is assigned to an emergency services call for the duration of the call. The ESRK is used to identify (for example, route to) both the emergency services provider and the switch that is currently serving the emergency caller. During the lifetime of an emergency services call, the ESRK also identifies the calling subscriber. ESRD and ESRK can be of two different types: NA for North American and EU for European, plus Other for other expected formats.

Encapsulation In object-oriented programming, data can be encapsulated in an object, which means all access to the data and manipulation of the data occurs through the object's methods. Legacy software or data can be encapsulated by giving it an interface that is compatible with object software.

Encryption Manipulating data (using a mathematical transformation) to make it unreadable to anyone who does not possess the decryption key.

E-OTD Enhanced Observed Time Difference, a way of calculating the position of a mobile station related to several base stations.

Ephemeris The predictions of current GPS satellite positions that are transmitted to the user in the data message.

Epoch A reference frame for a related set of time measurements. All measurements with respect to a given epoch are comparable. The structure of an epoch is important only to a system making time comparisons or calculations.

ESS Enhanced Signal Strength.

ETA Estimated Time of Arrival.

ETACS Extended Total Access Communications System.

Ethernet A type of local-area network used for high-speed communication among computers. IEEE 802.11 is a standard for wireless Internet.

ETSI European Telecommunications Standards Institute; the European telecommunications standardization body.

EUREF 89 (European Reference Frame 1989) Spatial reference system, which is a refinement of ITRF 89.

European Terrestrial Reference System (ETRS89) A reference frame for the Earth which is centered on Europe. Consistent within centimeters with the WGS'84 reference system.

Exactness The precision of the reported value.

External location entity Entity in the network or in the terminal that can provide a location as a response to a given request information. "External" means not specified by the WAP Forum.

Fast switching channel A single channel that rapidly samples a number of GPS satellite ranges. "Fast" means that the switching time is sufficiently fast (2 to 5 milliseconds) to recover the data message.

FCC (Federal Communications Commission) A governmental agency regulating the communication technology of the USA.

FDMA (Frequency Division Multiple Access) A radio transmitting system where one carrier frequency is allocated to each mobile station.

Feature A digital representation of a real-world entity or an abstraction of the real world. It has a spatial domain, a temporal domain, or a spatial/temporal domain as one of its attributes. Examples of features include almost anything that can be placed in time and space, including desks, buildings, cities, trees, forest stands, ecosystems, delivery vehicles, snow removal routes, oil wells, oil pipelines, an oil spill, and so on. Features are usually managed in groups as feature collections.

Feature collection A set of related features managed as a group.

Federated databases Separate databases that are structured, perhaps with middleware or special database access software, in such a way that they can be queried as a single database.

FOMA Freedom of Mobile Multimedia Access; the trade name for NTT DoCoMo's 3G network combining i-mode and W-CDMA.

FPLMTS Future Public Land Mobile Telecommunications System. Early name for the ITU effort to standardize 3G/IMT-2000, and you can see why it changed.

Frequency band A particular range of frequencies.

Frequency spectrum The distribution of signal amplitudes as a function of frequency.

GAIT GSM ANSI-136 Interoperability Team. A project promoted by the UWCC, the GSMA, and TDMA and GSM operators, to standardizse a multitechnology handset providing interoperability between TDMA and GSM networks.

Gateway A point on a network that enables data traffic in one format to change format (the original name for router was gateway). Today, it is considered to operate at application level. You need a gateway to convert information held in a WML format by a WAP site into HTML formats (for standard Web pages) or POP3 format for standard e-mail messages.

GCS Global Coordinate System.

GDC Geodetic Datum Coordinates coordinate system.

GDOP Geometric Dilution of Precision.

GEI Geocentric Equatorial Inertial coordinate system.

Geocentric With respect to, or centered upon, the center of gravity of the Earth.

Geocentric Coordinates (GCC) Used to define three-dimensional positions. They define the three dimensions with respect to the center of mass of the reference ellipsoid (in other words, the center of the Earth). The Z axis goes from the center through the North pole; the X axis is defined by the intersection of a plane defined by the prime meridian and the equatorial plane; and the Y axis is a plane 90 degrees east of the X axis, with its intersection through the equator. The values in the coordinate system are all in meters, and the coordinate represents an (x, y, z) offset from the center of the planet, based upon the WGS84 ellipsoid.

Geodata Information that identifies the geographical location and characteristics of natural or man-made features and boundaries of the Earth. Geodata represent abstractions of real-world entities, such as roads, buildings, vehicles, lakes, forests, and countries.

Geodata model A formalized system for representing geodata. The OpenGIS Specification defines geographic data types in the Open Geodata Model.

Geodetic datum A geodesy term often used synonymous with the term spatial reference system, SRS, referring to an earth model fixed to the ground for mapping purposes.

Geodetic distance The length of the shortest curve between those two points along the surface of the earth model being used by the spatial reference system.

Geographic application Applications that pertain to the Earth and Earth phenomena, with known spatial and temporal reference systems. Expressed in a human context versus a computer context.

Geographic relationship A member of a closed set of possible relationships between geographic entities: disjoint, meet, overlap, coincide, contains, inside, part_of, has_part.

Geoid An ellipsoid defined not by a length measurement between the center and the circumference, but by a gravitational constant.

Geomancy Also called Feng Shui in Chinese. The discipline of modeling, interpreting, and using data about force fields and lines on and in the Earth.

Geomatics The discipline of modeling, interpreting, and using data about the Earth.

Geometric Dilution of Precision (GDOP) See Dilution of Precision.

Geoprocessing application Computer applications that model, interpret, and use Earth information. The implementation of a Geographic Application on a computer. The terms “geoprocessing,” “geomatics,” and “geotechnology” mean approximately the same thing, although some groups make minor distinctions among them.

Geospatial function (or process) A function or process that handles or operates on geodata.

GERAN (GSM/EDGE Radio Access Network). The "air interface" (radio system) for GPRS and EDGE.

GIS Geographic Information System. GISs are special-purpose digital databases in which a common spatial coordinate system is the primary means of reference. GISs contain subsystems for: 1) data input; 2) data storage, retrieval, and representation, 3) data management, transformation, and analysis; and 4) data reporting and product generation. It is useful to view GIS as a process rather than a thing. A GIS supports data collection, analysis, and decision-making and is far more than a software or hardware product. Other terms for GIS and special-purpose GISs include: Land-Base Information System, Land Record System, Land Information System, Land Management System, Multipurpose Cadastre, and AM/FM (automated mapping and facilities management) system.

Global Navigation Satellite System, GLONASS Russian satellite system, similar to GPS but using different codes.

Glyph The graphic representation of a character.

GML (Geography Markup Language) A markup language in the XML family developed for representation of geographic data. The language is primarily developed for data interchange and storage.

GMLC Gateway Mobile Location Center.

GMPC Gateway Mobile Positioning Center is the interface to the Location-Based Application that receives location requests and provides final location estimates. The GMPC can request routing information from the HLR. After performing registration authorization, it sends positioning requests to and receives final location estimates from the *Visiting MSC* (VMSC).

GMT Greenwich Mean Time.

GPRS General Packet Radio Services. A system devised to enable higher data throughput speeds over GSM networks. It breaks data down into small packets and is similar to X.25 and IP. The crucial point about GPRS is that it provides a constant "always-on" connection as opposed to using standard data connections over GSM (based around ISDN) that requires the user to dial into a particular service. GPRS connections can typically yield between 20 and 40 kbit/s at present.

GPS Global Positioning System. Developed for the U.S. military for navigation and surveying, the GPS relies on satellites (and ground stations) for precise determination of location. Although GPS can be used to determine location very precisely (within centimeters, given the correct controls and proper use), it does not solve all the problems of locational determination in GIS databases.

GPS time The GPS system time, operationally defined as UTC time offset by a small integer number of seconds due to the irregular addition of "leap seconds" to account for the variable (generally slowing) rotation rate of the planet.

Graphic element A point, polyline, polygon, display text, or texture intended for graphic display.

Great-circle A circle on the surface of the earth with the same radius as the earth.

GSE Geocentric Solar Ecliptic coordinate system.

GSM Geocentric Solar Magnetospheric coordinate system.

GSM Global System for Mobile communications. Originally developed in Europe as Groupe Speciale Mobile in ETSI, it was intended as a common standard for the digital mobile telephone network. It is now by far the most dominant of second-generation digital telephony standards.

GSM Association Industry body promoting GSM worldwide.

GSM1800 GSM operating at 1.8 GHz; formerly DCS1800.

GSM1900 GSM operating at 1.9 GHz; formerly PCS1900.

GsmSCF GSM Service Control Function.

GTD Geometric Time Difference.

GUI Graphical User interface.

Guidance route Refers to a route from the start point to the destination point.

Guided journey Travel of a vehicle equipped with a guidance capability based on the prior specification of a destination and (optionally) intermediate waypoints.

Handle An index entry or unique name in software that identifies a catalog entry or other resource so that it can be found and utilized by another software facility.

Handover The action when the radio connection to a mobile station is handed over from one base station to another.

Hardover word The word in the GPS message that contains synchronization information for the transfer of tracking from the C/A to P code.

HDML HandHeld Markup Language; a system devised by Unwired Planet (now Openwave, formerly Phone.com) for defining information to be fed to mobile communication devices. Originally based on the same ideas as the Hypercard product from Apple. Products developed originally for HDML have frequently been modified to work with WAP.

Hierarchical database A database that stores related information in terms of pre-defined categorical relationships in a "tree-like" fashion. Information is traced from a major group, to a subgroup, and to further subgroups. Much like tracing a family tree, data can be traced through parents along paths through the hierarchy. Users must keep track of the hierarchical structure in order to make use of the data. The relational database provides an alternative means of organizing datasets.

HLR Home Location Register; the HLR contains location-based service subscription data and routing information. The HLR is accessible from the GMPC. For roaming mobile stations, the HLR might be on a different mobile network than the current SMPC. Functional unit in the GSM network.

Home MPC The MPC that is associated with the target mobile station's subscribers' Home System and to which the MS's user identity (e.g., directory number, MSID) is assigned for location-based services.

Horizontal error With a point expressed in terms of latitude and longitude, an expected error in the horizontal direction is represented by a circle, and the length of its radius serves as a horizontal error of its latitude and longitude.

HPLMN Home Public Land Mobile Network.

HSCSD High-Speed Circuit-Switch Data. A technology for improving data speeds over GSM. It actually relies on combining two existing GSM channels together. At top speed (43.2 Kbit/s) with HSCSD, the user is making the equivalent of three standard voice calls. Often proposed as a suitable way to broadcast video over mobile systems.

HSM Host Security Module (or Hardware Security Module); a hardware device used for storing keys and performing cryptographic functions under control of a host computer.

HTML HyperText Markup Language. A means of controlling and displaying information in the form of pages. HTML was created by Tim Berners-Lee as the means of creating WWW (World Wide Web) pages on the Internet and over private intranets. It has now become the favored means of creating Web sites on the Internet.

HTTP HyperText Transfer Protocol. An Internet Engineering Task Force (IETF) protocol for Web-based file transfer, defined in RFC2616.

HTTPS HTTP Secure.

iCal A calendar interoperability standard established by the Internet Mail Consortium.

iDEN Integrated Dispatch Enhanced Network. An enhanced version of TDMA used in the United SAtates, Canada, Mexico, Israel, and parts of Asia and South America. A Proprietary to the manufacturer Motorola, it is a rival to Europe's TETRA. In the United States, the main iDEN carriers are Nextel and Southern Linc.

IETF Internet Engineering Task Force. The body that is responsible for the design and evolution of the Internet's architecture. Its standards, developed by Working Groups within the IETF, are published as RFCs (Request For Comment).

IL Initial Location.

ILR Immediate Location Request.

IMEI International Mobile Equipment Identity. Unique identifier for digital handsets, given at Type Approval.

Immediate Position Request A position request where a single position estimate response is required immediately.

Immediate Query Service. In the WAP Location Framework, it allows an application to query WAP Location Query Functionality for the location of a WAP client and get an immediate response.

i-Mode Information mode. i-Mode is similar to WAP but proprietary to NTT DoCoMo at present. It utilizes cHTML (compact HTML) as its markup language instead of WML (like WAP).

Implementation A software package that conforms to a standard or specification; a specific instance of a more generally defined system.

IMSI International Mobile Subscriber Identity. The unique (numeric) identifier of a subscriber, carried on the SIM card. Typically only exposed in transactions with the network.

IMT 2000 International Mobile Telecommunications 2000; the ITU's standard for 3G mobile networks. Intended to provide global framework for intelligent mobile data networks. UMTS (Universal Mobile Telecommunications System) was Europe's proposal for IMT 2000.

IN Intelligent Network.

Information Community A community of geodata producers and users who share a common set of feature definitions and other semantics that structure their data. Organizations in an Information Community need to agree on metadata standards.

Interface A shared boundary between two functional entities. A standard specifies the services in terms of the functional characteristics and behavior observed at the interface. The standard is a contract in the sense that it documents a mutual obligation between the service user and provider and assures stable definition of that obligation.

Intermediary A service that provides functions by which to interconnect, adapt, and facilitate services offered by other parties, components, or environments. Common forms of intermediaries include agent, broker, mediator, and trader services.

Interoperability The ability for a system or components of a system to provide information portability and interapplication, cooperative process control.

Introductory Point Refers to an end point through which to go to the target; an appropriate point is set by an information creator. It is recommended to set a point that serves as an eventual mark, such as a station, intersection, or stop.

Ionosphere The band of charged particles 80 to 120 miles above the Earth's surface.

Ionospheric refraction The change in the propagation speed of a signal as it passes through the ionosphere.

IP Internet Protocol. See TCP/IP.

IPv4 IP version 4. A comprehensive standard to address and encode data for transmission over networks.

IPv6 IP version 6. A comprehensive standard to address and encode data for transmission over networks. Ipv6 differs from IPv4 (there is no IPv5) in many respects, notably in the addressing.

IRTS International Terrestrial Reference Service, a body managed by the university of Paris that is responsible for making sure that the Earth and the clocks we use keep the same time (by inserting leap seconds, among other things).

IS-136 Digital mobile cellular communication standard. Used in North America, Latin America, Asia Pacific and Eastern Europe.

ISDN Integrated Services Digital Network. The standard developed by the ITU (International Telecommunications Union) for digital telephone networks on which GSM is heavily based. When you make a data connection over GSM, it is in fact an ISDN call. Because each "service" is identified separately over GSM, most handsets default to only a single service: voice. Thus, when making a WAP call, you might find that the "data" service has not been enabled (so you need to make a call to customer services to "enable" it). The same principle applies to roaming.

ISO International Organization for Standardization (yes, the abbreviation is correct). The main ISO standard relating to geoprocessing is ISO/TC 211. The ISO is a de jure standards body, consisting of national standardization agencies.

ISO 8601 A standard that describes a large number of date/time formats.

ISO/TC 211 The main ISO standard working group (TC means Technical Committee) related to developing geograpical data processing standards.

ITRF International Terrestrial Reference Frame; the body that is responsible for making sure that the Earth and the clocks we use keep the same time, by inserting leap seconds, among other things.

ITRF 89 (International Terrestrial Reference Frame 1989) Global spatial reference system built on the reference ellipsoid GRS 80.

ITU International Telecommunications Union. The international body responsible for telecommunications co-ordination since 1861. A United Nations body since 1963, it includes ITU-T, the successor body to CCITT (Committee Consultatif Internationelle des Telegraphes et Telephones) that does technical work. ITU is also concerned with regulatory and political aspects of telecommunications.

Java A high-level object-oriented language that enables applets (applications) to be written once, and run anywhere (whatever the platform is). The aim is to help simplify application development. Developed by Sun Microsystems, and not a standard (but free implementations exist).

Java 2 Micro Edition (J2ME) "The mobile edition" of the program environment for wireless devices.

Java Script Scripting language, developed by Netscape, used on the Web.

Javacard The JavaCard Forum has developed specifications for running a subset of the Java language on a smart card.

kbps kilobits per second Measurement unit for the speed (bit rate) of transference of data.

Keys In a modern encryption system, the algorithm is generally assumed to be known, and what is kept secret is the key. There are many different forms of keys, each of which can be regarded as a string of meaningless bits until it is used to encode or decode a message.

LAC Location Area Code.

LAN Local-area network; a system for connecting computers so that they can communicate with one another.

Landsat A system of satellites that scan the Earth at a variety of wavelengths. The satellites return information that can be used to inventory and analyze a variety of natural and human resources.

Language-independent Describes a standard or specification that is not specified in terms of a specific programming language, but is implementable in a variety of languages.

Last-Known Position The current position estimate (and its associated time stamp) that is stored for the target mobile station in the MPC is referred to as the "last-known position" until replaced by a later position estimate and its time stamp.

Latency The time delay between making a request and obtaining a response.

Lat-Lon A position coordinate system using latitude and longitude coordinates with a reference ellipsoid such as the WGS-84 ellipsoid.

Layer A collection of geographic objects of a similar type. In raster structures, a set of information combined with other data sets covering the same geographic area.

L-band The group of radio frequencies extending from 390 MHz to 1550 MHz. The GPS carrier frequencies (1227.6 MHz and 1575.42 MHz) are in the L band.

LBB Location-Based Billing.

LBI Location-Based Information.

LBS Location-Based Services.

LCC Lambert Conformal Conic PCS coordinate system.

LCC Location Country Code.

LCCF Location Client Control Function.

LCCTF Location Client Coordinate Transformation Function.

LCF Location Client Function.

LCS Client An entity (e.g., service control function) that interacts with an MPC for the purpose of obtaining position information for one or more target mobile systems within a set of specified parameters (such as PqoS). LCS clients subscribe to LCS in order to obtain position information for the purpose of providing location-based applications. The LCS client is responsible for formatting and presenting data and managing the user interface.

LCS Location Service Service that makes use of the subscriber's location. Synonymous with Location-dependent Service and Location-based Service.

LDR Location Deferred Request.

LDT Location Determination Technology.

LE Location Estimate.

LEC Local exchange carrier.

Legacy system Software or database components inherited from a previous computing model which that do not fit into an open system environment without some modification.

LFS Location Fixing Schemes.

Line string A set of coordinate points and the lines that join them.

LIR Location Immediate Request.

LIR Location Information Restriction.

LKL Last-Known Location.

LMMF LMU Mobility Management Function.

LMU Location Measurement Unit, measurement unit utilized for certain positioning solutions.

Local time UTC time shifted according to the local time zone.

Localised Name The name of a geographic entity, expressed in human natural language. A localized name has an explicitly specified natural language, a specific writing system (script), and a specific character encoding used to structure the character codes in a binary data stream.

Location Reference to a position.

Location Attachment Service In the WAP Location Framework, a method that includes the location of the user in the request so that the response can be returned within the scope of the same transaction.

Location Information Information related to a position, including both various location formats (different coordinate systems and datum), and other types of location information (such as geo-codes, velocity, altitude, etc.).

Location Query Service Position request from a Location Enabling Server to a Location Server in the LIF specifications.

Location Services Client Subscriber Profile A collection of subscription attributes of parameters related to location services that have been agreed upon for a contractual period of time between the client and the provider of location services.

LOP Line of Position.

LSAF Location Subscriber Authorization Function.

LSB Location Sensitive Billing.

LSBcf Location System Broadcast Function.

LSBF Location System Billing Function.

LSCF Location System Control Function.

LSOF Location System Operations Function.

LSR Local Space Rectangular coordinate system.

LSR Location Service Request / Response.

LTU Location Determination Unit.

MAGIC API Mobile/Automotive Geo-Information Services Core Application Programming Interface.

Maidenhead Grid Squares A coordinate system designed to help radio amateurs designate geographical position. The grid identifies "Fields" consisting of area 20 degrees of longitude by 10 degrees of latitude with two alphabetic characters. An additional set of two numeric digits locates a specific two-degrees of longitude by one-degree of latitude" grid square" area within the field. Two additional alphabetic characters can be used to refer to a 5.0 minutes of longitude by 2.5 minutes of latitude "Sub-Square" within the grid. In each case, the longitude character precedes the latitude designator.

Manageability The ability to configure, reconfigure, manage, and control a system or the components of a system.

MAP Mobile Assisted Positioning.

Map distance The distance between the points as defined by their position in a coordinate projection (such as on a map when scale is taken into account).

Map matching The synchronization of information from navigation sensors with a database model of the driveable roadways, using the constraints of that model to improve the knowledge of the vehicle position with respect to the stored geometric representation. Map matching is often done for purposes of route guidance, where it is essential that the motion of the vehicle be related to the modeled road network.

MBP Mobile Based Positioning.

MCC Mobile Country Code. A number that is part of the global numbering plan, and used to identify the country where the call from the mobile station originates.

M-commerce Mobile commerce. A name for a number of ways and representations of commercial transactions using mobile technology.

Mediator A service which that acts as an intermediary capable of impartially negotiating between a service requester and a service provider regarding aspects of a service to be provided. The mediation function often follows broker, trader, or agent functions.

Metadata Graphical or textual information about the content, quality, condition, origins, and characteristics of data ("data about data").

Middleware Software (such as remote procedure calls) that enables applications to access data and computing resources distributed across networked computers, regardless of incompatible operating systems and networks.

MIN Mobile Identification Number.

MLC Mobile Location Center.

MLP Mobile Location Protocol. The API and transmission method for the data and processing instructions defined in the Location Interoperability Forum specifications.

MLS Mobile Location Services.

MMR Mobile Measurement Report.

MMS Media Messaging Service.

MNC Mobile Network Code. A number that is part of the global numbering plan and used to identify the network where a call from a mobile station originates.

MO Mobile Originated.

Mobile Positioning Center (MPC) The MPC serves as the point of interface to the wireless network for the position determination network. The MPC serves as the entity which that retrieves, forwards, stores, and controls position information within the position network. It can select the PDE(s) to use in position determination and forwards the position estimate to the requesting entity (e.g., LCS client) or stores it for subsequent retrieval. The MPC can restrict access to position information for a Target MS (e.g., release position information to authorized entities).

Mobile Station A mobile telephone system terminal (i.e., a mobile terminal and a valid SIM card). Main component of the GSM network.

MO-LR Mobile Originating Location Request.

MPS Mobile Positioning System.

MPTP Mobile Phone Telematics Protocol.

MPTP terminal GSM phone with Mobile Phone Telematics Protocol.

MSC Mobile Services Switching Centre. Functional unit in the GSM network.

MSC/VLR Mobile Switching Center/Visitor Location Register. The MSC contains functionality responsible for mobile station subscription authorization and managing circuit switched data as well as call-related and non-call related positioning requests of the location based service.

MSID Mobile Station Identifier.

MSIN Mobile Station Identification Number.

MSP Multiple Subscriber Profile.

MT Mobile Termination.

MT-LR Mobile Terminating Location Request.

Multi-channel receiver A GPS receiver that can simultaneously track more than one satellite signal.

Multipath error Errors caused by the interference of a signal that has reached the receiver antenna by two or more different paths; usually caused by one path being bounced or reflected.

Multiplexing channel A channel of a GPS receiver that can be sequenced through a number of satellite signals.

Natural language A written, spoken or signed language used by humans. Examples are Japanese, Dutch, Hindi, and English.

Natural language (NL) reference A human-understandable specification of a real-world entity, usually one with a localized name and a geographic entity class or other unique identification.

NAVSTAR The 24 GPS satellites.

Network Destination Code (NDC) Number in the global numbering plan that specifies where the call from the mobile station is directed.

Network latency The time it takes for a data packet to move across a network connection.

NI-LR Network Induced Location Request.

NMR Network Measurement Results.

NMT Nordiskt MobilTelefon System. Analog predecessor of GSM in the Nordic countries (Sweden, Norway, Finland, and Denmark), as well as several other European countries. Uses either 450 MHz or 900 MHz band.

Non-Personally Identifiable Data Defined as information not uniquely and reliably linked to a particular person, including but not limited to activity on a wireless network such as anonymous location or aggregate location statistics. Information identifying the geographic origin of one or more signals is considered Non-Personally Identifiable Data provided that information is not linked to or associated with any Personally Identifiable Data.

Northings The base latitudes in the UTM (Universal Transverse Mercator) system. In the northern hemisphere, northings are read in meters from the equator (0 meters). In the southern hemisphere, the equator is given the false northing of 10 million meters.

NSS Network Support Subsystem.

OBEX Object Exchange Protocol.

Obfuscate To intentionally make the measurement less accurate by adding randomness.

Object-oriented (OO) Software in which data and processing functions are packaged into small, discrete, interoperable modules offering advantages such as portability and easy maintainability.

Object technology Software scheme in which data and processing functions are packaged into small, discrete, interoperable modules offering advantages such as portability and easy maintainability.

OGC See "Open GIS Consortium."

OGC Abstract Specification A document (or set of documents) containing an OGC consensus computing technology independent specification for application programming interfaces and related technology based on object-oriented concepts that describes an application environment for interoperable geoprocessing and geospatial data products.

OGC Implementation Specification A document containing an OGC consensus computing technology dependent specification for application programming interfaces and related technology based on the Abstract Specification or domain-specific extensions to the Abstract Specification provided by domain experts (usually as a result of activity in the Domain Task Force).

OGM See "Open Geodata Model."

Omnicell A cell with the base station located in the center.

Open Geodata Model (OGM) That part of OpenGIS Abstract Specification which defines a general and common set of basic geographic information types that can be used to model the geodata needs of more specific application domains, using object-based and/or conventional programming methods.

Open GIS Consortium (OGC) Not-for-profit trade association founded in 1994 to provide the geoprocessing industry with a consensus process for developing interoperability standards and industry relationships that will hasten the commercial success of distributed geoprocessing. OGC envisions the full integration of geospatial data and geoprocessing resources into mainstream computing and the widespread use of interoperable, commercial geoprocessing software throughout the global information infrastructure.

Open system A system that implements open interface specifications and standards that promote application portability, scalability, interoperability, diversity, manageability, extensibility, compatibility with legacy components, and user portability.

OpenGIS Implementation Specification Detailed software specifications for implementing parts of the OpenGIS Abstract Specification on particular distributed computing platforms, such as OLE/COM and CORBA. The OGC Technical Committee issues *Requests for Proposals* (RFPs), and in response to these, members team up to submit OpenGIS Implementation Specification for review by the Technical Committee and the OGC Management Committee. In addition to enabling interoperability with each DCP, these groups endeavor to provide maximum interoperability between DCPs.

OpenGIS Service A software component providing object management, access, manipulation, interchange, or human-technology services for Open Geodata Model features.

OpenGIS Specification A software interface standard that enables interoperable geoprocessing which includes: real-time data sharing between GIS systems from different vendors; interoperation between dissimilar types of geoprocessing systems (GIS, Earth imaging, desktop mapping, navigation, etc.); and efficient discovery and access to remote geodata and geoprocessing resources in distributed computing environments.

Opt-Out Means by which the consumer subscriber takes action to withdraw permission whether or not he or she has previously opted in.

Orientation Heading or bearing on or near the surface of the Earth, with respect to true north.

Orthorectification Use of photogrammetric techniques to adjust and correct distortions in images.

OSI Open System Interconnection. A reference model developed by the ISO to describe network systems. Implemented in DECnet, but not used other than as a descriptive model today.

OSS Server The Operator Support System allows enables the operator to operate and maintain the Location Based Services Solution.

OTA Over-the-air (actually used as an abbreviation for Over-the-Air Provisioning). The method used to remotely manage applications on a subscriber handset.

OTD Observed Time Difference. The time difference between the reception of signals from two base stations, observed by the mobile station.

Packet-switched A kind of network technology where the data is sent in small packets (not necessarily electronic).

PAP-USER A WAP Push user defined identity; for example, john.doe@wapforum.org/TYPE=USER@ppg.carrier.com.

Parlay Standard API for intelligent network (IN) functions.

PCF Positioning Calculation Function.

PCM Pulse Code Modulation.

PCN Personal Communication Network; GSM with 1800Mhz frequency.

P-code The Precise code. A very long sequence of pseudo-random binary biphase modulations on the GPS carrier at a chip rate of 10.23 MHz, which repeats about every 267 days. Each one-week segment of this code is unique to one GPS satellite and is reset each week.

PCS Personal Communications Service; a term for 2G networks commonly applied to the 1900 MHz networks in the United States.

PDA Personal Digital Assistant. Term invented by Apple (for its Newton handheld computer) and now used as a generic label for handheld computers. Particularly associated with pocket computers with touch screens that accept stylus input, such as the Palm Pilot and Handspring Visor.

PDC Personal Digital Cellular. A digital mobile standard adopted in Japan that is based on TDMA technology.

PDE Position Determination Element.

PDE Position Determining Entity.

PDE Positioning Determining Equipment.

PDOP Position Dilution of Precision.

PDT Positioning Determination Technologies.

Personalization Using data about an individual to create a presentation that is adapted to that individual.

Personally-Identifiable Data Defined as information that can be used to identify a person uniquely and reliably, including but not limited to name, address, telephone number, e-mail address and account or other personal identification number, as well as any accompanying data linked to the identity of that person. WLIA regards the subscriber's wireless telephone number as a clear link to that person's identity. Information identifying the geographic origin of a wireless signal is considered Personally Identifiable Data when and if it is linked to or associated with any other Personally Identifiable Data as defined herein.

Photogrammetry Use of aerial photographs to produce planimetric and topographic maps of the earth's surface and of features of the built environment. Effective photogrammetry makes use of ground control, by which aerial photographs are carefully compared and registered to the locations and characteristics of features identified in ground-level surveys.

PHS Personal Handyphone System. A microcellular system connecting the handset to the ISDN network (similar to DECT, but deployed only in Japan, China, and Thailand).

PIN Personal Identification Number.

PIUA Positioning Information User Application.

PKI Public Key Infrastructure.

Platform Another term for computer hardware, including microcomputers, workstations, and mainframe computers, or for underlying software, such as an operating system that provides services to layered software. When discussing software, platform independence implies that the software can be run on any computer.

PLMN Public Land Mobile Network. A network served by one operator. Used to define a phone number in the WAP Location Framework.

PLS Personal Location Systems.

PLT Personal Location Terminal.

POI Point Of Interest. A model element corresponding to some geographic feature of potential interest to a mobile human.

POI Privacy Override Indicator.

Polygon A shape defined by lines connecting points.

POP3 Post Office Protocol 3. The standard defined means of providing an electronic mail mailbox over the Internet.

Position Determining Entity (PDE) The PDE determines the estimated geographic position of the Target MS. The input to the PDE for requesting the position is a set of parameters such as PQoS requirements. Multiple PDEs may serve the coverage area of an MPC and multiple PDEs maymight serve the same coverage area of an MPC utilizing different position determining technologies.

Position Estimate The geographic position of a Target MS as determined by a PDE. The reference system for the coding of the Target MS position is the World Geodetic System 1984, (WGS-84).

Position Quality of Service (PQoS) A set of attributes associated with a request for the geographic position of a Target MS. The attributes include the required horizontal accuracy, vertical accuracy, response time, priority, and maximum age of the Target MS position.

Position Coordinates in a reference system.

Position Location-based service Service that uses information about the location of clients.

Possessor The possessor of a particular piece of location information is the person/entity/organization that generated the information. Examples are if the device contains a GPS receiver, then the device or the owner of the device is the possessor of the GPS based position information; and if the location of a device is determined using network resources, e.g. an GMLC, then the carrier is the possessor of the location information; or if the location of a device is determined using assisted GPS with assistance provided by the carrier, then the device or the owner of the device is the possessor of the raw GPS data, whereas the carrier might be the possessor of the improved data; and if the location is determined using an external entity outside the carrier network, then the external entity is the possessor of the location information.

Post Meridian (P.M.) Time which occurs after noon.

POTS Plain Old Telephony Service.

PPS Precision Positioning Service.

PQoS Position Quality of Service.

PRAF Positioning Radio Assistance Function.

Precise Positioning Service (PPS) The most accurate dynamic positioning possible with standard GPS, based on the dual frequency P-code and no SA.

Precision Relative error; number of digits to which the measurement is accurate. In terms of data quality, refers to the level of measurement and exactness of description in a GIS database. Precise locational data can measure position to a fraction of a unit. Precise attribute information may specify the characteristics of features in great detail. It is important to realize, however, that precise data—no matter how carefully measured—might be inaccurate. Surveyors might make mistakes or data may be entered into the database incorrectly. Therefore, a distinction is made between precision and accuracy.

Priority The MPC can process requests for the position of a Target MS with different levels of priority. A request with a higher priority might be given faster access to position determining resources than a lower-priority request.

Privacy flag SS7 function to set a flag in the Home Location Register, which determines if the user is willing to give out his position or not.

Profile A collection of standards, with parameters, options, classes, or subsets, necessary for building a complete computer system, application, or function.

Protocol A set of semantic and syntactic rules that determine the behavior of entities that interact.

Proxy A server that works both as a client and a server.

PS Polar Stereographic PCS coordinate system.

PSAP Public Safety Answering Points.

Pseudo-random code A signal with random noise-like properties used by the GPS satellites. It is a very complicated but repeating pattern of ones and zeroes.

Pseudolite A ground-based differential GPS receiver that transmits a signal like that of an actual GPS satellite and can be used for ranging.

Pseudorange A distance measurement based on the correlation of a satellite transmitted code and the local receiver's reference code that has not been corrected for errors in synchronization between the transmitter's clock and the receiver's clock.

PSMF Positioning Signal Measurement Function.

PSTN Public Switched Telephone Network.

Public key A public key encryption algorithm is one in which one key is published and the other is kept secret.

QoP Quality of Position, which can mean any of the following: Age of the location information; in other words, when the information was actually collected. Accuracy of the location information, i.e., how accurately the information can be technically measured. Confidence in the accuracy information; for example, "with 65 percent probability."

QoS Quality of Service.

Raster Originally, a scan line in an electronic display (such as a television or computer monitor). In geoprocessing, raster refers to digital geographic databases built from "grid cells" in a matrix. A raster display builds an image from pixels, small square picture elements of coarse or fine resolution. A raster database maintains a "picture" of reality in which each cell records some sort of information averaged over the cell's area. The size of the grid cell can range from centimeters to kilometers. Many satellites, like Landsat and SPOT, transmit raster images of the Earth's surface. Reflectance of sunlight at a certain wavelength is measured for each cell in an image.

RDF Resource description framework. A standard from the W3C for metadata describing (digital) objects.

Real-time Refers generally to systems that respond immediately or synchronously to external events.

Reference Model Provides the complete scientific and engineering contextual framework for a technology area. The underlying elements, rules, and behaviors.

Relational Database Stores data in such a way that it can be added to, and used independently of, all other data stored in the database. Users can query a relational database without knowing how the information has been organized. Although relational databases have the advantages of ease-of-use and analytical flexibility, their weakness can be slower retrieval speed. SQL (structured query language) is an interface to a relational database.

Relative positioning GPS positioning using data from at least two GPS receivers. A method used in order to increase the accuracy by compensating for signal and satellite errors.

Remote Procedure Call (RPC) An API for remote execution of detailed functions.

Representative location Representative location of a target; an information creator sets an appropriate point in the boundary of the target. It is recommended to set a point in the center of the institutional or administrative boundary. For a mobile entity, the point where it actually is set.

Request for Comment (RFC) The name for the standards set by the IETF.

Request for Information (RFI) A general request to the industry to submit information to a Working Group in an organization in anticipation of an RFP.

Request for Proposals (RFP) An explicit request to the industry to submit proposals to a Working Groups in a standards organization for a technology.

Resource Record (RR) Data field in the DNS database, describing characteristics of the named IP address, e.g. location (as described in RFC 1876).

Rhumb line Curve with a constant geodetic bearing.

Roaming Technical name for the capability of a single handset to work in conjunction with more than one mobile network. In practice roaming means that the handset will work when its owner travels abroad (since the operator can "lock" the handset into the home network in the country where the user has the subscription).

Route A path to a destination starting either at the current position of a vehicle or person or at a designated starting waypoint and possibly passing through some intermediate waypoints.

Route coordinate point string The route from the introductory point to the terminal point is represented by a string of coordinate points. The points are numbered in ascending order from the introductory point to the terminal point. The introductory and terminal points are excluded from the numbering of points. It is recommended to set coordinate points at branches of the route or points where the route changes in shape significantly between the introductory and terminal points.

RRLP Radio Resource LCS Protocol.

RSA The Rivest-Shamir-Adleman algorithm is the form of public-key encryption most widely used today, particularly for digital signatures and key exchange.

RSVP Resource Reservation Protocol.

RTD Real-Time Difference. Timing offset between two base stations.

RTP Real-Time Transport Protocol.

RxLev A measurement of the effect needed to connect to the mobile stations to the base stations near it. This measurement can be used to measure the distance from the base station to the mobile station.

Satellite constellation The arrangement in space of a set of satellites.

SC Service Center.

Scalability The ability to change the component configuration of a system to fit desired application contexts.

Scale The relationship between distance on a map and the corresponding distance on the Earth's surface. Map scale is often recorded as a representative fraction such as 1:1,000,000 (1 unit on the map represents a million units on the Earth's surface) or 1:24,000 (1 unit on the map represents 24,000 units on the Earth's surface). The terms "large" and "small" refer to the relative magnitude of the representative fraction. Because 1/1,000,000 is a smaller fraction than 1/24,000, the former is said to be a smaller scale. Small scales are often used to map large areas because each map unit covers a larger earth distance. Large-scale maps are employed for detailed maps of smaller areas.

Schema A description of a feature's attributes, or more specifically, the specific attribution model for a feature in terms of primitive data types and constraints on these types.

SCP Service Control Point.

SCP Signaling Control Point.

Sectorcell A cell served by a directed base station (not covering a full circle).

Selective Availability (SA) A policy adopted by the US Department of Defense to introduce some intentional clock noise into the GPS satellite signals thereby degrading their accuracy for civilian users. This policy was discontinued as of May 1, 2000, and now SA is turned off.

Semantics The interpretation of the content of a representation, such as the human-centred natural language reference on the one hand and a canonical storage representation of a model element on the other hand. Semantic translation is the conversion between one representation and another, preserving as much meaning as possible.

Server Entity that supplies the location of the target to the client; one end of the protocol.

Service A computation performed by an entity on one side of an interface in response to a request made by an entity on the other side of the interface.

Service Providers Service providers are components, within a client-server context, that provide a response to a specific client request for service. Even though service providers can also be clients in the client-service provider model, the fundamental aspect of service providers in this discussion is that they are suppliers of some service.

Serving Mobile Positioning Center (SMPC) Manages the overall coordination and scheduling of resources required to perform positing of a mobile station. It also calculates the final location estimate and accuracy. In one mobile network, there may be more than one SMPC. The SMPC and GMPC functionality can be combined in the same physical node, or reside in different nodes.

SIM Subscriber Identity Module. Smart card holding the user's identity and telephone directory and SMSs. Applications can reside on the SIM. Provides all pertinent information about the user including airtime creditworthiness (for prepaid). Known as USIM in 3G (the Universal Subscriber Identity Module). Without the SIM, the terminal can only be used for emergency calls.

SIM Application Toolkit (SAT) A GSM standard (11.14) providing mechanisms (or tools) that network operators can use to produce SIM controlled features (an application) and services able to function with any SIM Toolkit Mobile. In WAP, there is alsoS@T (SIM @lliance Toolbox), the specifications for interoperable systems and products for adding WML-based (WAP) services to SIM Application Toolkit enabled (GSM Phase2+) handsets.

Slow switching channel A sequencing GPS receiver channel that switches too slowly to allow the continuous recovery of the data message.

SLPP Subscriber LCS Privacy Profile.

SM Solar Magnetic coordinate system.

Smart card A plastic card with an embedded microchip (integrated circuit) that enables the storage, addition, and processing of information.

SMLC Serving Mobile Location Center.

SMS Short Message Service. Facility for sending text messages in GSM (and some other mobile systems). On GSM networks, the maximum length of the message is 160 characters, but multiple messages can be chained. Other mobile network technologies, however, such as TDMA, implement SMS in a slightly different manner. The crucial point here is that many TDMA handsets could not send (Mobile

Originate) messages, although they could receive (Mobile Terminate) messages. SMS is used to provision terminals with configuration data (such as WAP settings) and was originally developed for that purpose (using the control channel), although its widest acceptance today is as person-to-person text messaging system.

SMTP Simple Mail Transfer Protocol.

SOAP Simple Object Access Protocol. A messaging protocol defined in XML.

Space segment The part of the whole GPS system that is in space; in other words, the satellites.

Spatial Reference System (SRS) A geodetic system for interpreting geographic or grid coordinates.

Specification A document written by a consortium, vendor, or user that specifies a technological area with a well-defined scope, primarily for use by developers as a guide to implementation. A specification is not necessarily a formal standard.

Spread spectrum A system in which the transmitted signal is spread over a frequency band much wider than the minimum bandwidth needed to transmit the information being sent. This process is done by modulating with a pseudo-random code for GPS.

SPS Standard Positioning Service.

SQL Structured Query Language. A relational database query and manipulation language.

SS Signal Strength.

SS7 Signalling System No. 7. The system used for internal signaling in ISDN.

SSL Secure Sockets Layer. A form of data encryption used in computer-based transactions and standardized under the name TLS or Transport Layer Security.

Standard A document that specifies a technological area with a well-defined scope by a formal standardization body and process.

Standard Opt-In Defined as a process that requires active choice on the part of the wireless subscriber to express permission or consent. For example, pushing the "locate me" button on a wireless handset is an expression of permission to be located.

Standard Positioning Service (SPS) In GPS, the normal civilian positioning accuracy obtained by using the single frequency C/A code.

State Plane Coordinate System (SPC) A locational reference system developed in the United States in the 1930s that provides positional descriptions accurate to 1 foot in 10,000. The SPC system divides the United States into 125 zones (5 cover Texas) and employs both Lambert conformal and Transverse Mercator projections (depending upon a state's size and shape). Within any given SPC zone, X-Y coordinates are given in eastings and northings. A central meridian passes each zone and is given a false easting of 2 million feet. A false northing of 0 feet is established below the southern limit of each zone.

Static positioning Location determination when the receiver's antenna is presumed to be stationary on the Earth. This allows enables the use of various averaging techniques that improve accuracy by factors of over more than 1,000.

String of locus coordinate points In order to indicate a route up to the current point for a mobile entity, the past locus from the current point is represented by a string of coordinate points. Points are indicated in reverse historical order starting from the current point, which is excluded from the numbering of points. In order to represent the locus properly, it is recommended to set coordinate points at branches of the route or points where the route changes in shape significantly.

Subscriber The entity or customer that pays the subscription for the client. The user and the subscriber need not be the same person; for example, a company (the subscriber) might supply terminals to its employees (the users).

Surface Configuration Model Defines the geometric characteristics of the Earth's surface, exclusive of features that fall upon the surface; defined in terms of elevation, shape, roughness, slope, and aspect with the later properties possibly derived from elevation.

Surface Feature An entity that lies on the Earth's surface or is referenced to the Earth's surface.

SVG Scalable Vector Graphics Markup language in the XML family developed for presentation of vector graphics.

SyncML Synchronization Markup Language. XML language to describe object synchronization processes.

System Internal Interface (SII) An interface between components within an application platform.

T1P1.5 Committee in TIA that has been active in defining standards for positioning. The work has been taken over by 3GPP.

TA Timing Advance.

TACS Total Access Communications System (old mobile telephony system).

Target The entity whose location is known by the server and desired by the client. The protocol does not specify how the server learns the location of the target.

Target MS Privacy Profile The profile detailing the location privacy information for the mobile station (e.g., LIR restriction mode, LCS Client exception list).

Target MS The target mobile station is the object to be positioned by the MPC. For network based positioning methods, no support for LCS is required by the Target MS. For mobile assisted and mobile based positioning methods, the Target MS actively supports LCS. For all positioning methods, the ability to control privacy for the MS subscriber is required.

TCP/IP Transmission Control Protocol/Internet Protocol. A system devised for coordinating the transfer of data over networks. It has grown to become the foundation for the global communications network that we know now as the Internet.

TDMA Time Division Multiple Access. Digital mobile telephony system deployed in the United States based on the same principles as GSM. EDGE will work on both TDMA and GSM networks.

TDOA Time Difference of Arrival.

TDOP Time Dilution of Precision.

TD-SCDMA Time Division—Synchronous Code Division Multiple Access. Successfully combines two leading technologies—an advanced TDMA system with an adaptive CDMA component to present yet another 3G technology. Championed by CATT (China Academy of Telecommunications Technology).

Terminal Point Refers to an eventual destination point for the target. An information creator sets an appropriate point within the boundary of the target. It is recommended to set a point where an eventual arrival is completed such as a parking lot, entrance, or gate.

Terminal Also called a mobile terminal or mobile station, a device that holds the WAP client typically used by a user to request and receive information.

Terrain distance A distance measurement which takes the local vertical displacements into account. In other words, you also take the elevation into distance and measure the real distance that a mobile station travels (a line on a sloping plane is longer than a line on a flat surface, since it also has a vertical distance). Terrain distance can be based either on a geodetic distance or a map distance. In practice, this measurement is usually measured on the ground (e.g. by wheel rotations).

TETRA TErrestrial Trunked RAdio. Digital standard developed by ETSI to satisfy the user requirements for PMR (Private Mobile Radio) /PAMR (Public Access Mobile Radio) where GSM is not applicable. Mostly used for emergency services in Europe.

TIA Telecommunications Industry Association. A U.S.-based telecommunications standards organization.

TIGER Topologically Integrated Geographic Encoding and Referencing file. A type of digital map developed by the U.S. Bureau of the Census to support the 1990 population census. Census maps in TIGER format succeed the previous DIME format. TIGER files are available for every county in the United States and for the millions of census blocks in urban areas. Although the accuracy of TIGER files varies from county to county, partly for reasons beyond the control of the Bureau, they are likely to improve in coming decades. The TIGER files are a particularly important resource for many urban GIS.

Time Elapsed days and seconds within a day with respect to an epoch.

Timeslot The unit of time in a TDMA system. Each timeslot can carry information for one connected subscriber.

Timestamp A precise relative time assigned to navigation data, internally consistent for a single mobile device.

TLS Transport Layer Security (formerly Secure Sockets Layer or SSL). Technique for secure (encrypted) transmission over the Internet.

TM Transverse Mercator coordinate system.

TOA Time of Arrival. A way of measuring the distance from the mobile station to the base station.

Tool A software component, sometimes called an application object, which can act as either a service provider or service requester within an application platform.

Topology Properties of geometric forms that remain invariant when the forms are deformed or transformed by bending, stretching, and shrinking. Among the topological properties of concern in GIS are connectivity, order, and neighborhood.

Trader A kind of intermediary service which acquires services from one or more providers for “resale” to a service requester. The trader service insulates requester and provider services from having to interact directly with one another. The trader is responsible to the requester for all aspects of the requested service.

Traffic channel A well-defined channel that is used to carry speech or data information to a specific subscriber.

Transaction Rate The frequency of position information requests for a location-based service.

Translation The process of converting data or commands from one computer format to another or from one computer language to another.

Transparency The ability of systems or components of systems to hide the details of their implementations from other client or server systems or components of systems.

TTA Telecommunications Technology Association. Korean telecommunications body.

TTC Telecommunication Technology Committee. Japanese telecommunications body.

TTFF Times to First Fix. The time to obtain position information after a position request.

Tuple An ordered set such a set of coordinates that define a point.

UAPROF User Agent Profile. Specification for descriptions of capabilities in wireless devices that uses the CC/PP model.

UDP User Datagram Protocol. A companion protocol to TCP for services that does not require the same type of transmission control that TCP provides.

UL User Location.

ULC User Location Camel.

ULE User Location Emergency.

UL-TOA Uplink Time of Arrival. Network based positioning solution.

UMTS Universal Mobile Telecommunications System. The standard for 3G networks formulated by ETSI for use throughout Europe and merged with Japanese proposals in the scope of 3GPP to form a global standard in IMT-2000.

UMTS Forum Industry body formed to promote the interests of UMTS/3G. www.umts-forum.org.

Unstructured Supplementary Service Data (USSD) Protocol for data transport within GSM. USSD is actually a low-bit rate channel that uses the GSM signaling system. It is an alternative to using the standard dial-up GSM channel. Both USSD and SMS are possible alternatives for transmitting WAP data.

Uplink The transmission from the mobile station to the BTS.

URI Uniform Resource Identifier. A string referring to an object (created following a set of design rules).

URL Uniform Resource Locator. A string identifying resources in a network by an address. A subset of URI.

US User Status.

User Agent User agent (or content interpreter) is any software or device that interprets WML, WMLScript, or other content. This agent can may include textual browsers, voice browsers, search engines, and so on.

User interface The way a receiver conveys information to the person using it (the controls and displays).

User segment The part of the whole GPS system that includes the receivers of GPS signals.

User A user is a person that interacts with a user agent to view, hear, or otherwise use rendered content.

UTC Coordinated Universal Time (the abbreviation used to mean Universal Time Coordinates, hence UTC). Promulgated by the Bureau International des Poids et Mesures (http://www.bipm.fr/) and the International Earth Rotation Service (http://www.hpiers.obspm.fr/).

UTM Universal Transverse Mercator (coordinate system).

UTM Coordinate System A planar locational reference system that provides positional descriptions accurate to 1 meter in 2,500 across the entire Earth's surface except the poles. Based on the Universal Transverse Mercator map projection. At the poles, the Universal Polar Stereographic projection is used. The UTM system divides the earth's surface into a grid in which each cell, excluding overlap with its neighbors, is 6 degrees east to west, and 8 degrees north to south (with the exception of the row from 72-84 degrees north latitude). For any position in the UTM grid, X-Y coordinates can be determined in eastings and northings. Eastings are in meters with respect to a central meridian drawn through the center of each grid zone (and given an arbitrary easting of 500,000 meters). In the northern hemisphere, northings are read in meters from the equator (0 meters). In the southern hemisphere, the equator is given the false northing of 10 million meters.

UTRAN The radio system for Universal Telephony Radio Access Network (WCDMA).

UWCC Universal Wireless Communications Consortium. Industry body representing those interested in TDMA (D-AMPS) and AMPS networks; promoted its version of 3G called UWC-136, which is now part of EDGE.

Validation The process of testing an application or system to ensure that it conforms to a specification.

VAS Value-Added Services. Typically information or entertainment services delivered by an operator using voice or SMS.

vCal A calendar inter-operability standard by the Internet Mail consortium; superseded by iCal.

vCard A standard for business card inter-operability by the Internet Mail Consortium.

VDOP Vertical Dilution of Precision.

Vector displays and databases Databases that build all geographic features from point, that is, from discrete X-Y locations. Lines are constructed from strings of points, and polygons (regions) are built from lines that close.

Vector methods In geoprocessing, methods of representing geographic features from points, lines, and polygons as opposed to raster techniques which record geographic features within a matrix of grid cells. The choice between vector and raster GIS has much to do with the application being considered, because both methods have strengths and weaknesses. Many current GIS permit transformation between vector and raster input and output.

Velocity Speed and direction.

Vertical error With a point expressed in terms of altitude, an error expected upward or downward with the indicated altitude as a starting point, serve as a vertical error. If values differ between upward and downward errors, the greater of the two is used.

View SQL "Select" statement used to provide temporary information about a given table(s) of a Database Management System without actually creating a subset or new table.

Virtual reality Refers generally to interactive multimedia environments that present users with a sensory experience similar in some ways to our experience of the real world.

VLR Visitor Location Register. A database maintained by a mobile telephone system to keep track of the current users active in the system.

W3C World Wide Web Consortium. An industry consortium developing recommendations for markup languages and protocols for the World Wide Web.

WAE WAP Application Environment.

WAP Wireless Application Protocol. A standard created by Ericsson, Motorola, Nokia and OpenWave (formerly Phone.com and Unwired Planet) for providing Internet access via mobile communications devices. Part of the standard defines how information should be displayed—using the *Wireless Markup Language* (WML), while another part defines how information should be sent and received via WAP gateway servers.

WAP Client In the context of push, a WAP client is a device (or service) that can receive push content from a server. In the context of pull, a WAP client is a device that can initiate requests to a server for content. See also "device."

WAP-FEP/PEP WAP Feature Enhancing Proxy/Performance Enhancing Proxy (formerly WAP Gateway).

WCDMA Wideband CDMA (radio transmission technology). A higher capacity version of CDMA which also includes features from GSM. This interface was picked by Europe and Japan as the air interface for 3G networks based on UMTS/IMT 2000.

Web Services Architecture under definition in the W3C to transport data and processing instructions encapsulated in XML as SOAP messages and WSDL descriptions.

WGS 84 World Geodetic System 1984 The global spatial reference system which is used for GPS navigation. There is also a reference ellipsoid and a geoid height system called WGS 84.

WIM Wireless Identity Module. This module corresponds to SIM in WAP.

Wireless Pull Advertising (Pull Messaging) Defined as any content sent to the wireless subscriber upon request shortly thereafter on a one-time basis. For example, when a customer requests the local weather from a WAP-capable browser, the content of the response, including any related advertising, is Pull Messaging.

Wireless Push Advertising (Push Messaging) Defined as any content sent by or on behalf of advertisers and marketers to a wireless mobile device at any time other than when the subscriber requests it. Push Messaging includes audio, *Short Message Service* (SMS) messages, e-mail, multimedia messaging, cell broadcast, picture messages, surveys, or any other pushed advertising or content.

Wireless Spam Push Messaging that is sent without Confirmed Opt-In.

WLL Wireless Local Loop. Popular term for fixed mobile networks which replace fixed line (copper) networks with wireless connections in order to provide local telephony. Often WLL networks re-use existing cellular technologies.

WLS Wireless Location Services.

WML Wireless Markup Language. The chosen method for creating pages of information on WAP sites. It is an XML application.

World Geographic Reference System (GEOREF) Coordinate system used in aircraft navigation. It is based on latitude and longitude and divides the world into 12 bands of latitude and 24 zones of longitude (each 15 degrees wide). The grid makes it possible to designate which grid square you are in by two letters, and the grid can be further subdivided by more letters and numbers.

WSDL Web Services Description Language. A language to describe web services.

WSP Wireless Session Protocol.

WTA Wireless telephony application.

WTLS Wireless Transport Layer Security. The wireless equivalent of SSL.

WTP Wireless Transaction Protocol. A level in the WAP protocol.

XHTML eXtended Hypertext Markup Language. The XML version of HTML, and the base for the new markup language in WAP.

XML eXtensible Markup Language. Created as a kind of "super language" for structured information (actually a rule set to define markup languages).

XSLT eXtensible Stylesheet Language Transformations. Language developed for transformation of the structure of XML documents.

Index

A

B

C

D

M

N

O

P

Q

T

U

V

Y–Z

CUSTOMER NOTE: IF THIS BOOK IS ACCOMPANIED BY SOFTWARE, PLEASE READ THE FOLLOWING BEFORE OPENING THE PACKAGE.

This software contains files to help you utilize the models described in the accompanying book. By opening the package, you are agreeing to be bound by the following agreement:

This software product is sold as is without warranty of any kind, either express or implied, including but not limited to the implied warranty of merchantability and fitness for a particular purpose. Neither Wiley nor its dealers or distributors assumes any liability for any alleged or actual damages arising from the use of or the inability to use this software. (Some states do not allow the exclusion of implied warranties, so the exclusion may not apply to you.)